10.00

Vibrational Spectroscopy
of
Trapped Species

Infrared and Raman Studies of Matrix-Isolated Molecules, Radicals and Ions

Edited by

H. E. Hallam

University Reader in Chemistry,
University College of Swansea

A Wiley–Interscience Publication

JOHN WILEY & SONS

London · New York · Sydney · Toronto

Library of Congress catalog card number
72-8601

ISBN 0 471 34330 7

Set on Monophoto Filmsetter and printed by
J. W. Arrowsmith Ltd., Bristol, England

Foreword

The matrix isolation method has now been in active development for almost two decades. Its original aim was to permit leisurely spectroscopic study of highly reactive molecules, molecules that would be transient or non-existent under normal conditions. Independently, it was proposed for use with ESR, visible ultra-violet (electronic) and infrared (vibrational) spectroscopy. The method has undoubtedly had its greatest impact in the infrared spectral studies, the subject of this book. There are several reasons for this. First, the only alternative means for recording the infrared spectrum of a transient species, flash photolysis with rapid scanning, has proved to be quite difficult. Second, matrix shifts of infrared bands are almost always quite small, so that the spectrum closely resembles that of the gas phase species. This facilitates both identification and interpretation. Third, matrix absorption features are extremely sharp. This enhances sensitivity and permits resolution of closely spaced bands, including small spectral shifts due to heavy isotopes.

These advantages were not won easily, however. Though the first matrix studies began prior to 1954, it was several years before the first infrared spectra of positively identified transient molecules were reported: HNO by H. W. Brown and G. C. Pimentel (1958) and HCO by G. E. Ewing, W. E. Thompson, and G. C. Pimentel (1960). These prototype experiments were successful only after reliable low temperature cells had been developed in which systematic study could be conducted of the effects of concentration, deposition conditions, and temperature upon isolation efficiency and diffusion rate.

Now, the matrix isolation method has come into full flower. The acceleration of application is indicated by the statistics. Prior to 1958, the *only* free radicals whose infrared spectra had been recorded were those of the stable molecules NO and NO_2. Then, in the 1961–1965 period, the infrared spectra of some 30 additional diatomic and triatomic transient molecules were recorded with the matrix method. In the subsequent five year period, this number had risen to about 70, more than double the earlier five year figure.

Today, a very large number of laboratories are equipped to conduct matrix infrared studies and the variety of applications is increasing. The fairly recent addition of laser Raman has considerably increased the power of matrix isolation spectroscopy. Many interesting studies of high temperature species have been reported and the use is growing. Weak complexes can be fruitfully studied. Perhaps one of the richest potentialities yet to be exploited is the method as an adjunct to the synthesis of new compounds. Prototypes that furnish examples are KrF_2 and HOF, both of which were prepared in gram amounts following their first discovery and identification in matrix infrared experiments. All of these developments are well described in the chapters that follow here.

Matrix isolation need no longer be regarded as a highly specialized and quite difficult technique. The instrumental developments of the past few years and the large store of experience that has been accumulated should and will attract new practitioners. They will, in turn, show us new possibilities. Both this increasing use and the broadening area of application will be encouraged and aided by this valuable book.

GEORGE C. PIMENTEL
University of California

Berkeley, April 1973

Preface

This monograph is intended to provide a survey of studies of the vibrational spectra of species trapped in solid matrices. In a sense it can in part be considered as a follow-up of the standard text *Formation and Trapping of Free Radicals* edited by A. M. Bass and H. P. Broida, Academic Press (1960) and covers the work produced in the past decade.

It is only the widespread availability of liquid helium refrigerant and the commercial production of miniature Joule–Thompson cryocoolers since the mid nineteen-sixties that the so-called matrix isolation technique has become generally accessible to most spectroscopic laboratories. The result has been a great proliferation of matrix studies so that the method is rapidly becoming a routine spectroscopic technique. The greatest effort has been concerned with free radicals and unstable molecules but attention is increasingly being directed towards stable molecules. The book could have been devoted entirely to species trapped in inert matrices at cryogenic temperatures but it was felt that great benefit could be achieved by encompassing two other fields of trapped species, namely clathrates and ion impurities in alkali-metal halide matrices.

To attain an authoritative account it was decided to invite contributions from active leaders in each of the fields covered, despite the disadvantages of a multi-author work. The editor wishes to record his thanks to the authors for their co-operation and especially for limiting themselves to their particular briefs and avoiding major overlaps. The minor amount of overlap in some chapters has been purposely included by the editor in the belief that it enhances the usefulness of the book, and, if necessary, allows chapters to be read in isolation. Authors were not briefed to be completely exhaustive in their literature coverage but have in fact approached this and each chapter covers its particular field in a most comprehensive manner. Most of the important work published up to the end of 1972 is included and also some papers which appeared early in 1973.

I am most grateful to Professor Howard Purnell for his support and encouragement and to the Spectroscopic Panel of the Hydrocarbon Research Group of the Institute of Petroleum for their interest and support which enabled cryogenic studies to be inaugurated in my research group, and to the Science Research Council for its continued support. I am greatly indebted to my research collaborators over the years for their painstaking work and ideas. Their names appear frequently in the text but this is little recompense for their efforts; however, it has been a particular pleasure to be able to call upon two of them as contributors. Dr Austin Barnes is specially thanked for his interest and detailed criticisms and for his meticulous attention and speed in preparing the indexes.

I take great pleasure in thanking Professor George Pimentel for his kindness in writing the Foreword and to whom all matrix isolation spectroscopists are deeply indebted. I am grateful for the permission of authors and copyright owners to reproduce various figures and tables and I also wish to thank Mrs Sue Howells for the excellence of her secretarial assistance. Finally, I am deeply indebted to my wife, Joan, for her encouragement, understanding and forbearance.

Swansea, April 1973   H. E. HALLAM

Contributors

L. ANDREWS
Associate Professor of Chemistry, University of Virginia, Charlottesville, Virginia 22901, U.S.A.

A. J. BARNES
I.C.I. Fellow, Department of Chemistry, University College of Swansea, Singleton Park, Swansea SA2 8PP, Wales.

H. E. HALLAM
University Reader in Chemistry, University College of Swansea, Singleton Park, Swansea SA2 8PP, Wales.

D. C. MCKEAN
Senior Lecturer in Chemistry, University of Aberdeen, Meston Walk, Old Aberdeen, Scotland.

G. F. SCRIMSHAW
Research Demonstrator in Chemistry, University College of Swansea, Singleton Park, Swansea SA2 8PP, Wales.

G. A. OZIN
Associate Professor of Chemistry, University of Toronto, 80 St. George Street, Toronto 5, Ontario, Canada.

A. SNELSON
Senior Chemist, Illinois Institute of Technology Research Institute, 10 West 25 St., Chicago, Illinois 60616, U.S.A.

W. F. SHERMAN
Lecturer in Physics, Wheatstone Physics Laboratory, King's College, Strand, London WC2R 2LS, England.

G. R. WILKINSON
Professor of Physics, Wheatstone Physics Laboratory, King's College, Strand, London WC2R 2LS, England.

Contents

1 Introductory survey

H. E. HALLAM

Matrix-isolation is a technique for trapping species as isolated entities in an inert solid, or matrix, in order to investigate their properties, usually by spectroscopic methods. A suitable matrix must, at the temperature of the experiment, be a solid which is inert, rigid with respect to diffusion, and transparent in the spectral region of interest. In addition to noble-gas crystals, reactive-gas crystals, ionic crystals and molecular solids can all be used as matrices.

The technique of trapping at low temperatures in an inert solid matrix was originally developed as a means of studying unstable molecular species. Under these conditions the lifetime of a trapped species is considerably increased. The matrix cage severely restricts the occurrence of bimolecular collisions, whilst the cryogenic temperatures (usually 4 K, liquid helium, or 20 K, liquid hydrogen) employed, effectively prevent any reaction with an activation energy greater than a few kilojoules per mole. Thus the trapped species can be studied at leisure using conventional spectroscopic techniques, as opposed to time-resolved spectroscopic observation employed in flash photolysis, the second major technique for the spectroscopic study of reactive species.

The matrix isolation approach was first applied by G. N. Lewis[1] (1941), who studied the phosphorescence of low concentrations of various aromatic molecules suspended in a rigid glassy medium at low temperatures. The use of argon and nitrogen as matrix supports for the isolation of molecules and the photoproduction of free radicals was simultaneously proposed (1954) by G. Porter[2] and by G. C. Pimentel.[3] The subsequent development of the technique, particularly in its adaptation to infrared studies, has been largely pioneered by Pimentel and his associates.

The advantages to be gained from applying matrix isolation to the spectroscopy of stable molecules are now becoming more appreciated, and the technique is rapidly becoming a standard one for studying stable molecules. The isolation of monomeric solute molecules in an inert environment reduces intermolecular interactions, resulting in a sharpening of solute absorptions compared with other condensed phases. With the exception of a few small hydrides, such as the hydrogen halides or ammonia, rotation does not occur in matrices thus much narrower bands are obtained than in the vapour phase. The matrix technique now finds wide applications in most fields of spectroscopy, particularly IR, UV, and ESR. This monograph brings together all of the IR studies and the very recent, and thus at present limited, Raman studies.

It is convenient to divide up the study of reactive species in inert low temperature matrices into separate chapters, one on free radicals and intermediates (Chapter 5), and one on vaporizing molecules (Chapter 6).

Radicals can be produced by photolysis or discharge in the gas phase, then diluted with an excess of inert gas and frozen onto a suitable optical support. *In situ* production of radicals from suitable precursors trapped in a matrix can also be employed. Reactive matrices (e.g. carbon monoxide, carbon dioxide) are utilized to produce radicals as secondary products, e.g.

$$H_2S(CO) \overset{h\nu}{\to} H + SH$$

$$H + CO \to HCO$$

Another technique, originated by Andrews and Pimentel[4] (1966), for the production and stabilization of free radicals and other reactive species, involves the reaction of a molecular beam of lithium with another species which has been co-condensed with an excess of argon.

$$Li + CH_3I \to CH_3 + LiI$$

The use of the MI technique for the study of reactive species remains its largest and perhaps most significant application and is reviewed (Chapter 5) by L. Andrews, a former associate of Pimentel.

At high temperatures the population of excited rotational and low frequency vibrational levels is so high that the spectrum of a vaporizing molecule is necessarily a superposition of many hot bands and thus is extremely complex to interpret. The application of the MI technique to the study of the composition of molecular beams produced by a Knudsen effusion cell was first suggested by Pimentel[5] (1960) and the first successful experiment was reported by Linevsky[6] (1961). Linevsky demonstrated how the problem could be overcome by simultaneous condensation of a beam of vaporizing molecules (lithium fluoride), effusing from a Knudsen cell, and a stream of noble gas to form a low-temperature matrix. The technique has found many valuable applications since. For example, the bending vibrations of the group IIB halides have been the subject of considerable study. At first it was believed that these bending modes have frequencies about $400 \, cm^{-1}$, but these assignments could not be reconciled with the thermodynamic properties, in particular, with the entropy of the MX_2 molecules. Low temperature matrix isolation studies showed that these frequencies are appreciably lower than $400 \, \text{cm}^{-1}$ and therefore the statistically calculated entropies are higher than obtained with the former assignments. These determinations brought the spectroscopic observations in line with the thermodynamic investigations. This considerable field is reviewed in Chapter 6 by A. Snelson, one of Linevsky's early collaborators.

For a complete interpretation of the spectra described above it is important to establish that the molecular parameters obtained from MI spectra are not significantly perturbed by the matrix environment. This can be achieved only by fundamental studies of stable molecules under matrix conditions; such studies are described in Chapter 3. It is found that vibrational frequencies in matrices generally occur at lower frequencies than those observed in the gas phase. In noble-gas matrices, however, the frequency shifts are relatively small, typically less than 0·5 per cent. Thus thermodynamic functions and force constant values derived from matrix data will differ only slightly from gas-phase values. Relationships such as the isotope sum and product rules, which hold well for gas-phase frequencies, are thus equally valuable for matrix frequencies. Detailed comparisons of gas-phase and matrix spectra for some simple linear triatomic XYZ molecules have recently been made by Overend et al.[7] These indicate that the small shifts in the vibrational frequencies are determined largely by small perturbations of the quadratic terms in the intramolecular potential energy; the anharmonic terms (cubic and quartic) are considered to be the same as those in the gas phase.

Most entrapped species will be so tightly held as to prevent rotation, which entails the loss of structural information from rotational line spacing and symmetry designations from gas-phase vibrational-rotational band envelopes. However, the resultant sharpening of vibrational absorptions allows near-coincident fundamentals or isotopic splittings to be resolved. Two recent examples are hydrogen sulphide[8] and sulphur dioxide.[9] For hydrogen sulphide, because of the near coincidence of the two stretching frequencies v_1 and v_3, the antisymmetric stretching fundamental v_3 has not been observed in either the gas, liquid or solid phases. In the gas phase v_1, the symmetric stretch, is observed at $2614 \cdot 6 \, \text{cm}^{-1}$ and v_2, the bending fundamental, at $1182 \cdot 7 \, \text{cm}^{-1}$; v_3 is estimated from combination band frequencies to lie at $2627 \cdot 5 \, \text{cm}^{-1}$. When hydrogen sulphide is isolated in an argon matrix at 20 K two sharp bands are observed at $2628 \cdot 3$ and $2581 \cdot 8$ with an intensity ratio of c. 1 : 9, which are assigned to v_3 and v_1 respectively. For sulphur dioxide a sample consisting of 60 per cent. ^{16}O and 40 per cent. ^{18}O with sulphur isotopes in natural abundance has been examined[9] in a krypton matrix at 20K. All of the 15 fundamentals of the species $^{32}S^{16}O_2$, $^{34}S^{16}O_2$, $^{32}S^{16}O^{18}O$, $^{34}S^{16}O^{18}O$, $^{32}S^{18}O_2$ and one of the fundamentals of $^{34}S^{18}O_2$ were resolved and observed in a single scan. Analysis of the data yields values of force constants and anharmonic corrections which are in excellent agreement with the values derived from the gas-phase data. The bond angle which leads to the best fit of the experimental matrix data is 119° 37′, in excellent agreement with the value of 119° 19′ determined from microwave data. This illustrates the negligible effect of the solid matrix on the bond angle and the considerable precision with which the value can be deduced from the nearly pure vibrational spectra of the matrix-isolated molecule. It is now arguable that parameters obtained from IR matrix data are more reliable than those obtained from gas-phase IR spectra, in that the sharp absorption bands obtained in an inert matrix at cryogenic temperatures allow superior resolution to that in gas-phase spectra.

The narrow band-widths obtained with the technique have been utilized by Barnes and Hallam[10] to introduce a valuable method for studying conformational isomers. The monomer O—H stretching absorption of ethanol in dilute solution in carbon tetrachloride at ambient temperatures, displays very slight asymmetry the interpretation of which has roused considerable controversy. In an argon matrix at 20 K, two monomer absorptions are revealed,[10] at 3657·6 and 3662·2 cm^{-1} which are attributable to the two conformers of ethanol, with the OH group *trans* or *gauche* to the methyl group. Other modes which might be expected to be sensitive to the conformation of the molecule are split into similar doublets. The study provides a complete vibrational assignment for the *trans* molecule, the more stable conformer.

The technique is also finding extensive usage in molecular association studies. In these instances, at low concentrations monomeric entities are isolated but at high concentrations dimers and large multimers are isolated and give rise to characteristic absorptions which are also sharper than obtained in any other phase, and thus can be studied individually. The association process can be controlled and followed in a very elegant manner by depositing a low concentration mixture and then allowing the matrix to warm up slightly (say from 20–35 K for argon) and thus soften; the trapped monomers then diffuse to form dimers and high multimers. The association behaviour of the hydrogen halides and of alkanols in particular have now been extensively studied by the MI method and they are far more informative than their counterpart studies in other phases.

Early use of the technique for the synthesis and structural investigation of the xenon fluorides has recently been extended by Turner and by others to the preparation of unstable or metastable transition metal carbonyls under cryogenic conditions. This has been done both by starting with the parent metal carbonyl and successively photolysing off carbon monoxide ligands in an inert matrix[11] and by synthesizing the metal carbonyls directly[12] from the metal and carbon monoxide in an argon matrix at 4 K and annealing, for example, $Ni(CO)_{1 \to 4}$ and $Ta(CO)_{1 \to 6}$. Such studies offer interesting possibilities for the synthesis of novel inorganic and organometallic compounds and a new and exciting field is opening up which can be predicted to show a considerable rate of growth. It is clear from kinetic studies that the mechanism of solid-state photolytic processes is significantly different from those in the gas phase. Thus we find novel reactions occurring under cryogenic matrix conditions which may well be able to be exploited on an industrial scale.

Despite the generalization that the matrix had little influence on the trapped species, upon which the utilization of the MI technique for structural studies is dependent, even in the early matrix studies, it soon became apparent that the environment even of a noble gas 'inert' matrix had several effects on the spectral features of the trapped guest species. For example, the stretching frequency for lithium fluoride occurs as a singlet at $898 \, cm^{-1}$ in the gas phase and shifts to $867 \, cm^{-1}$ in a neon matrix, to $830 \, cm^{-1}$ in krypton, and appears as a doublet at $837/842 \, cm^{-1}$ in argon and $814/823 \, cm^{-1}$ in xenon. The frequency shifts are interpreted in terms of different electrostatic interactions between the trapped molecule and the surrounding matrix, and the doublet in terms of more than one trapping site in the matrix. Other perturbations may arise due to the site symmetry of the trapped species which may cause a splitting of any degenerate vibrations. The influence of the matrix on the frequencies and intensities of the vibrational modes of a molecule has in itself become an interesting field of study and this has been the underlying fundamental theme of the author's research group. Studies of this sort in solid solutions have the advantage over similar work in liquid solutions that the environment of the molecular unit under consideration is generally known with much more certainty. Such studies of matrix environmental effects are essential for a complete interpretation of results obtained by the MI method but are also of considerable interest in their own right since they provide knowledge concerning intermolecular forces in the solid state. These experimental studies embrace inert and reactive matrices and are described in Chapter 3 whilst A. J. Barnes, in Chapter 4, reviews the theories which have been proposed to account for the perturbations exerted by the matrix on the solute.

Small molecules such as the hydrogen halides, ammonia and methane have been extensively studied in a variety of matrices with particular regard to their rotational motions. In noble-gas matrices discrete rotational lines are observed whose reversible variation of intensity on temperature cycling allows them to be assigned to transitions corresponding to quantized rotational motion of the trapped solute in the host lattice. It is of considerable interest to compare these phenomena with similar ones observed for the trapped molecular guest in clathrates (Chapter 8) and the impurity ion in many ionic lattices (Chapter 7).

Chapter 3 also provides a brief account of some applications of matrix isolation for the quantitative analysis of multicomponent gas mixtures. The method, developed by Rochkind,[13] involves condensing the gas mixtures, diluted 100 : 1 with nitrogen at 20 K by controlled-pulse deposition. This results in the isolation of individual species and the spectra are characterized by extremely sharp absorptions which contrast markedly with gas phase spectra with their rotational features. Nitrogen is utilized as a matrix since, to date, no trapped molecule in this medium has been found to exhibit rotational features. The technique appears to be superior to gas chromatography for the quantitative analysis of μ mole quantities of isotopic isomers such as deuterated ethylenes.

The increasing availability of liquid helium refrigerant resulted in the gradual growth of matrix studies in the early 1960's and double Dewar cryostats which utilize liquid refrigerants were well described by Mauer[14] in 1960. The commercial production of miniature liquefier systems, stimulated by development in cryogenic detectors for the USA satellite programme, ushered in the rapid expansion in IR matrix isolation studies which has occurred in the past few years. These cryostats are described in Chapter 2. In these cryocoolers refrigeration is achieved by the open-cycle Joule–Thomson expansion of high pressure cylinder gases under a throttling process. Closed-cycle cryocoolers which involve continuous fluid circulation and thus avoid consumption of gas are also described; several models specially produced for matrix isolation spectroscopy have recently been marketed and these will add further impetus to MI studies. This chapter also gives an account of the ancillary equipment required for MI experiments and provides a comprehensive compilation of the physical properties and solid-state spectra of a wide variety of matrix materials.

An ionic crystal, usually an alkali-metal halide, functions as a matrix when it forms a dilute solid solution with an impurity species, usually another ion, of interest. The physics of the interactions in these solid lattices, usually at room temperature but also at low temperatures, is treated together with a number of covalent molecular matrices in Chapter 7 by W. Sherman and G. Wilkinson. This chapter is the longest in the book as it provides for the first time a basic account aimed at the physical chemist spectroscopist rather than the physicist. With the upsurge in interest in solid-state chemistry in recent years this account should prove most opportune. The overlap of interests, such as rotational phenomena and environmental effects, with low-temperature molecular matrices is seen to be considerable and greatly contributes to the unifying theme of the monograph.

Many molecular solids, both covalent and ionic, have natural cavities in their structures which allow the inclusion of foreign molecules. For example, crystallization of solutions of quinol in the presence of small molecules such as oxygen, carbon monoxide and methane produces crystals which are different from ordinary α-quinol. These crystals, which are stable, are non-stoichiometric compounds in which guest molecules are entrapped within cavities in the host lattice and are known as clathrates. These cage-like complexes thus provide special examples of matrix-isolated molecules. They are of considerable interest since the structure of a number of them have been elucidated by X-ray diffraction studies, which gives us exact knowledge as to the size and shape of the cavity occupied by the trapped species, hence we can describe the environment and the interactions of the guest molecule fairly precisely. They have the advantage that their crystal stability allows them to be studied at ambient temperatures, when the absorbing molecular guest will possess considerable thermal energy. The advantage of stability is, however, greatly outweighed by the serious disadvantage, as compared with a noble-gas or nitrogen matrix, that the absorption spectrum of the host lattice is usually extensive and may altogether obscure that of the guest species. There are clearly also restrictions on the size of the guest molecule that can be accommodated.

All of the IR studies of clathrates that have been reported are discussed by D. McKean (Chapter 8) together with the first Raman studies which were published as this monograph was going to press. Again it is seen that there is an overlap of interest with the themes described in the previous chapters in that clathrate studies are not restricted to ordinary temperatures but often utilize cryogenic conditions as a means of studying the rotational motion of the enclathrated guest.

Recent developments in laser-Raman spectroscopy[15] have resulted in this technique taking its place alongside IR spectroscopy as a complementary way of observing vibrational spectra. Solids are now amenable to study and several Raman investigations of clathrates and ionic matrices have been reported and are discussed in the appropriate chapters. Preliminary attempts have also been made to study low temperature matrices by laser Raman, the problems and successes of which are described by G. Ozin in Chapter 9. This chapter was added at a very late stage in the preparation of the monograph when it became known in spectroscopic circles that several workers, including Ozin, had, after much painstaking endeavour, overcome the formidable experimental problems and were producing some successful Raman spectra of cryogenic matrices. For example, Shirk and Claassen[16] obtained a Raman spectrum of SF_6 in argon at 4·2 K at a mole ratio of 1 : 500. Ozin (see references in Chapter 9) has applied the technique to reactive species such as Br_3, $XeCl_2$ and SiF_3, and high temperature species such as SeO_2 and $(SeO_2)_2$. Nibler and Coe[17] have also reported success with the technique including measurement of depolarization ratios of methane, carbon tetrachloride and carbon disulphide in nitrogen at 12 K. Professor Ozin very kindly agreed to write, in the space of a few weeks, an account of these efforts, which are presented in the hope of stimulating growth in this almost virgin field. It is now clear that Raman matrix isolation spectroscopy is a viable technique although Raman MI investigations are unlikely to become as routine as infrared MI studies. They are so necessary to complete our knowledge of small molecular species and larger molecules of high symmetry that many developments in this area can be expected. Work is now under way to further optimize the experimental parameters which will allow the technique to be applied to further systems of interest.

1.1 References

1. G. N. Lewis, D. Lipkin and T. T. Magel, *J. Amer. Chem. Soc.*, **63**, 3005, (1941); G. N. Lewis and D. Lipkin, *J. Amer. Chem. Soc.*, **64**, 2801, (1942).
2. I. Norman and G. Porter, *Nature*, Lond., **174**, 508, (1954).
3. E. Whittle, D. A. Dows and G. C. Pimentel, *J. Chem. Phys.*, **22**, 1943, (1954).
4. W. L. S. Andrews and G. C. Pimentel, *J. Chem. Phys.*, **44**, 2361, (1966).
5. G. C. Pimentel, p. 109, in *Formation and Trapping of Free Radicals*, ed. A. M. Bass and H. P. Broida, Academic Press, New York, 1960.
6. M. J. Linevsky, *J. Chem. Phys.*, **34**, 587, (1961).
7. C. B. Murchison and J. Overend, *Spectrochim. Acta.*, **27A**, 1509, (1971); D. Foss Smith, Jr., J. Overend, R. C. Spiker and Lester Andrews, *Spectrochim. Acta.*, **28A**, 87, (1972).
8. A. J. Barnes and J. D. R. Howells, *J. C. S. Faraday II*, **68**, 729, (1972).
9. M. Allavena, R. Rysnik, D. White, V. Calder and D. E. Mann, *J. Chem. Phys.*, **50**, 3399, (1969).
10. A. J. Barnes and H. E. Hallam, *Trans. Faraday Soc.*, **66**, 1932, (1970).
11. A. J. Rest and J. J. Turner, *Chem. Comm.*, **375**, 1026, (1969).
12. A. Kaldor and R. F. Porter, *Inorg. Chem.*, **10**, 775, (1971).
13. M. M. Rochkind, *Analyt. Chem.*, **39**, 567, (1967); **40**, 762, (1968).
14. F. A. Mauer, 'Low temperature equipment and techniques' Chapter 5 in *Formation and Trapping of Free Radicals*, ed. A. M. Bass and H. P. Broida, Academic Press, New York, (1960).
15. P. J. Hendra and P. M. Stratton, *Chem. Rev.*, **69**, 325, (1969).
16. J. S. Shirk and H. H. Claassen, *J. Chem. Phys.*, **54**, 3237, (1971).
17. J. Nibler and D. Coe, *J. Chem. Phys.*, **55**, 5133, (1971).

The individual chapters contain extensive literature references where further details of their topics can be found. The listing below gives the review accounts of the vibrational spectroscopy of trapped species which have appeared up to December 1972.

R1 'Radical formation and trapping in the solid phase', G. C. Pimentel, Chapter 4, in *Formation and Trapping of Free Radicals*, ed. A. M. Bass and H. P. Broida, Academic Press, New York, (1960).
R2 'Solid solutions', Section VI, in review on 'Infrared spectra of crystals', W. Vedder and D. F. Hornig, *Adv. Spectroscopy*, **2**, 189, (1961).
R3 'Matrix technique and its application in the field of chemical physics', G. C. Pimentel, *Pure Appl. Chem.*, **4**, 61, (1962).
R4 'IR spectral perturbations in matrix experiments', G. C. Pimentel and S. W. Charles, *Pure Appl. Chem.*, **7**, 111, (1963).
R5 'Molecular interactions in clathrates: a comparison with other condensed phases', W. C. Child, jun., *Quart. Rev. Chem. Soc.*, **18**, 321, (1964). (I.R. studies of clathrates published before 1963 are discussed on pp. 342–346.)

R6 'The matrix isolation of vaporizing molecules', W. Weltner, jun., *Proc. of the International Symposium on Condensation and Evaporation of Solids*, 243, (1962), published 1964.

R7 'Thermodynamic properties of vaporizing molecules', W. Weltner, jun., *Proc. 1st Meeting Interagency Chem. Rocket Propulsion Group Thermochem.*, *New York, 1963*, **1**, 27, (1964).

R8 'Recent advances in infrared spectroscopy of matrix-isolated species', H. E. Hallam, in 'Molecular spectroscopy', *Proc. of 4th Institute of Petroleum Hydrocarbon Research Conference*, Institute of Petroleum, London, p. 329, (1968).

R9 'Infrared studies of matrix-isolated species', A. J. Barnes and H. E. Hallam, *Quart. Rev. Chem. Soc.*, **23**, 392, (1969).

R10 'IR spectroscopy at sub-ambient temperatures I. Literature Review', T. S. Herman and S. R. Harvey, *Appl. Spectroscopy*, **23**, 435, (1969). 'III Molecules and Molecular Fragments within Matrices', T. S. Herman, *ibid*, 461. 'IV Sample Cells', T. S. Herman, *ibid*, 473.

R11 'Spectra of radicals', D. E. Milligan and M. E. Jacox, Chapter 5 in *Physical Chemistry*, **Vol. 4**, Academic Press, New York, (1970), and M. E. Jacox and D. E. Milligan, *Appl. Optics*, **3**, 873, (1964).

R12 'Infrared and Raman spectroscopy of trapped species', H. E. Hallam, *Ann. Repts. Chem. Soc.*, 117, (1970).

R13 'The study of free radicals and their reactions at low temperature using a rotating cryostat', J. E. Bennett, B. Mile, A. Thomas and B. Ward, pp. 1–77, in *Advances in Physical Organic Chemistry*, **Vol. 8**, Academic Press, London and New York, (1970).

R14 'Matrix isolation', J. S. Ogden and J. J. Turner, *Chemistry in Britain*, **7**, 186, (1971).

R15 'Infrared spectra of free radicals and chemical intermediates in inert matrices', Lester Andrews, *Ann. Rev. Phys. Chem.*, **22**, 109, (1971).

R16 'Applications of matrix isolation infrared spectroscopy', A. J. Barnes, *Reviews in Anal. Chem.*, **1**, 193, (1972).

2 Experimental techniques and properties of matrix materials

H. E. HALLAM and G. F. SCRIMSHAW

Contents

2.1 Introduction

In matrix isolation experiments, the temperature required depends upon the matrix being used. Not only does the material have to be condensed, but the temperature has to be sufficiently low to prevent diffusion of the solute in the matrix lattice and subsequent recombination of radicals or polymerization and aggregation of monomeric stable species. Whilst liquid-air or liquid-nitrogen temperatures may be adequate for a few systems it is desirable to use as low a temperature as possible, and to employ liquid hydrogen (20 K) or liquid helium (4 K) as the refrigerant.

There are several basic problems in the design of suitable cells for obtaining the infrared spectrum of a solute dispersed at high dilution in a solid matrix at cryogenic temperatures. The cell must be an integral part of a cryostat capable of achieving the low temperature necessary for the solute, mixed with the gaseous material, to be condensed. The cell must be evacuated, ideally to 10^{-5} mmHg, to reduce heat leaks to a minimum. There must also be an injection inlet as a means of supplying the gaseous sample mixture to the cold sample support window within the evacuated cryostat. A variety of types of cells in present use will be detailed in the following sections.

In addition to the basic cell, a variety of ancillary equipment is needed. This includes a suitable vacuum system for sample handling and for evacuation of the cryostat. Some means of measuring the temperature achieved in the cell, ideally at the sample site, and a means of following the pressure in the gas handling line and in the cryostat evacuation system, are also required. For certain types of cryostats (§2.2.5) a system for handling high pressure refrigerant gases is also needed.

The choice of matrix material will of course be governed by the nature of the solute and the experiment. However a knowledge of the properties of matrix materials is necessary, both for the reasons which apply to choice of solvent in liquid solution IR studies and for considerations peculiar to low temperature matrix isolation spectroscopy. The properties of a number of potential matrix materials are listed in later sections of this chapter.

2.2 Apparatus

2.2.1 Ancillary equipment

It is not intended to discuss ancillary apparatus exhaustively, as it is dealt with very fully elsewhere; readers are referred to the detailed chapter by Mauer[1] for a review of matrix isolation equipment.

There are many standard reference books on vacuum technique and on temperature and pressure measurement which should be consulted for experimental detail, and a few of these are listed in the references.[2-5] The Science Research Council publishes a list[6] of equipment, materials and suppliers of cryogenic equipment. The list contains all known UK suppliers of items; some foreign suppliers are also included where no equivalent equipment is manufactured in the UK, or if the information is likely to be particularly helpful to users of the booklet.

An outline of the ancillary equipment required, namely the vacuum system, and pressure and temperature measuring apparatus will be given.

(i) Vacuum system

This normally consists of a glass line, made wherever possible of wide bore tubing to enable high pumping speeds to be achieved. The complexity of the system will depend upon the use to be made of it, but a reasonable degree of versatility is desirable. A typical line is shown in Figure 2.1.

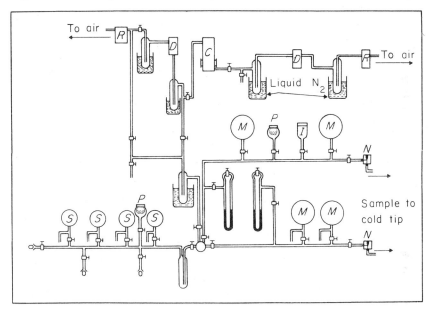

Figure 2.1 Schematic diagram of vacuum line for double injection matrix isolation experiments: S = storage bulb, M = mixing bulb, R = rotary pump, D = diffusion pump, I = ionization gauge, P = Pirani gauge, N = needle valve, C = cryostat cell

Obvious requirements include several bulbs, (S), each with a suitable freeze-out tip and capable of isolation from the line, in which to store gases, and at least one 2 litre or larger bulb (M) to act as a reservoir for the gaseous mixture to be studied. Preferably two sample lines each with at least one such reservoir, and each with its own manometer and needle-valve exit, should be provided, to allow for 'sandwich deposition' of two different mixtures, in reaction studies for example. It is necessary for the reservoir to be large to avoid a drastic fall-off in pressure of the mixture supply as the sample is used. This would cause a concomitant fall-off in deposition rate for a given needle-valve setting, especially when depositing fairly large amounts of dilute gas mixture. Matrix gas is best added discontinuously from an over-pressure so as to create turbulence and thus promote mixing within the bulb.

In order to follow the rate of deposition and the amount deposited, a mercury manometer is very useful, a typical experiment showing a pressure fall, in a sample reservoir plus adjacent line volume of 2·5 litres, from 700–400 mmHg for example, at a rate of 1·5 mmHg per minute.

Sufficient outlets ending in sockets (or cones) should be included, to enable additional apparatus such as sample cylinders or bulbs, drying tubes, cold traps, etc., to be plugged into the system. There should be enough taps to isolate the different parts of the system from one another and from the pumping units.

Use of a controllable leak, such as a micrometer needle valve (N), on the sample outlet line allows the rate of flow of sample mixture to the cryostat cell (C) to be varied accurately as required. Not only will too high a flow rate tend to give badly scattering matrices, and possibly aggregation of solute species in the matrix due to local temperature rise, but if excessive it may raise the pressure in the cryostat sufficiently to cause an appreciable heat leak, resulting in a temperature rise in the cell with a consequent warming of the matrix.

It is sometimes useful to have a section of the vacuum system capable of being heated above the ambient temperature, possibly by means of heating tapes wound around the glass line. This will enable solutes having a low vapour pressure at room temperature to be mixed with the matrix gas for examination. In general, however, such work is better carried out using a specially designed furnace (§2.2.8).

The whole apparatus, both sample handling line and cryostat, is usually evacuated by one or more mechanical fore pumps (R), normally of the rotary oil type. These are used for the initial pumping-down of the system and for any further rough pumping, i.e. from atmospheric pressure. This will arise each time a new unit, such as a sample gas cylinder, is connected to the line thereby letting that section up to atmospheric pressure. They also act as backing pumps to diffusion pumps (D), usually water-cooled mercury diffusion pumps, which are needed in order to reduce the pressure below 10^{-3} mmHg to the value approaching 10^{-5} mmHg usually necessary.

Liquid nitrogen cold traps are used, on each side of the mercury pumps, to prevent back-diffusion of vapours into the vacuum system and to hold pumped-off samples.

(ii) Pressure measurement

Some method of measuring the pressure in the line is necessary, and since the pressure range being used varies during an experiment, more than one means of measurement must be available.

A mercury manometer built into the line is used for higher pressures up to 1 atmosphere. It is used for rough pumping work, for measuring high pressures of gas present such as matrix material, and for monitoring the gas flow through the needle valve to the cryostat during deposition.

A more sensitive means is required for low pressure work, such as measuring the vacuum achieved in the cryostat to see whether it is sufficiently good to commence cooldown. There are several possible methods of measuring very low pressures. A Vacustat connected to the line is always useful and will register down to 10^{-3} mmHg. Ionization gauges (I) cover a considerable range (10^{-9}–10^{-3} mmHg), and have the advantage that the cold-cathode gauge is resistant to corrosive gases. Pirani gauges (P) have a range of 10^{-4}–0.3 mmHg but have to be calibrated against a MacLeod gauge (range 10^{-6}–1 mmHg). A Tesla coil is a convenient accessory for spot checks in a given section of line, to give an idea of the degree of vacuum present. An excellent review of low-pressure measuring gauges is given by Steckelmacher.[7]

(iii) Temperature measurement

For obtaining the temperature of a localized area in a small cell, enclosed within a vacuum, with sufficient accuracy, sensitivity and speed of response, thermocouples are the most suitable devices. Negligible heat leak occurs since one is using long thin wires, and rapid response is obtained to changes in local temperature. They are sufficiently accurate for most purposes, the reference junction normally being kept at 0 °C in an ice/water bath or in liquid nitrogen at 77.3 K and a potentiometer used to measure the e.m.f. produced. Continuous recording can of course be used if so desired.

Copper/constantan and chromel/constantan thermocouples are satisfactory for most purposes, especially at 20 K and above. However, the thermoelectric power, dE/dT, of thermocouples approaches zero at absolute zero, and so in many cases sensitivity is inadequate in the 4–20 K range. Better sensitivity is obtained with a gold (cobalt doped) alloy/copper thermocouple[8] in this temperature range. Mauer[1] mentions the possibility of variations in calibration sometimes being encountered with this thermocouple, and it may require spot calibration and correction. An alternative is the gold (iron doped)/silver thermocouple, which also has a high sensitivity in the lower temperature range. Representative values of e.m.f. for these thermocouples at 4 K, 20 K and 77 K, together with their sensitivity are given in Table 2.1. Detailed tables of temperature/e.m.f. values for a wide variety of thermocouples are given in the comprehensive series of National Bureau of Standards publications.[9]

An excellent handbook of low temperature measurements[9a] has been provided by Cryophysics SA of Geneva which contains material drawn from papers presented at the 5th Symposium on Temperature Measurement. The handbook briefly reviews all the methods available particularly for temperatures in the range below 30 K and gives a detailed table for a typical calibration of a Johnson–Matthey wire Au (0.03 atm per cent. Fe)/Chromel-P thermocouple. This review also emphasizes that such figures should be used

Table 2.1 Thermocouples: selected values of e.m.f. $E(\mu V)$ and sensitivity $dE/dT(\mu V/K)$ at various temperatures

Thermocouple	4 K	20 K	77 K	Ref. temp. (K)
Cu/Constantan	—	6160	5470	273
	—	8	15	
Cu/Constantan	2·62	60·40	723·5	0
	1·295	5·765	16·290	
Chromel/Constantan	—	9690	8640	273
	—	12	25	
Au(0·07 at. per cent Fe doped)/Ag	982	768	378	273
	13	12	4	
Au(0·07 at. per cent Fe doped)/Cu	1716	1500	923	273
	13	13	8	
Au(2·1 at. per cent Co doped)/Cu	8930	8750	7279	273
	5	15	33	
Au(2·1 at. per cent Co doped)/Cu	8·22	179·6	1777·2	0
	4·044	16·428	35·205	

as a guide only, since variations can occur from batch to batch of material. However, it appears that data from one batch can be scaled, after making a few spot checks on another batch, to give an adjustment and thus provide a working calibration table for a particular thermocouple. Another recent general review on cryogenic thermometry has been given by Rubin.[9b]

2.2.2 Single-Dewar cryostats

This is the simplest design of cryostat, and is suitable for use with either liquid air (90 K) or, preferably, liquid nitrogen (77 K) as refrigerant. Such a cell can of course be used equally well with cooling mixtures such as solid carbon dioxide/acetone.

The basic requirements of the cell are:
 (i) A sample window, on which the material to be studied is condensed.
 (ii) A tubular reservoir, or 'cold finger', connected to the sample window and in good thermal contact with it, to hold refrigerant.
(iii) An outer jacket, or vacuum shroud, in which the outer infrared-transparent windows are fixed; the space between the inner refrigerant tube and the outer jacket must be evacuable, to minimize heat transfer into the cell and to prevent atmospheric water from freezing on to the matrix support.
(iv) A means of introducing the gaseous sample into the enclosed, cold, cell and spraying it on to the cold window.
 (v) A temperature-measuring device attached to the sample area.

Such a cell is therefore basically a standard low-temperature IR cell, modified for the admittance of the gaseous sample after evacuation and cooling. These cells can be constructed of either glass or metal. Pyrex glass has low thermal expansion and high strength and is generally easier to work with than metal; thus a glass cell can be of simpler construction and cheaper. It is advisable to have the inner surfaces of the vacuum jacket silvered as in a Dewar flask to help minimize heat leakage.

A metal cryostat although more difficult to construct has the advantage that it is possible to incorporate other mechanical features more readily. Stainless steel is the preferred metal since it has lower thermal conductivity than brass or copper, and also has high resistance to chemical attack by the mixtures used in the experiments.

Various accessories and modifications can be incorporated depending on the design and the requirements of the experiment. Obviously the cryostat must be reasonably easy to assemble and dismantle for repairing, cleaning, and polishing sample windows. Standard vacuum techniques apply, and use is made of rubber, Teflon, or neoprene O-rings in suitably shaped grooves, and standard cone-and-socket ground-glass joints. In all-metal systems, if it is desirable to bake to remove adsorbed substances, one cannot use rubber, etc., and metal gaskets are preferred. These are also better for high-vacuum systems, for example 10^{-7} mmHg, to avoid contamination from any constituents of the O-rings having an appreciable vapour pressure.

Alkali halide windows can be screwed up against an O-ring or, if preferred, cemented into place, when they are for use as outer cell windows. In the case of sample window plates which get cold and must have good thermal contact in order to enable the sample to condense efficiently at the required temperature, probably the most satisfactory way is to hold the window in a metal (copper or brass) block by a ring and washers secured by screws, or by a screw-in annular face-plate and washers. The metal block is attached to the end of the refrigerant tube either through graded seals (for a glass tube) or by screwing into the threaded tip of the metal tube.

Electrical leads can be led into the cryostat through glass-to-metal seals in the cell wall of glass units; thermocouple wires can be passed through a small hole and sealed with wax, as it is better to avoid junctions in the lengths of wire. Mechanical linkages can be made in several ways. A shaft can be rotated slowly using standard cone-and-socket ground-glass joints. Sylphon bellows can be used to transmit linear motion. Shafts through O-ring seals can be made vacuum-tight, and used to both slide and rotate.

2.2.3 Double-Dewar cryostats

Although single-Dewar cryostats are adequate for working down to liquid nitrogen temperatures, it is necessary to use double-Dewar cryostats with liquid hydrogen or helium. Otherwise loss of the expensive refrigerant by evaporation would be intolerable, even if the required temperature could be maintained.

The design is basically simple, namely the liquid helium or liquid hydrogen (primary refrigerant) Dewar and its vacuum jacket is surrounded by a liquid nitrogen (secondary refrigerant) Dewar and its vacuum jacket. This reduces heat flow into the primary refrigerant and is very efficient. In spite of this it has been shown that, although storage Dewars can be designed with an evaporation rate for helium as low as 1–5 per cent. per day[10,11] depending upon material of construction (metal or glass), for Dewars used in spectroscopic studies evaporation rates are usually several times this, due to heat leakage through openings in the radiation shield and, to a lesser extent, to conduction, and can be 50 per cent. or more per day.

There are various examples of suitable cells, both glass and metal, in the literature.[12,13,14] This type of apparatus is again essentially a conventional low-temperature cell modified to include a means of spraying the gaseous sample mixture on to the cold window. An essential requirement of such a cell is that of having the liquid helium reservoir and its attached sample cell rotatable when assembled, in order that the angle of sample deposition may be varied.

Details of construction of such a cell are standard and again are covered by Mauer,[1] who also describes a method[15] for siphoning liquid helium when it is not required to recover the gas. If the gas is to be recovered, the method of Broom and Rose-Innes[16] can be used. The explosive nature of hydrogen/air mixtures requires extra care[17] when using it as the refrigerant although the same technique of handling is employed. A typical design[14] of cell is shown in Figure 2.2. The basic construction of this cryostat consists of a series of concentric stainless steel tubes welded together in such a way as to provide four compartments. An outer jacket evacuated to c. 10^{-2} mmHg surrounds a liquid nitrogen jacket which in turn is separated from the helium reservoir by a vacuum space at 10^{-5}–10^{-6} mmHg. The copper thermal insulation shield surrounds the inner vacuum space and reduces heat transfer to a minimum. Prior to assembly of the cryostat an O-ring annulus of indium wire is made to seal the thermal shield to the base of the liquid nitrogen chamber. Pre-cooling of the inner chamber is effected by pouring liquid nitrogen directly into the inner chamber. After this has cooled to c. 90 K the nitrogen is withdrawn prior to transferring the liquid helium. Provided the system is leak free, the helium cryostat assembly can be used continuously over periods in excess of 6 hours without any replenishment of liquid helium.

2.2.4 Continuous flow cryostats

Very recently a number of liquid helium cryostats have been developed which do not employ secondary refrigerant but depend upon the continuous circulation of liquid helium through a heat exchanger. Figure 2.3 shows the general design of a continuous flow cryostat. The upper part which carries the co-axial refrigerant input and exit tubes, vacuum vent valves, sample stage and radiation shield, is the refrigeration unit, and the lower part is a rotatable and demountable vacuum shroud fitted with appropriate optical windows. The refrigerant is circulated by a rotary pump and the boiled-off

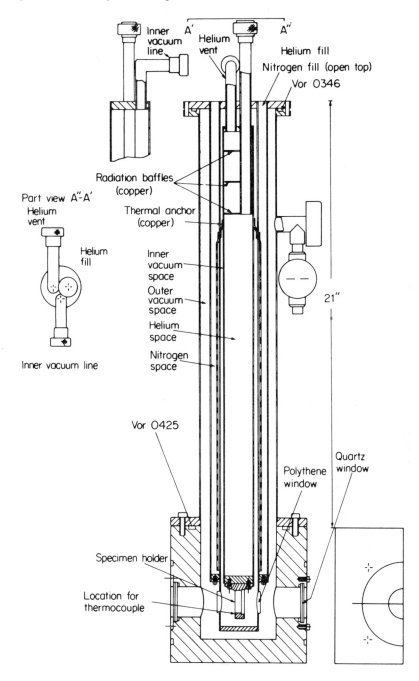

Figure 2.2 Typical metal double-Dewar cryostat [reproduced by permission from P. Rowland Davies, *Disc. Faraday Soc.* **48**, 181 (1969)]

Figure 2.3 Cutaway drawing of a continuous flow cryostat (model CF 104) reproduced by permission from Oxford Instruments Co. Ltd, Osney Mead, Oxford). The upper part carries the refrigerant input and exit tubes, vacuum vent valves, an additional entry port, and a 10-way electrical connector feed for the built-in resistance thermometer, heater, and any other required facility. The lower part consists of the heat exchanger and sample holder surrounded first by a radiation shield and then by the rotatable vacuum shroud (19 × 6·3 cm) with outer windows. Total height *c.* 45 cm and weight *c.* 1·5 kg

exhausting gas cools the radiation shield thus obviating the need for a secondary refrigerant. The flow is controlled by a micrometer needle valve which allows a temperature range of 3–300 K to be obtainable with a stability of 0·5 K between 10–70 K and 0·1 K between 3–10 K. These stabilities are based on manual settings and are more than adequate for matrix isolation studies. Finer control can be achieved by means of a temperature controller provided as an additional accessory.

Cooldown time is of course dependent upon experimental conditions, but is approximately 10 minutes, and will consume about 1 litre of liquid helium. Consumption thereafter again depends upon conditions but is typically *c.* 900 cm³ h⁻¹ at 4·2 K and *c.* 500 cm³ h⁻¹ at 17·5 K. Since the refrigerant is used for the duration of the experiment only, the use is far more economical than double-Dewar liquid cryostats. The refrigeration capacity (Table 2.2a)

Table 2.2(a) Continuous flow cryostats

Cyostat	System	Temp. range (K)	Refrigeration capacity[a]
Oxford Instruments			
CF104	liq. N_2	65–300	
CF104	liq. He	3–80	360 mW at 4·2 K and 900 cm³ h⁻¹

a. Depends upon flow rate.

depends upon the flow rate but even at the above rates it is perfectly adequate for infrared MI experiments. At high flow rates they should also provide sufficient cooling power for Raman experiments (see Chapter 9) and models are available (e.g. CF 105 of Oxford Instruments Co. Ltd) with an exit window at the base of the vacuum shroud.

All of these continuous flow models are extremely compact, lightweight and portable, and can be operated in any orientation; furthermore, they require no servicing.

2.2.5 Joule–Thomson open-cycle cryostats

Using the types of cell previously discussed, one is faced with the necessity of using liquid refrigerants. For liquid nitrogen and liquid air this is not a serious drawback, since they are usually readily available and not over expensive, and many laboratories have their own equipment for preparing them. The use of liquid hydrogen and liquid helium is however a different matter. Although these refrigerants are available, they are also very expensive and running a low temperature programme can be costly, even when provision is made for recovering the helium gas. A further minor complication is that the low temperature cell required is usually bulky and unwieldy to handle.

An experimental advance has been achieved in the design[18] and commercial manufacture[19,20] of miniature cryostats of novel design based on the miniature Joule–Thomson refrigeration systems developed by Geist and Lashmet[21,22] for the 15–200 K temperature range. A variety of these cryostats are now available commercially, covering the range 4–200 K, and as they operate principally on high-pressure gases they avoid the necessity of handling the low temperature liquids. Their compactness makes them ideal for fitting into restricted spectrometer sampling areas but their small size limits their wattage of refrigeration. Their cooling power is, however, quite adequate for all IR spectroscopic applications (see Chapter 9, however, for discussion of their use in laser-Raman spectroscopy). Some of these cryostats will be described briefly below and the basic details listed in Table 2.2b.

Table 2.2(b) Commercial Joule-Thomson cryostats

Cryostat	System	Temp. range (K)	Refrigeration capacity
Air Products			
AC-1	N_2 gas	68–300	7 W at 77 K
AC-2	N_2 gas/H_2 gas	16–200	4 W at 23 K
AC-2L	N_2 liq./H_2 gas	16–77	6 W at 22 K
AC-3L	N_2 liq./H_2 gas/He gas	3·6–77	4 W at 20 K
			500 mW at 4·4 K
Hymatic			
SES-227	N_2 gas/H_2 gas	20–77	1·5 W at 21 K
SES-237	N_2 liq./H_2 gas	20–77	4.W at 20 K

The model in use in our own laboratories will be dealt with in greater detail.

Air Products 'Cryo-Tip'

Air-Products and Chemicals Inc. of Allentown, Pennsylvania,[19] have been responsible for the commercial development of these miniature cryostats and they are marketed under the trade name of Cryo-Tip. They can be obtained covering a range of temperatures and supplying various wattages of cooling (Table 2.2b).

Model AC-2. This cryostat requires no liquid refrigerants. Cooling is supplied by a built-in miniature two-fluid open cycle Joule–Thomson liquefier operating from cylinders of high-pressure nitrogen and hydrogen. This model delivers from about 1–2 W at the cold tip depending on the gas pressures used, but it is possible to get special nozzling giving *c.* 4 W cooling if required. This hydrogen/nitrogen cryostat is capable of reaching 15–16 K, at which temperature it normally supplies up to 2 W of refrigeration. It can be operated so as to hold the sample window from this temperature up to 200 K, with a temperature control of ± 0.3 K. It is compact, and easy to handle (it can be held in one hand) and, being metal, is not fragile; it can also be operated in any orientation.

The basic components are analogous to the liquid refrigerant cells described earlier, namely, a sample support window connected to a cooling tube, a radiation shield around the sample area, a vacuum jacket around the whole, and a means of introducing the gaseous sample mixture. The cold tip is cooled by Joule–Thomson expansion hence ancillary equipment to handle and control the high-pressure gases is also needed. A diagram of the cell is shown in Figure 2.4, and details of the nitrogen-hydrogen cryostat are as follows.

The heat exchanger providing the refrigeration has two separate but intertwined circuits of coiled-finned tubing, with a series of constrictions, one for the nitrogen gas and one for the hydrogen gas. In the Joule–Thomson cycle, cooling occurs when a non-ideal gas expands without doing work. Before cooling will occur, however, the gas must be pre-cooled to below its inversion temperature. The inversion temperature of nitrogen is above room temperature, and so liquefaction of the high-pressure nitrogen occurs on expansion at ambient temperatures. Since hydrogen has an inversion temperature of 160 K, pre-cooling by the liquid nitrogen allows hydrogen liquefaction to take place. Only the copper tip of the cooling tube is at liquid hydrogen temperature. A radiation shield is attached to the cooling tube at the liquid nitrogen level so that it achieves about 80 K, thus greatly reducing heat leak at the tip of the heat exchanger.

It is possible to vary the exact temperature at the tip, since this is a function of the gas pressure above the pool of liquefied gas. This gas pressure is

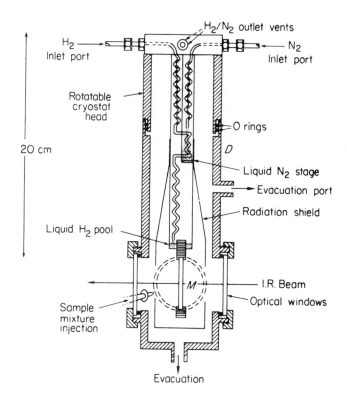

Figure 2.4 Liquid hydrogen Cryo-Tip (Air Products and Chemicals Inc.) and matrix-isolation cell (*M*)

referred to as the hydrogen back pressure. The back pressure is increased above, or lowered below, atmospheric pressure to achieve higher or lower temperatures respectively than that normally existing at the cold tip. It should be noted that to lower back pressure, it is necessary to pump on the gas above the liquefied hydrogen pool. Thus it is necessary to vent the pump exhaust out of the room to avoid risk of explosion. To achieve a sample temperature of 77 K, it is only necessary to maintain the nitrogen gas pressure and reduce the hydrogen gas pressure below the level required to maintain liquefaction. This provides hydrogen gas at liquid nitrogen temperature at the sample area. If it is desired to run the cryostat as a liquid ntrogen cell (i.e. not going below 66 K, with pumping), this can be done by reconnecting the high-pressure nitrogen gas supply to the hydrogen inlet on the heat exchanger. This supplies liquid nitrogen to the sample area. Cooldown to 20 K from room temperature takes about 45 minutes.

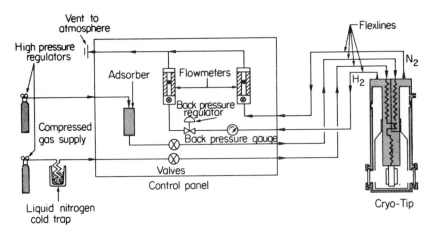

Figure 2.5 Flow diagram of two stage H_2/N_2 Cryo-Tip and ancillary equipment (reproduced and redrawn by permission from Air Products and Chemicals Inc.)

The ancillary equipment (Figure 2.5) for handling and controlling the high-pressure gases consists of a control panel, gas clean-up units to protect the Joule-Thomson liquefier, and flex-lines for the gases. The cryostat is connected to the cylinder gas supplies through valves and piping mounted on to the control panel which carries pressure regulators for the hydrogen and nitrogen and flowmeters for monitoring the flow rate during operation. The high pressure flex-lines into the cell, carrying 1500 psi of nitrogen and 1000 psi of hydrogen during cooldown, are bronze metal bellows type with a woven cable reinforcement. For safety purposes it is advisable to strap these down as far as possible within the flexibility required. The outgoing (low pressure) gases are combined at the control panel to dilute hydrogen, and are vented out of the laboratory. Venting through to roof level is preferable but through a window quite adequate due to the rapid diffusion rate of hydrogen.

The gas clean-up units contain an adsorber such as molecular sieve at ambient temperature, to remove water and carbon dioxide from the nitrogen, and silica gel in a liquid nitrogen cooled trap to remove nitrogen from the hydrogen, to avoid blocking the cryostat Joule–Thomson coils. These units require periodical (c. 1000 hours operation for high purity hydrogen) regeneration by heating for 48 hours in a flow of argon or nitrogen. The nitrogen trap is heated at 300 °C with c. 200 psi nitrogen flow; the nitrogen trap is heated at 135 °C with c. 3 psi argon flow. The gas flow lines are purged at low pressure before using high pressure, to remove any air from the lines and from the Joule–Thomson coils.

The cell shroud has four optical ports of which two are required for the infrared beam. The other two are normally blanked off by metal plates, unless wanted, for example, for photolysis experiments. There are two sample injection inlets at 45° to the infrared beam ports. Thus one can either spray from one port or both ports at 45° angle to the sample support plate, or by rotating the Cryo-Tip and attached sample plate through 45°, spray at 90° to the plate. It is usually preferable to spray at 90° to the cold window, since the area covered by a cone from a given point at a given distance is appreciably larger than at a 45° surface, and material outside the beam area is wasted. A considerable amount of the sprayed sample will bounce off the cold window and either condense on to other cold surfaces or be pumped away. It has been estimated[23] that c. 60 per cent. of the molecules in a stream of gas hitting a surface will condense on to the surface, provided that the surface is held at a temperature well below the melting point of the gas. For monitoring the temperature during cooldown and deposition, a chromel-constantan thermocouple is embedded in the matrix support.

The evacuation system ideally consists of a double pumping line, each being a mercury diffusion pump backed by a mechnical rotary pump. One system pumps on the Cryo-Tip from the outlet provided at the top, whilst the other pumps from the bottom of the shroud. A vacuum better than 10^{-4} mmHg is needed for good performance.

The refrigeration systems can be bought with a variety of vacuum shrouds each designed for a specific type of application. The infrared shroud does not have injection nozzles supplied to fit the sample inlet ports, but a satisfactory system can be made by soldering a hypodermic needle through the port, and connecting the outer coupling end to fine-bore copper tubing leading to the sample system needle-valve. In general, shrouds will be designed and constructed by the user for the specific matrix experiment envisaged.

Operation of the AC-2 Cryo-Tip. The whole system, that is vacuum line and Cryo-Tip, is pumped down to approximately 10^{-4} mmHg or better from both top and bottom vacuum connections simultaneously and pumping is continued throughout operation. Cylinders of nitrogen and hydrogen are attached to the gas-handling manifold. The high-pressure nitrogen and hydrogen inlet valves on the control panel are opened, so that in subsequent steps high pressure build-up will not occur in the line. The manifold valves are opened for the same reason, and the cylinders opened. After closing the valves which isolate the cylinders from the manifold, thus trapping gas at full cylinder pressure in the section of line, all joints are tested for leaks with a bubbler. If no leaks are found the cylinders can be closed until required. After the gas in the manifold and line has bled off, all valves (including high-pressure inlet valves on the control panel) are closed.

The sequence of operation is then as follows. Open the high-pressure inlet valves on the control panel. Hydrogen and nitrogen gas cylinder pressures are set to approximately 250 psi by means of the appropriate manifold valves, to purge out the whole gas handling system and Cryo-Tip. The indicators on the control panel gas-flow meters can be seen to be slightly raised and steady. This low-pressure purge is carried out for about 20 minutes, after which cooldown is commenced. Raise the nitrogen pressure in the system to 1500 psi. The hydrogen pre-purifier unit Dewar is about three-quarters filled with liquid nitrogen, and five minutes after this, the hydrogen pressure is increased to 1000 psi. It normally requires 45 minutes from this point to achieve 20 K, when the nitrogen flow-meter should read 7 and the hydrogen flow-meter 3·5 (both on a scale of 10). The temperature registered by the thermocouple is checked, using the potentiometer. Upon reaching 20 K, the gas pressures can be reduced to about 1000 psi nitrogen and 700 psi hydrogen. (The exact values will depend on the rate of heat leakage, etc.).

For deposition, the top part of the Cryo-Tip and the connected sample deposition plate are rotated through 45° to the direction of the beam. This puts the face of the sample plate at right-angles to the injection needle. The needle-valve on the sample line is then opened to allow the gas mixture to be deposited at the desired rate. If the deposition rate is high, or an especially thick layer of sample is to be laid down, it may be advisable to increase the refrigerant gas pressure to, say, 1150 psi nitrogen, 800 psi hydrogen, to help temperature control. It can be reduced once deposition is finished. The deposited sample is then rotated back into the infrared beam in order to run the spectrum.

It is sometimes desired to run the spectrum at higher temperatures or to raise the sample temperature for diffusion. A temperature of *c*. 30 K can be achieved by use of the appropriate valve on the control panel, raising the hydrogen back-pressure to approximately 200 psi. By cutting the hydrogen supply from the cylinder the temperature can be further raised, up to 77 K. A careful watch on the temperature should be kept, using the thermocouple, as it is easy to get into an irreversible cycle, due to evaporation of the matrix.

After running the spectrum, the temperature of the cell must be raised to evaporate the sample. Close both gas cylinders, and leave all valves open. When the gas pressures have fallen to 1 atmosphere in the system, as shown by the flow-meter balls and zero reading on the manifold gauges, close all valves including the high-pressure hydrogen and nitrogen inlet valves on the control panel. The temperature in the cell will slowly rise, the sample normally evaporating off the salt plate in about 30 minutes and being pumped away.

Other Air-Products models available are as follows:

(i) *AC-1*. This is a simplified, single gas, system, using commercial purity nitrogen gas to achieve a temperature range of 68–300 K with a stability of ±0·1 K at 68–110 K, ±0·5 K at 110–200 K. The temperature obtained is varied by varying the liquid nitrogen vapour pressure, gas flow, or by using an electric heater. The cooldown time is 30 minutes depending on the sample to be cooled. Refrigeration is conserved by countercurrent heat exchange between the high and low pressure gas.

(ii) *AC-2L*. A modified version of the AC-2, using hydrogen gas and liquid nitrogen, for precooling. Its temperature range is 15–77 K, and it can deliver up to 6 W of cooling at 22 K.

(iii) *AC-3L*. This is a three-stage Cryo-Tip, for operation at liquid helium temperatures. Helium and hydrogen gases are used, with liquid nitrogen as pre-coolant. A schematic diagram of the refrigeration unit is shown in Figure 2.6. The unit operates in the range 3·6 to 70 K, giving from 500 mW at 4·4 K to 4 W at 20 K, temperature stability varying from ±0·1 K to ±0·5 K with temperature. Cooldown time is 40–70 minutes, depending upon specimen mass and shielding. The normal model must

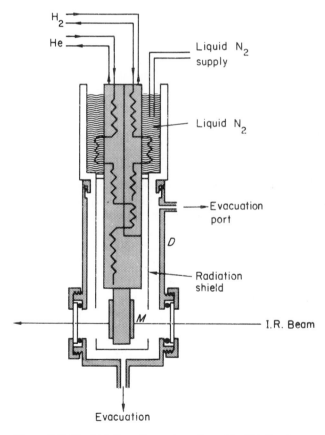

Figure 2.6 Liquid helium three-stage Cryo-Tip (Air Products and Chemical Inc.)

be operated within 30° of the vertical, although models for use in horizontal positions are available. The unit is rather heavier than the others and also more cumbersome, weighing 10 kg as against the 0·5–1 kg of the AC-1 and AC-2 models.

These refrigerators are designed to use relatively inexpensive commercial grade compressed gas which is consumed at a rate of 3 to 6 hour per standard cylinder. This time is usually sufficient for the completion of a typical matrix-isolation experiment but by use of a bank of cylinders connected to a common manifold, experiments of longer duration can be performed. Rochkind[24] has described a gas control system involving two batteries of four cylinders and has run a matrix experiment continuously for 30 hours. Whatever the planned length of experiment it is advisable to use high purity gases in order to reduce the risk of blockage; the cost is little more than commercial grade and is partly offset by the increased life of the adsorber units.

Hymatic Engineering Company Limited, 'Minicoolers'

This Company[20] manufactures two liquid hydrogen temperature cryostats, SES 227 and 237, which incorporate Minicoolers as refrigeration units. The Hymatic Minicoolers are miniature gas liquefiers with two coaxial stages combining the Joule–Thomson cooling effect with a counterflow heat exchanger. They will deliver up to 10 W of useful cooling at temperatures down to 77 K (Table 2.2b). A variety of other Minicoolers, of various physical sizes, wattages and temperature ranges, are available.

SES-227. This is a two-gas system, incorporating a two-stage Minicooler type MAC.215, and the necessary gas control and cleaning equipment and pipework. High purity commercial hydrogen gas and nitrogen, air or argon are required, together with a small amount of liquid nitrogen used in the hydrogen and gas cleaning units. The MAC-215 produces 1 W of cooling down to 21 K. The nitrogen air or argon pre-cooler cools the hydrogen gas below its inversion temperature. Using nitrogen/hydrogen combinations useful cooling power up to 1·5 W can be obtained.

SES-237. The unit uses a supply of liquid nitrogen as pre-coolant, and a single-stage Minicooler type MC-8 operating on hydrogen gas. It is claimed to produce liquid hydrogen within 15 minutes from start of operations. Temperatures within the range 77–20 K can be maintained accurately by control of the gas pressure. Cooling power provided is up to 4 W at 20 K. This unit can also be used from 77 K up to room temperature, using nitrogen gas in the MC-8 unit and no precooling. It must be used vertically.

The systems mentioned above are not a comprehensive listing, but cover perhaps the main commercial Joule–Thomson cryostats at present available.

2.2.6 Closed-cycle cryostats

An alternative system of refrigeration is the closed system, staged expansion, Stirling refrigeration cycle. Cooling takes place in an enclosed space, and involves the out-of-phase operation of two pistons during which the refrigerant fluid is passed to and fro between five interconnecting sections (Figure 2.7): the compression space, the cooler, the regenerator, the freezer and the expansion space. A very good account of this system is given by Kohler.[25]

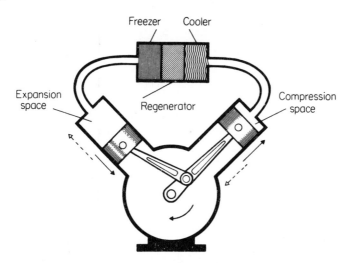

Figure 2.7 Schematic diagram of Stirling Refrigeration Cycle

The cycle can best be visualized as taking place in four successive stages. Although these are considered as separate phases for clarity, they tend to merge together in practice. The cycle begins (1) with compression (by an electric motor) of a gas at environmental temperature; next (2) the gas is transferred to a cold space, and passes in turn through the cooler, which dissipates the heat of compression, the regenerator, which cools the gas nearly to the refrigerating temperature, temporarily storing its heat, and the freezer. Then (3) the gas is expanded into an expansion space, and becomes still cooler as a result of the work it does against the expansion piston. In phase (4) the piston moves the gas back to the compression space. On the way through the freezer the cold gas absorbs heat from whatever the system is designed to cool, and then, passing through the regenerator, the gas picks up the stored heat and is returned to the environmental temperature for the commencement of a new cycle.

Closed-cycle cryostats are extremely efficient and are much lower in running costs than open-cycle systems since there is no continued expenditure for gas; generally, the only utility required to operate these units is electrical power. They provide more refrigeration power and also allow operation, in any orientation, for indefinite periods of time. The capital cost of a closed-cycle machine is however greater than an open-cycle system.

Commercial cryostats employing the Stirling refrigeration cycle are marketed by Philips[26] in the USA under the name Cryogem refrigerators. They are available in Europe through NV Philips, Eindhoven, Holland and in the UK through Pye Unicam Limited, Cambridge. The Norelco double stage Cryogem uses air cooling and helium gas and covers the temperature range 20–300 K with a cooling power of 1 W at 20 K.

The first closed-cycle cryostats were designed to meet the needs for cryocooled parametric amplifiers and in general provided greater cooling power than required for cryospectroscopy. The increasing utilization of matrix-isolation spectroscopy has been accompanied by rapid developments in closed-cycle systems. Several different cycles, operating on helium gas, are being utilized in commercial cryostats which have been specifically designed for cryospectroscopy and at the time of going to press the cost of these units are about half that of the early closed-cycle cryocoolers and about double that of the open-cycle Joule–Thomson systems. Choice of refrigerating cycle is based on an operating compromise between a reasonable cooldown time and a low operating temperature, in addition to factors such as freedom from vibration and a comfortable sound level.

Cryogenic Technology Incorporated[27] produce a range of models (Table 2.3) under the name Cryodyne which use the Gifford–McMahon cycle. For example, the Cryodyne 20 operates over 18–300 K with a cooldown time of c. 25 minutes and 1 W cooling power at 19 K. In an assessment trial of this unit we have obtained satisfactory matrix spectra at temperatures approaching 15 K. Air Products Limited have recently introduced a Displex series based on a modified Solvay cycle. Their Model CS-202 is a two-stage displacer/expander with an operating range of 12–300 K, a cool-time of c. 45 minutes to 17 K and 1·5 W cooling power at 20 K. Matrix experiments currently reported in press claim to be operating this unit in the 12–14 K range. British Oxygen Company Limited also have a two-stage unit, operating on the Taconis cycle: the CRS 1R-16 mk 2 provides cooling down to 10 K with a cool-down time of 40 minutes and 1·5 W cooling power at 16 K. The cold-head is however substantially larger and more cumbersome than the other models described above.

Each of the above systems makes use of a separate compressor connected by flexible precharged gas lines to the cold head thus ensuring cold-source flexibility and freedom from vibrations. Daily running costs are negligible and operational times of about 3000 hours are claimed without operator attention. Maintenance after this length of usage is simply a matter of changing items such as absorber units, valve discs and seals.

Table 2.3 Closed-cycle cryostats

Cryostat	Temp. range (K)	Refrigeration capacity
Cryogem		
Single stage model	30–300 K	15 W at 77 K
		1.5 W at 30 K
Double stage model	20–300 K	3 W at 30 K
		1 W at 20 K
Cryodyne		
Model 70	35–300 K	5 W at 50 K
		1 W at 34 K
Model 20	18–300 K	1·5 W at 22 K
		1 W at 19K
		0 W at 15 K
Air Products Displex		
Model CS-102	30–300 K	5 W at 38 K
(single stage)		0 W at 30 K
Model CS-202	12–300 K	2 W at 25 K
(double stage)		1·5 W at 20 K
		0 W at 12 K
BOC-CRS		
1R-16 mk 2	10–300 K	1·5 W at 16 K
		2·2 W at 20K

Air Products Limited have also recently marketed a closed-cycle plug-in refrigeration system which utilizes their open-cycle Cryo-Tips as basic units; a single such compressor can operate multiple Cryo-Tip refrigeration units in separate laboratories.

2.2.7 Rotating cryostat

A novel development in apparatus for low temperature studies is the rotating cryostat. Developed by Thomas,[28] and modified by Bennett and Thomas,[29] the apparatus was designed as a means of mixing two solids together intimately at the molecular level in a controlled manner for free-radical work.

The apparatus allows for alternate layers of the two (or three) reactants to be sprayed on top of one another on to the surface of a rotating drum. The deposit thus obtained consists of the required thickness of one reactant sandwiched between required thicknesses of the other reactant. This is repeated to give a multilayer deposit of *c*. 0·1–0·5 mm thick. Examination of the material is carried out, either *in situ* or following its removal.

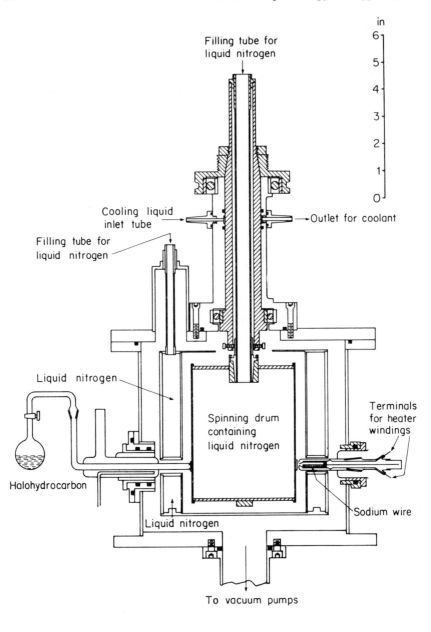

(i)

Figure 2.8 Rotating cryostat [reproduced by permission from J. E. Bennett and
A. Thomas, *Proc. Roy. Soc.* **A280**, 123 (1964) and Shell Co. Ltd]

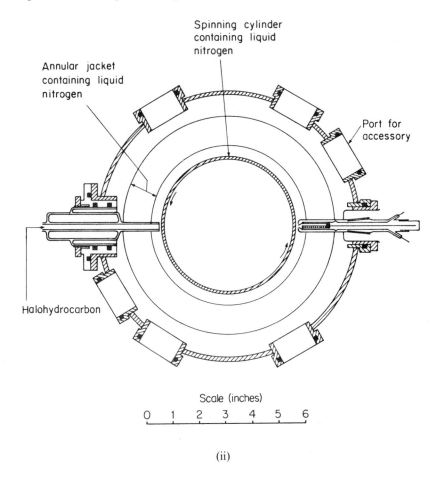

Scale (inches)

0 1 2 3 4 5 6

(ii)

The cryostat is shown in Figure 2.8. The central stainless-steel drum is filled with liquid nitrogen and spins in a high vacuum (up to 10^{-5} mmHg) at a controlled speed of 2300 rpm. The gaseous sample beam effuses from a slit near to the moving surface, and so collides with the drum. The vapour freezes immediately, forming a solid layer several monolayers thick on the outside of the drum. This layer is carried around with the drum, and gas from the second sample beam is deposited on top of it. The flow rates of the two sample beams are adjusted individually to vary the thickness of the layers from a small fraction of a monolayer to several monolayers. In this way the desired ratio of the two samples can be achieved. For their free radical studies, Bennett and Thomas,[29] and Mile[30] deposited one atom of sodium for every 25 or so molecules of halocarbon. They thus had very few sodium atoms next to one another and radicals formed by the reaction are trapped on the surface of the solid, and are covered by a fresh layer of halo-carbon on the next revolution of the drum.

It is possible to study the deposited samples *in situ* by reflection spectroscopy, and this has been done in the ultraviolet/visible region by Bennett *et al.*[31] Alternatively, the sample can be removed and transferred to an ESR spectrometer. It is of course necessary to keep the sample in a high vacuum and at liquid nitrogen temperature throughout handling. This is achieved by scraping the deposit off the drum with the blade of a stationary scraping tool, the drum being slowly turned by hand. A ramrod is then used to push the removed deposit down the hollow tube of the tool into one of several nylon capsules. This capsule is then allowed to fall down a chute which leads out of the base of the cryostat into a glass tube surrounded by liquid nitrogen in a Dewar vessel. The Dewar vessel can then be lowered sufficiently to enable the glass tube to be sealed with a small flame. This sealed tube containing the sample is then quickly transferred to a small, specially designed, Dewar vessel containing liquid nitrogen, which is placed in the cavity of an ESR spectrometer. No IR spectrum has been reported using the rotating cryostat which, although a difficult operation, is quite a feasible one.

The apparatus can be used with liquid nitrogen in the sample cylinder and liquid nitrogen in the annular cooling jacket.[29] It can also be used with liquid helium in the sample cylinder and liquid nitrogen in the jacket,[28] depending upon the temperature of operation required.

Certain problems of design had to be overcome in producing this cryostat. For example, the spinning parts of the cryostat have to spin safely at high speeds in a high vacuum and heat generated by the spinning shaft has to be removed by circulating silicone fluid. Good thermal insulation is also required. Ports are provided around the circumference for observation, and for admission of reactants, vacuum sealing with accessories being made by an O-ring in a machined groove in the port. Temperature measurements are made by means of two small carbon resistors cemented into very small copper blocks which are soldered to the central reservoir. Resistance measurements are only made when the drum is stationary.

2.2.8 Special techniques

There are various specialist applications of the matrix isolation technique, which are discussed in later chapters while the practical techniques involved are outlined here.

(i) *Knudsen effusion*

There are considerable experimental difficulties in studying the spectra of vapour in equilibrium with materials at high temperatures. These arise mainly from the chemical reactivity of the high-temperature species with the containing cell and optical windows and the difficulty in obtaining a high enough concentration of the desired species in the optical beam. Furthermore, high temperature spectra are usually very complex due to the presence of features arising from transitions involving populated high rotational and vibrational levels. These difficulties are largely overcome by the combination of Knudsen effusion and the matrix isolation technique first successfully achieved by Linevsky.[32] The method is now so widespread as to warrant a full chapter (Chapter 6) in this monograph but it is more convenient to describe the experimental technique in this section.

Knudsen effusion is a convenient method of obtaining the vapour species from high-temperature materials; it is easier than the furnace method, and one can control reaction with the container more readily by using a variety of container materials and liners. Since the Knudsen cell is small, it is easily mounted into a vacuum system, and can be heated by induction or electron bombardment. The cell is attached at the bottom of the cryostat, and the inlet for the matrix gas is next to the inlet from the cell, so that both beams will intermingle. A molecular beam of the vapour species is produced, and condensed together with a stream of matrix gas on to the cold sample window. The spectrum of the high-temperature species can then be run in the usual matrix isolation fashion. A typical two chamber Knudsen cell is shown in Figure 2.9. In *chamber 1* the solid is evaporated to produce an equilibrium two phase system at temperature T_1. The vapour leaks through tube T into *chamber 2* held at temperature T_2. By choosing appropriate temperatures, T_1 and T_2, a wide range of conditions for chemical equilibrium may be achieved. Superheating in *chamber 2* allows investigation of gas equilibrium by examining the spectra of the vapour effusing into the matrix. To achieve uniform temperature conditions an equilibrium *chamber 2* is provided with a demountable system of shields S.

(ii) Pseudo matrix isolation (PMI)

A novel application of the matrix isolation technique to quantitative analysis of multicomponent mixtures has recently been developed by Rochkind.[33] Although basically the same as the standard method, PMI involves condensation of the sample at 20 K in a nitrogen matrix by controlled-pulse deposition, instead of by continuous flow. Thus only small quantities of the gaseous solute-matrix mixture impinge on to the cold plate at a time, each pulse of mixture condensing before the next one arrives at the surface. This technique appears to allow more rapid deposition but without any extensive interference by solute-solute interactions, and pulse annealing appears to dispel multiple site effects. Details of the modifications are given by Rochkind.[24,33] Beer's-Law-plots of the bands obtained with mixed solutes yield good analytical curves and satisfactory quantitative analyses (§3.7).

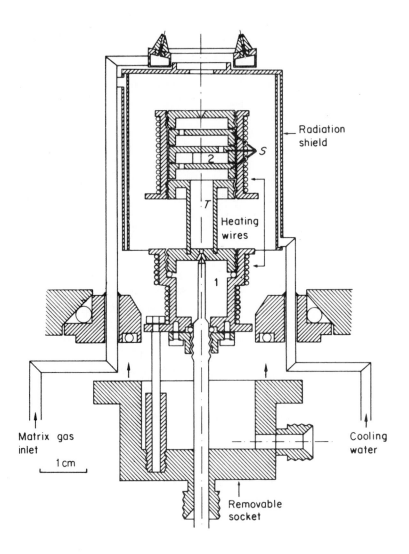

Figure 2.9 Knudsen furnace and matrix isolation cell for examining high temperature species [reproduced by permission from R. A. Frey, R. D. Werder and H. H. Gunthard, *J. Mol. Spectroscopy*, **35**, 260 (1970)]

2.3 Matrix materials

2.3.1 Requirements

The number of substances which are of potential use as matrix materials is very large. Suitability of particular materials will depend to some extent upon the type of experiment envisaged and the capabilities of the sample-handling line and cryostat available. There are however several properties that a substance must have if it is to be a successful matrix material.

(i) *Purity.* As in other fields of solid-state chemistry it is essential that the matrix materials should be of a high degree of purity. Several non-reproducible spectra have been attributed to the presence of impurities in the lattice, e.g. atmospheric nitrogen from even[34] minute leaks in the appratus.

(ii) *Vibrational spectrum.* As with liquid solvents, absorption bands in the region of the solute absorption will mask the bands of interest. With the high M/A ratios necessary for solute isolation fairly thick films have to be deposited, thus fundamental vibration modes of the matrix species cause a blackout over a wide spectral region and modes normally referred to as weak may interfere severely.

The IR spectrum of a fairly thick solid layer will often show appreciable scattering of light. These reflection losses limit the total thickness of matrix which may be deposited and hence sets a limit to the dilution ratio achievable. From this point of view, a glassy matrix is to be preferred to a crystalline one. A matrix which appears to be completely opaque to visible radiation will often allow sufficient IR radiation to pass through to give a spectrum. It is necessary to use an attenuating comb in the reference beam and to adjust the spectrometer's operating parameters (slit, gain, etc.) to enable a reasonable spectrum to be obtained from the small fraction of transmitted energy. Since scatter is also a function of wavelength, it will sometimes be necessary to run the spectrum in sections, with different settings of the attenuating comb, slit, etc., for each section.

The temperature at which a matrix is deposited often has a pronounced effect upon the scattering produced by the solid film. An example is xenon, which forms a reasonably transparent matrix when deposited at 66 K, whereas if the same amount of xenon is condensed below 50 K, a highly scattering film results.[35] The deposition temperature can be varied between the lower temperature limit of the particular cryostat being used and the temperature at which diffusion occurs. An improvement can sometimes be achieved by gently annealing the matrix subsequent to deposition. A very slow rate of deposition may reduce scattering due to the production of a smoother surface; high rates can cause 'splash patterns' to be frozen in.

The problem of background scattering is different in Raman MI spectroscopy because the Raman technique depends on specular reflexion from the matrix rather than transmittance through it. This is discussed in Chapter 9.

(*iii*) *Inertness.* Of necessity the matrix must not react with the solute. The limitations of this criterion will depend upon the nature of the solute and the experiment. The possibility of planning a matrix to react with a given solute constitutes a special type of experiment (see Chapter 5). The exact limitations of this criterion will naturally depend upon the nature of the solute. Thus the noble gases can be regarded as inert for all solutes, whereas hydrocarbons might be suitable if the purpose of the experiment was to be a study of hydrogen-bonding. If free radical studies are being undertaken, and the matrix being photolysed, hydrocarbons or even nitrogen might be classified as reactive. However *all* matrices perturb the energy levels of the trapped species and thus to some extent modify vibrational frequencies (see Chapter 4).

(*iv*) *Rigidity.* Since the aim of the MI technique is to study isolated species, rigidity of the matrix is of prime importance. If this requirement is not fulfilled, diffusion of the solute molecules within the matrix will occur, leading to recombinations in the case of free radical studies, and to aggregation and the formation of multimeric species in studies of stable molecules.

The rigidity of a matrix is highly temperature dependent, the effect of temperature on diffusion being very marked in most cases; the first criterion for assessing the suitability of a matrix material is thus the melting point of the substance.

The problem of diffusion in solid matrices has been discussed by Pimentel.[36] In the absence of actual diffusion experiments on a system he gives the rough guide that the melting point, T_m, in K, of the matrix material must be at least double the temperature at which the matrix is to be used. This limiting value is well known by investigators of solid-state studies of non-stoichiometric compounds as Tammann's rule, which relates the temperature, T_d, at which diffusion first becomes appreciable, to the absolute m.p., for salts, oxides, etc., $T_d \simeq 0.57 T_m$, for covalent compounds $T_d \simeq 0.90 T_m$. This is only a rough guide since clearly diffusion does not suddenly occur at a given temperature but gradually increases as the number of lattice defects increases and as progressively more solute molecules have sufficient thermal energy. Recent studies, however, indicate that in some systems appreciable diffusion can occur below the Tammann temperature. For example, in free radical experiments, Mile[30] found diffusion occurring at temperatures well below this limit and radicals beginning to disappear when the temperature rose to between one-tenth and one-third of T_m. Literature values of the estimated diffusion temperature (T_d) for the common matrix materials are given in Table 2.4. These considerations limit the range of matrices that can be used with liquid nitrogen cryostats, if isolation of the solute species is of importance to the experiment being undertaken. The temperature rule should therefore be that it is advisable always to operate at the lowest temperature attainable.

Table 2.4 Thermal properties of matrix materials

	T_d (K)	m.p. (K)	b.p. (K)	Vapour Pressure = 10^{-3} mm at (K)[a]	L_f (Jmol^{-1})	$-U_0$ (Jmol^{-1})	$\lambda(20\,K)$ (Wm^{-1} K^{-1})
Ne	10	24·6	27·1	11	335	1874	0·4
Ar	35	83·3	87·3	39	1190	7724	1·3
Kr	50	115·8	119·8	54	1640	11155	1·2
Xe	65	161·4	165·0	74	2295	16075	2
CH_4	45	90·7	111·7	48	971		0·1
SF_6		222·7	209·4(sub)				
N_2	30	63·2	77·4	34	721	6904	0·4
O_2	26	54·4	90·2	40	444		
Cl_2		172·2	239·1	118	6820	32400	
CO	35	68·1	81·6	38	836	7950	
NO		109·6	121·4	66	2310		
CO_2	63	216·6	194·6(sub)	106	8339	26987	
N_2O		182·4	184·7	99	6535	24267	
SO_2		197·6	263·1	138			
H_2S		187·7	212·5	104			
C_2H_6		89·9	184·5(sub)				
C_2H_4		104·0	169·5				
C_2H_2		191·4	189·6(sub)				
CF_4 (14)[b]		89·2	145·2				
CHF_3 (23)		118·0	191·2				
$CClF_3$ (13)		92·2	191·8				
$CBrF_3$ (13B1)		105·2	215·4				
$CHClF_2$ (22)		113·2	232·4				
CCl_2F_2 (12)		115·2	243·4				
$CHCl_2F$ (21)		138·2	282·1		2736		
C_2F_6 (116)		172·6	195				
$CClF_2CF_3$ (115)		167·2	234·5				
$CClF_2CClF_2$ (114)		179·2	2770·0				
C_4F_8 (C318)		231·8	267·4				

a. R. E. Hornig and H. O. Hook, *R.C.A. Review*, **21**, 360, (1960).
b. Number given to Arcton, Freon, etc.

Of relevance to the question of rigidity is the ability of the matrix to accommodate the solute molecules in the lattice, either in lattice sites or in holes in the matrix, without undue distortion of the lattice being caused. This will be discussed later under crystal data.

(v) *Volatility*. The matrix material and the solute are normally mixed in the gas phase, and then sprayed on to the cold plate in the sample cell. This means therefore that the matrix material must have a sufficiently high vapour pressure at room temperature to allow it to be handled in the vacuum line. This line may sometimes have provision for heating above the ambient temperature, but the choice is obviously limited to materials having a boiling point within a moderately limited range. (See, however, Chapter 6.)

Having deposited the condensed layer in the cryostat, it has to remain there under a vacuum of $c.\ 10^{-4}$ mmHg. This requires a low vapour pressure at the temperature of the refrigerant being used. Conversely, it gives an upper limit to the temperature at which a given matrix material can be used. Too high a vapour pressure not only causes some of the layer to be lost, but in so doing impairs the vacuum, thereby causing an increased heat leak and possibly raising the temperature of the deposited layer. The upper limit to the temperature range in which a matrix can be used is thus given by the temperature at which its vapour pressure is about 10^{-3} mmHg. These values (Table 2.4) are seen to be slightly higher than T_d values. Thus lattice rigidity rather than vapour pressure determines the upper temperature limit.

It can be seen that argon, krypton and xenon can be used with liquid hydrogen cryostats but that neon requires liquid helium temperatures (4 K).

(vi) *Latent heat of fusion.* The latent heat of fusion L_f is of importance since it is a measure of the amount of heat to be removed from the support area on condensation of the gaseous mixture to form a matrix. The amount of heat dissipated on condensation depends also upon the rate of deposition and must not exceed the cooling capacity of the cryostat, otherwise the sample area temperature will rise, possibly permitting diffusion to occur. Values of L_f are given in Table 2.4 together with m.p. and b.p. values.

(vii) *Lattice energy.* The lattice energy U_0 is important, since being the energy required to form a lattice at 0 K, it also relates to the energy required to remove a molecule from its place in the lattice, i.e. diffusion of a molecule in the lattice.

The lattice energy U_0 is related to the zero-point energy E_z and the cohesive energy E_0 (or latent heat of sublimation L_0) at 0 K by the equation $U_0 = E_0 + E_z$, and so U_0 can be calculated from values of E_0 (or $-L_0$) and E_z. Zero-point energy can normally be neglected, so that essentially $U_0 \simeq E_0$ at 0 K. However, values of E_z for the noble gases are[37] neon 619·2 Jmol^{-1}, argon 778·2 Jmol^{-1}, krypton 606·7 Jmol^{-1}, and xenon 514·6 Jmol^{-1}, which yield calculated ratios of zero-point energy to cohesive energy, $E_z/-U_0$, of 0·33 for neon and 0·032 for xenon. These values are appreciable and so it appears[37] that quantum effects are important for all but the heaviest of the noble gases at 0 K.

(viii) *Thermal conductivity.* The thermal conductivity of the matrix material is of considerable practical importance. Gas at a temperature of say 295 K is impinging upon and condensing on the surface of a plate at low temperature, for example 20 K, being cooled by a boiling refrigerant. Upon condensation, heat must be removed from the sample area. As the layer of matrix grows thicker, conduction of the heat being supplied by fresh condensing gas must take place through this layer. If the thermal conductivity of the matrix material is poor, local heating will occur, having deleterious effects; a poor matrix may result which is opaque and causes scattering. Moreover local diffusion of solute may take place, with concomitant loss of monomer due to multimerization, or with recombination of radicals in photolysis studies.

Related to this is the deposition rate. Since the experiment depends on obtaining an IR or Raman spectrum, it is obviously necessary to lay down sufficient solute to give a spectrum of adequate intensity. The amount deposited will depend on several interlinked factors. These are:
(i) The concentration of the solid solution, i.e. M/A ratio.
(ii) The rate of deposition.
(iii) The total deposition time.
The concentration is usually determined by the nature of the experiment, but usually a dilute mixture is preferred, for example 1000 : 1, for effective monomer isolation. The rate of deposition will be limited by the thermal conductivity of the matrix material (and cooling power available). Obviously the better the conductivity, the faster the deposition rate which can be tolerated without localized heating ensuing. The time of deposition is therefore the factor most amenable to control, though often limited by the refrigerant supply per experiment.

Thermal conductivity, λ, is defined as the time rate of transfer of heat by conduction, through unit thickness, across unit area, for unit difference of temperature. Values are given in Table 2.4 in units of $W\,m^{-1}\,K^{-1}$ but since various units appear in the literature, conversion factors are given below:

$$1\,cal\,s^{-1}\,cm^{-1}\,K^{-1} = 418\cdot4\,W\,m^{-1}\,K^{-1}$$

$$1\,J\,cm^{-1}\,s^{-1}\,K^{-1} = 0\cdot2390\,cal\,s^{-1}\,cm^{-1}\,K^{-1}$$

$$1\,Btu\,ft^{-1}\,h^{-1}\,^{\circ}F^{-1} = 0\cdot01730\,J\,cm^{-1}\,s^{-1}\,K^{-1}$$

Two further points should be noted. Firstly, the thermal conductivity of a film condensed from the vapour phase may be different from that of a film obtained by solidification from a melt, due to either a different crystal lattice being formed or a less compact crystal. Secondly, the thermal conductivity of molecular solids tends to be low as compared with metals and ionic substances, the magnitude depending on the amount of anharmonicity of the lattice vibrations and on the temperature.

Noble gases

The main cause of thermal resistance is anharmonic interactions between lattice vibrations.[38] The variation of thermal conductivity of the noble

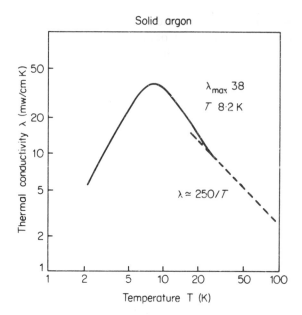

Figure 2.10 Thermal conductivity against temperature curve for argon [re-drawn from data in G. L.
Pollack, *Rev. Mod. Phys.* **36**, 748 (1964)]

gases with temperature follows a graph of the type shown in Figure 2.10
and such graphs, compiled from various data including that of White and
Woods[38] are given by Pollack.[39] The values of the λ/T curve maxima are:
neon 4·7 W m^{-1} T$_K^{-1}$ at 3·4 K; argon 3·8 W m^{-1} T$_K^{-1}$ at 8·2 K; and krypton
1·7 W m^{-1} T$_K^{-1}$ at 11·5 K. No experimental data are available for xenon,
but the calculated maximum is 2·4 W m^{-1} T$_K^{-1}$ at 16·1 K. The values in
Table 2.4 for neon, argon and krypton at 4 K, and for argon and krypton
at 20 K, are taken from the graphs. The value for neon at 20 K is taken
from the equation,[39] given for the tailing-off portion of the curve at higher
temperatures, $\lambda = 7·5/T$ above 12 K. Values for argon and krypton can be
obtained from the analogous equations,[39] argon $\lambda = 25·0/T$ above 25 K,
and krypton $\lambda = 27·0/T$ above 60 K, both in W m^{-1} T$_K^{-1}$. Some data given
in the literature at temperatures below 15 K have been considered unreliable[37]
due to severe disorder introduced during the preparation and cooling of the
polycrystals.

Nitrogen

   Roder[40] studied the thermal conductivity of solid nitrogen, and concluded
that the temperature dependence of its thermal conductivity is similar to
that of the noble gas solids and to that of hydrogen. The value given in the

table is the average of several values around the temperature quoted. Above c. 20 K, the thermal conductivity varies as B/T where $B = 8.0 \pm 0.6$ W m^{-1}.

Methane

Gerritsen and van der Star,[41] using a solid section of methane, found $\lambda = 0.12$ W m^{-1} T$_K^{-1}$ at 20 K. Clarke and Gordon[42] found $\lambda = 0.35$ W m^{-1} T$_K^{-1}$ at 84 K, and suggest that this is in reasonable agreement, since the conductivity of a hydrocarbon would be expected to increase with temperature.

Carbon dioxide

Gerritsen and van der Star[41] attempted to measure the thermal conductivity of carbon dioxide, and although they could not obtain a value, they estimate it to be much higher than methane. This apparently high value for carbon dioxide possibly results from more polar bonding in the carbon dioxide lattice and its lower molecular volume.

2.3.2 Choice of matrix

The obvious choice of matrix materials are the noble gases and nitrogen since they have no absorption in the near infrared and are for most purposes chemically inert. Helium is excluded, since it cannot be solidified in the cryostats; neon can only be used in liquid helium cryostats. Xenon forms notoriously bad matrices in terms of scatter, unless deposited above c. 50–60 K. Argon is the most commonly used matrix support; it forms extremely good matrices in liquid hydrogen cryostats, and, furthermore, is cheap.

A second group are the simple hydrocarbons, methane, ethylene and acetylene, which are of interest as reactive matrices.

A third group, of ten compounds, consists of nitrogen, oxygen, carbon monoxide, carbon dioxide, carbon disulphide, sulphur dioxide, sulphur hexafluoride, nitric oxide, hydrogen sulphide, and chlorine. These molecules provide a wide range of properties, and combinations thereof. All have been used occasionally in matrix studies, and their use is increasing.

A fourth series of compounds of potential interest as matrix materials are the halocarbons. These are readily available commercially and are listed under various trade names (Arcton, Freon, etc.) but the identifying number, as quoted in Table 2.4, is always the same. They provide an interesting group of similar compounds with a gradation of properties. Only those with a boiling point below room temperature (293 K) have been listed, an arbitrary choice to avoid the necessity of using a heated line to handle the matrix gas. A disadvantage of these compounds is that they have a vibrational spectrum, as of course do the second and third group of materials above. Furthermore, there is only a limited amount of physical data on them available in the literature.

2.3.3 Electrical properties

The various properties dealt with in this section are of importance in the consideration and calculation of the frequency shifts of solute absorption bands due to solute-matrix interactions, as detailed in Chapters 3 and 4 of this book.

(i) *Dielectric constant* (ε')

The Clausius–Mosotti (CM) relation is given by

$$\chi^{(e)} = \frac{3}{4\pi} \frac{M}{N\rho} \left(\frac{\varepsilon' - 1}{\varepsilon' + 2}\right)$$

where $\chi^{(e)}$ is the electric susceptibility per molecule, N = Avogadro's number, M = molecular weight, ρ = density, and ε' = dielectric constant (permittivity).

The dielectric constant of any real gas is determined both by the properties of isolated molecules and by the effects of molecular interactions. If the interactions are small and the molecules have no permanent dipoles, the interactions can be completely neglected, $\chi^{(e)} \equiv \alpha$ the average polarizability of the molecules, and the dielectric behaviour of the gas is represented well by the above CM equation with α replacing $\chi^{(e)}$.

If the molecules have permanent dipoles, $\chi^{(e)}$ contains other terms, which are temperature dependent, due to the alignment of the dipoles by the field. The effect of molecular interactions is usually fairly important, and the complete virial expansion of the CM function must be considered,[43] namely

$$\left(\frac{\varepsilon' - 1}{\varepsilon' + 2}\right) \frac{M}{N\rho} = \left(\frac{4\pi\alpha}{3}\right) + B\left(\frac{M}{N\rho}\right) + C\left(\frac{M}{N\rho}\right)^2 + \dots$$

The coefficients B, C, ... are called 2nd, 3rd ... dielectric virial coefficients. The second coefficient B incorporates the effects of pair interactions, the 3rd one C, that of the 3-body interactions and so on.

Only a few values of ε' for the solid state at low temperatures are available (Table 2.5).

Noble gases. To a first approximation, the CM function depends only on the polarizability of the molecules. If the polarizability is considered to be temperature independent (although it decreases with increasing pressure at high pressures), ε' is sensitive to changes in ρ, the density, as the temperature is varied. On lowering the temperature, the lattice contracts (**a** decreases) and the density ρ increases. As the density increases ε' increases, keeping the CM function constant. Therefore, $\Delta \mathbf{a}^3$ $(T_1 \rightarrow T_2)$ has been used to extrapolate a known ε' value at T_1 to give ε' at T_2.

Table 2.5 Electrical properties of matrix materials

	$\mu \times 10^{18}$ (esu cm)[a]	$Q \times 10^{26}$ (esu cm^2)[b]	α(Å^3)	ε' (20 K)[c]	n	IP (ev)[d]
Ne	0	0	0.395		1.23 (4 K)	21.56
Ar	0	0	1.641	1.63	1.29 (20 K)	15.76
Kr	0	0	2.480	1.88	1.28 (4 K)	14.00
Xe	0	0	4.044	2.19	1.49 (40 K)	12.13
CH$_4$	0	0	2.60	1.79	1.33 (90 K)	12.65
SF$_6$	0	0	4.52			
N$_2$	0	-1.4	1.767		1.22 (4 K)	15.5
O$_2$	0	-0.4	1.598		1.25 (4 K)	12.07
Cl$_2$	0	$+6.1$	4.61			11.48
CO	0.112	-2.5	1.977			14.01
NO	0.153	-1.8	1.74			9.26
CO$_2$	0	-4.3	2.63		1.22 (4 K)	13.78
N$_2$O	0.167	-3.5	3.00			12.91
SO$_2$	1.63	4.4	3.89			12.34
H$_2$S	0.97		3.78			10.42
C$_2$H$_6$	0	-0.8	4.47			11.55
C$_2$H$_4$	0	$+2.0$	4.22		1.44 (63 K)	10.46
C$_2$H$_2$	0	3.0	3.49			11.40
CF$_4$	0	0	3.86			14.6
CHF$_3$	1.65		3.09			13.86
CClF$_3$	0.50					12.91
CBrF$_3$	0.65					11.89
CHClF$_2$	1.42					12.45
CCl$_2$F$_2$	0.51					12.31
CHCl$_2$F	1.29					12.39
C$_2$F$_6$	0					
CClF$_2$CF$_3$	0.52					
CClF$_2$CClF$_2$	0.5					
C$_4$F$_8$	0					

a. NSRDS-NBS 10 (1967).
b. A. D. Buckingham, R. L. Disch and D. A. Dunmur, *J. Amer. Chem. Soc.*, **90**, 3104, (1968); ref. 52.
c. Extrapolated from data in ref. 47.
d. NSRDS-NBS 26 (1969).

On checking values of ε' (Ar) given in the literature, it was found that the values of Eatwell and Jones[44] are self-consistent, on basis of $\mathbf{a}^3$ estimation, but they are not in accord with those of Lefkowitz *et al.*,[45] being too large. Lefkowitz *et al.* give two values of ε' (Ar) at 77 K, one (the lower, 1·586) being for pure f.c.c. argon, and the other (the higher, 1·7036) being for solid argon which is only predominantly f.c.c. and so presumably $\varepsilon'_{\text{h.c.p.}} > \varepsilon'_{\text{f.c.c.}}$. The effect of impurity on stabilizing the h.c.p. form of solid argon has been examined by Jones and Woodfine[46] who have shown that small amounts (e.g. < 1 per cent.) of heavy impurities (as xenon) strongly influence ε' of solid noble gases. It is therefore possible that the relatively high values of ε' found by the other workers[44,47,48] at the melting point are due to the presence of traces of impurity.

It was assumed above that α is temperature independent. However, Smith and Pings[49] studied the polarizability of an argon atom in solid argon. They found a variation in α with **a** the lattice constant. Over the temperature range 20–80 K, there is an increase of α of ~ 1 per cent. with increasing **a**. The decrease in density with increasing **a** over this temperature range is c. 8 per cent. They are opposing effects therefore, and in calculating ε' by extrapolation from 80 K to 20 K the calculated $\Delta\varepsilon'$ should be appropriately reduced. This effect presumably operates in the other noble gases, although the magnitude of the effect is unlikely to be the same.

(ii) Refractive index (n)

The expansion for the refractive index of a substance, analogous to the CM relation for ε', is the Lorentz–Lorenz function

$$[R] = (M/\rho)(n^2 - 1)/(n^2 + 2)$$

where $[R]$ is the molar refractivity, M is the molecular weight, ρ is the density and n is the refractive index. In this expression, ε' is replaced by n^2. The Maxwell relation, $\varepsilon' = n^2$, is only valid under certain conditions. In the basic equation $n = \sqrt{\mu'\varepsilon'}$, where $\mu' = $ the permeability. Both μ' and ε' are unity in vacuum, and usually μ' can be considered to be very nearly unity, giving $n \simeq \sqrt{\varepsilon}$ or $n^2 = \varepsilon'$.

The Maxwell relation is only confirmed experimentally provided: (a) the substance contains no permanent dipoles, (b) the measurements are made with radiation of long wavelength (IR), and (c) n is not measured near a wavelength of absorption. The first condition depends on the fact that dielectric constants are measured at low frequencies (500–5000 kHz), whereas refractive indices are measured with radiation of frequency c. 10^{12} kHz. A permanent dipole cannot line up fast enough to follow an electric field alternating with such a high frequency. Therefore permanent dipoles contribute to ε' but not to n.

The second condition is due to the effect of high frequencies on the induced polarization. Using high-frequency radiation (visible region) only the electrons in the molecules can adjust themselves to the rapidly alternating electric field of the light waves. The more sluggish nuclei stay more or less in their equilibrium positions. Using IR radiation, which has lower frequency, the nuclei are also displaced.

Values of refractive index of solid phases are given in Table 2.5. They are all measured in the visible region, using either the sodium D line (589 nm) or the mercury green line (546 nm).

(iii) Polarizability

If the internuclear distances in a molecule are altered, changes occur in the electron distribution and in the forces binding the electrons, and so the value of the electron polarizability (α) changes accordingly. The mean value of the polarizability can be determined[50] from the refractive index n of the gas, using the equation

$$(n - 1) = 2\pi N\alpha$$

where N is the number of molecules per cubic centimetre.

For matrix purposes, the available values of polarizability are not entirely satisfactory—being determined in the gas phase at visible wavelengths, rather than in the low temperature solid at IR wavelengths. It should be noted that, as mentioned in section (i), Smith and Pings[49] found a variation of α with **a** (the lattice constant) for solid argon. Over the temperature range 20–80 K, an increase of α of c. 1 per cent. was found with increasing **a**.

Values of polarizability are given in Table 2.5. The mean polarizability $\alpha = \frac{1}{3}(\alpha_1 + \alpha_2 + \alpha_3)$, of the three principal components of the polarizability, is used. Molecular polarizabilities parallel and perpendicular to the symmetry axis are of current interest, for their relevance to the direct measurement of molecular quadrupoles, and the use of the parameter $(\alpha\| - \alpha\bot)$[51] in frequency shift calculations.

(iv) Ionization potentials

This parameter is required for the calculation of the dispersion contribution (Chapter 4) to solute-matrix interactions. The ionization potential, usually expressed in electron-volts, is the work necessary to remove a given electron from its atomic orbit to infinity. Values of 1st ionization potentials are listed in Table 2.5.

(v) Electric multipole moments

A set of electric multipole moments (charge, dipole, quadrupole, ... 2^n pole) is associated with any system of electric charges. The energy of interaction of the system with an external field, and the potential of the electric field at any point outside the distribution of charges and arising from it, is related to these moments. In order to study intermolecular forces, and to attempt an assessment of relative contributions of the various interactions in a system, knowledge of the multipole moments is required.

The definitions for molecular electric multipole moments usually used[52-55] are those of Buckingham[53] and Kielich.[54]

The multipole moment tensors used are, in the general case,

$$M_{\alpha\beta...\mu} = \frac{(-1)^m}{m!} \int \rho r^{2m+1} \frac{\partial m}{\partial r_\alpha \partial r_\beta ... \partial r_\mu}\left(\frac{1}{r}\right) d\tau$$

where μ_α ... are components of the dipole, quadrupole ... and general mth rank moment tensors respectively. The charge density (including the charge density of the nuclei) at the point $(x, y, z) = \rho$. The subscripted r is x, y, z depending on whether the subscripts have the value 1, 2, or 3; $d\tau = dx\,dy\,dz$. This equation has $\int \rho r \cdot d\tau$ equivalent to $\sum_i e_i r_i$ of Buckingham,[53] in which the summation is over all the charges in the molecule. With the coordinates axes taken as fixed to the frame of the molecule at its equilibrium configuration, rather than fixed in space, the multipole moments only apply to non-rotating molecules.

The multipole moments given in Table 2.5 are relative to an origin at the centre of mass of the molecule. All higher rank tensors depend on the position of the origin of the coordinate system, only the first non-zero molecular multipole moment tensor for a given molecule being independent of this position.

Dipole moments. The dipole moment tensor is $\mu_\alpha = \int \rho r_\alpha\,d\tau$. Dipole moments can be measured by several methods[56-59] which are all well established. The values of μ, given in Table 2.5, were obtained from the NBS compilation.

Quadrupole moments. The quadrupole moment tensor is given by

$$\theta_{\alpha\beta} = \tfrac{1}{2} \int \rho[3r_\alpha r_\beta - r^2\delta_{\alpha\beta}]\,d\tau$$

where if $\alpha = \beta$, $\delta_{\alpha\beta} = 1$; and if $\alpha \neq \beta$, $\delta_{\alpha\beta} = 0$. Quadrupole moments cannot be measured easily or accurately. Most of the methods used are indirect methods, and certain assumptions and approximations (of doubtful validity), have to be made about intermolecular forces, if θ is to be computed from the experimental data. The two most reliable techniques[52] are induced bire-fringence experiments[60-62] and measurements of anisotropy in diamagnetic susceptibility by molecular beam resonance methods.[63,64] Although present knowledge is somewhat limited and tentative, the relatively long range and angular dependence of the quadrupole field make θ an important parameter in the investigation of various physical properties. The topic was reviewed by Buckingham[53] and, more recently, by Krishnaji and Prakash,[55] and Stogryn and Stogryn.[52] These reviews give a detailed survey of the different methods available for the evaluation of θ, of which there are eight principal ones, and assessments of their relative merits and demerits. Data on quadrupole moments are also given by Orcutt[65] and Kielich.[66] Apart from theoretical calculations, the sign of quadrupole moments at present are only given by diamagnetic susceptibility, induced birefringence, and enthalpy measurements. Values in Table 2.5, unless given a plus or minus sign, refer therefore to absolute values.

Molecules with centres of symmetry (e.g. homonuclear diatomic molecules, hydrogen and nitrogen, symmetric linear molecules, carbon dioxide, carbon disulphide and acetylene, or planar molecules, boron trifluoride and benzene) have all odd-rank tensors identically zero. Thus they cannot have permanent dipole, octopole, or indeed $2^{(2n+1)}$ multipole moments; they may however have quadrupole or hexadecapole moments.[53] For tetrahedral molecules, for example methane or carbon tetrafluoride, the ground state dipole and quadrupole moments are zero, the principal multipole being an octopole. Octahedral molecules, for example sulphur hexafluoride, have hexadecapoles (the field of which varies as r^{-6} at long range) as their principal multipole.

In molecules whose leading multipole is a quadrupole, the lattice structure of the solid may be influenced by quadrupole-quadrupole forces. The structure of solid carbon dioxide and nitrogen (low-temperature form) is face-centred cubic, the molecular axes being orientated in such a manner that quadrupole-quadrupole interaction energy is a minimum.

2.3.4 Crystal data

In this section are given data on crystal structures and phase transition temperatures. This information is necessary for the interpretation of matrix spectra, and differences between spectra at various temperatures.

The common matrix materials, the noble gases, are spherical molecules. Solids whose molecules or atoms are essentially spherical in shape and are held by non-directional bonds adhere to the *principle of closest packing*. This tendency of spherical molecules to crystallize in one of the closest-packed structures is a consequence of the fact that, by so doing, they achieve the maximum number of nearest neighbours, thereby maximizing the inter-molecular van der Waals forces. Because of the importance of nearest-neighbour interactions on trapped species we shall outline these close-packed structures.

Simple close-packed structures

The most efficient way of packing together equal spheres, called closest packing, can be achieved in two ways, each of which utilizes the same volume (74 per cent.) of the total space. The *hexagonal close-packed structure* (h.c.p.) is built up as follows. Surround a sphere by six other equal spheres in the same plane (Figure 2.11a). Then add a second layer of similarly-packed

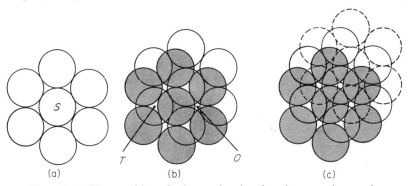

(a)   (b)   (c)

Figure 2.11 Close packing of spheres, showing three layers, a, b, c, and tetrahedral and octahedral holes

spheres staggered so that they nestle into the depressions formed by the first-layer spheres (Figure 2.11b). A third layer is then added with each sphere directly above a sphere of the first layer and so on in an alternating manner.

The *cubic close-packed structure* (c.c.p.) is built up with layers *a* and *b* as above but with the third layer *c* added so that its spheres are not directly above those of either layer *a* or *b*. The stacking sequence is thus *abcabc*... for c.c.p. and *ababab*... for h.c.p. In both of these arrangements, each sphere has 12 nearest neighbours—six in its own plane, three in the plane above, and three in the plane below. In a cubic close-packed lattice the unit cell is a face-centred cube (f.c.c.) of side a_0, the lattice parameter.

These structures are of importance for cryogenic matrix studies since all the noble gases crystallize in either c.c.p. or h.c.p. structures. The electron cloud in the nitrogen and carbon monoxide molecules are nearly spherical in shape and crystals of solid nitrogen and of carbon monoxide also have either c.c.p. or h.c.p. structures. Many of these solids are polymorphic, i.e. they crystallize in more than one structure. The details of the structures and phase transitions of these and other matrix molecules are listed at the end of this section.

Close-packed lattices can have three possible guest sites, *substitutional* (*S*), in which the guest molecule replaces a host molecule, and two types of *interstitial* sites. Interstices between layers of close-packed spheres are of two kinds, *tetrahedral* (*T*) and *octahedral* (*O*) (Figure 2.11). A tetrahedral hole has four spheres adjacent to it (Figure 2.12a) and an octahedral hole has six spheres adjacent to it (Figure 2.12b). The geometry of these holes

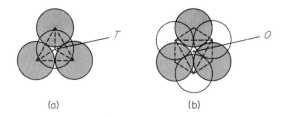

(a)   (b)

Figure 2.12 Tetrahedral and octahedral holes between layers of close-packed spheres

is of importance to matrix studies in relation to the perturbations which the trapped species will experience. A tetrahedral hole is in itself not tetrahedral in shape but is so named because a small guest atom trapped in such a hole would have four neighbouring matrix spheres arranged at the corners of a tetrahedron. For such a grouping it can readily be shown by simple geometry that the radius of the small sphere that occupies the site is given by $0.225r$

where r is the radius of the large sphere, or $0.159a_0$. Thus in order to occupy the tetrahedral site without disturbing the closest-packed lattice, the radius of a trapped spherical species should be no greater than 0.225 that of the matrix spheres. The tetrahedral hole indicated in Figure 2.12(a) is formed by one sphere of the top layer b and three spheres of the bottom layer a; there is such a hole directly under each sphere of the top layer. Additionally, there are tetrahedral holes from three spheres of the top layer and one from the bottom; thus there is also one tetrahedral hole above each sphere of the bottom layer. In the crystal as a whole there is one tetrahedral hole above and one below every close-packed sphere, and hence there are twice as many tetrahedral holes as close-packed spheres.

Figure 2.12(b) shows a grouping of six adjacent spheres, three from layer a and three from layer b, surrounding an octahedral hole. By simple geometry, it can be shown that the radius ratio of small (trapped) sphere to large (matrix) sphere is 0.414, or $0.293a_0$, so that an octahedral site is noticeably larger than a tetrahedral site (Table 2.6). If we follow one column of spheres vertically (best demonstrated by a three-dimensional model), octahedral sites and spheres alternate, for both c.c.p. and h.c.p. lattices. Thus there is one octahedral site for every sphere in the lattice, i.e. there are half as many octahedral sites in a closest-packed lattice as there are tetrahedral sites.

Table 2.6 Site diameters (spherical cavity sites)

	Subst. (Å)	O_h Int. (Å)	T_d Int. (Å)	Ref.
Ne (4 K)	3.156	1.31	0.71	a
Ar (4 K)	3.755	1.56	0.85	b
(20 K)	3.760			
Kr (4 K)	3.992	1.65	0.90	c
(20 K)	3.997			
Xe (4 K)	4.336	1.80	0.97	d
(20 K)	4.339			
CH_4 (4 K)	4.147	1.73	0.94	e
(20 K)	4.175			
CF_4 (20 K)g	c. 4.76	—	—	f

a. D. N. Batchelder, D. L. Losee and R. O. Simmons, *Phys. Rev.*, **162**, 767, (1967).
b. O. G. Peterson, D. N. Batchelder and R. O. Simmons, *Phys. Rev.*, **150**, 703, (1966).
c. D. L. Losee and R. O. Simmons, *Phys. Rev.*, **172**, 944, (1968).
d. D. R. Sears and H. P. Klug, *J. Chem. Phys.*, **37**, 3002, (1962).
e. S. C. Greer and L. Meyer, *Z. Angew. Phys.*, **27**, 198, (1969).
f. S. C. Greer and L. Meyer, *J. Chem. Phys.*, **51**, 4583, (1969).
g. Monoclinic crystal structure.

Lattice dislocations

The above discussion refers to an ideal crystal but all crystals deviate to some extent from the perfect state due to the presence of lattice flaws or imperfections. Just as many properties of real gases derive from their imperfections, so do many of the properties of real crystals. The subject of dislocations in the solid state is a vast one in its own right but only a brief mention of them can be given here.

The two simplest types of straight line dislocations are the *edge* and *screw* types. A pure *edge dislocation* is shown in Figure 2.13a, and involves essentially the insertion of an extra half-plane of atoms *X Y* into the lattice. That is, one of the planes of spheres terminates, resembling a knife stuck part way into a piece of cheese.

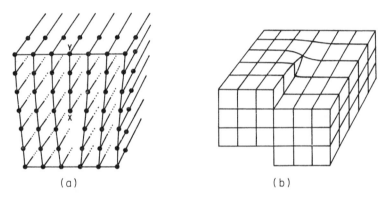

(a)   (b)

Figure 2.13 Illustration of formation of (*a*) a pure edge dislocation, (*b*) screw dislocation

In a *screw dislocation* part of the lattice is displaced with respect to the other part (Figure 2.13b). We may think of achieving this by first cutting through a perfect crystal, then forcing the material on one side of the cut to move down with respect to the material on the other side, by one or more lattice units.

Apart from the fact that imperfections provide additional trapping sites, dislocations offer a ready pathway for diffusion in the solid-state. The solid matrix formed in a typical matrix isolation experiment would be expected to be imperfect because of the nature of the deposition process, i.e. rapid condensation from ambient to cryogenic temperatures. It is known that argon deposited at 4 K has a high density of dislocations since the observed X-ray diffraction pattern (see Chapter 9 in ref. 1) shows considerable intrinisic line broadening. On warming this solid, a gradual exothermic re-ordering process occurs with crystallized unorientated argon being produced at about 30 K. This is accompanied by a rise of the vapour pressure to around

10^{-3} torr, suggesting that partial redeposition may occur. Electron diffraction of films of argon deposited at 7 K also show broad rings, sharpening on warming or on deposition at a higher temperature up to an 'evaporation temperature' of 30 K.[66a] Some of the other matrix materials behave in a similar manner and nitrogen deposited at 20 K is reported to show more disorder than that deposited at 4 K and subsequently annealed at 20 K. It has been estimated[66b] that the average size of the microcrystallites formed from a condensing gas is of the order of 100 Å which give rise to grain boundaries.

Multiple trapping sites

It follows that there are three main classes of trapping sites: substitutional, interstitial and dislocation sites. Bands observed in the infrared of matrix-isolated species are generally exceedingly sharp and spectra are highly reproducible. This suggests that the guest molecules are all in similar sites since different sites will offer different environmental perturbations and hence different frequency shifts. The most likely normal site is the substitutional site since it alone is large enough to accommodate a normal size diatomic guest. Dislocation sites may well be larger but they are non-uniform. The octahedral interstitial sites are much smaller than the substitutional sites but could conceivably accommodate the smaller diatomics whereas tetrahedral interstitial sites must be too small to be seriously considered. Larger guest molecules must occupy sites formed by removing two or more lattice molecules. In some cases (examples are given in several of the chapters) matrix-isolated species give multiplet spectral features which are not due to rotation or aggregation. These multiple bands have been explained in terms of multiple trapping sites, although to date there is no study which provides information as to the exact nature of the sites.

Structural parameters of matrix materials

The structures and known low temperature phase transitions of the most common matrix materials are listed below and structural parameters given in Tables 2.6 and 2.7.

Noble and spherically symmetrical molecules

All the noble gases except helium crystallize in c.c.p. structures. The lattice parameters have been determined by X-ray methods over a range of temperatures and are listed by Pollack[39] in his detailed review of the physical properties of the noble gases. More recent values of site diameters of the noble gases, together with those of methane and carbon tetrafluoride, at the two commonly used matrix temperatures of 4 K and 20 K are given in Table 2.6. Further values can be obtained from the key references listed which provide X-ray data from 0 K to T_m.

Table 2.7 Site diameters (non-spherical cavity sites)

	Crystal structure	Transition temp. (K)	Mean site diam. (Å)	Approx. site shape (Å)	Ref.
N_2 (4 K)	f.c.c.	35·6	3·991	4·52 × 3·42 × 3·42	a
(20 K)			4·004		
O_2 (23 K)	monoclinic	23·8	3·640	4·18 × 3·20 × 3·20	b
Cl_2	rhombohedral	—			c
CO (23 K)	f.c.c.	61·6	3·999	4·61 × 3·48 × 3·48	d
CO_2 (20 K)	f.c.c.	—	3·93	5·32 × 3·00 × 3·00	d
N_2O (20 K)	f.c.c.	—	4·00	5·38 × 3·07 × 3·07	e

a. Ref. 74.
b. C. S. Barrett, L. Meyer and J. Wasserman, *J. Chem. Phys.*, **47**, 592, (1967).
c. R. L. Collin, *Acta Cryst.*, **5**, 431, (1952); **9**, 537, (1956).
d. C. S. Barrett and L. Meyer, *J. Chem. Phys.*, **43**, 3502, (1965).
e. Site diameter calculated from density data.

Argon. Although the stable form is c.c.p., h.c.p. is a metastable phase at all temperatures below the m.p. (84 K) in high purity crystalline argon. In the presence of traces of impurities such as 1–2 per cent. of air, oxygen, nitrogen or carbon monoxide, argon crystallizes with h.c.p. structure[67] without any trace of the cubic phase. As the temperature is lowered, higher nitrogen content is needed to stabilize the h.c.p. phase and at low temperatures the c.c.p. is stable from 1 per cent. to *c.* 55 per cent. of nitrogen; for example, at 20 K *c.* 50 per cent. nitrogen is required to stabilize the h.c.p. phase.

Methane. X-ray studies indicate that solid methane (m.p. 89 K) is f.c.c. over the whole temperature range 4·2–75 K. However, if the lattice constant of the f.c.c. structure is plotted against temperature, a slight anomaly[68] but no discontinuity is apparent at 21 K. There is also an anomaly at 65 K which corresponds to that seen[69] in NMR measurements, and a broad specific heat anomaly is centred at 8 K. The inability of X-rays to detect any major changes at 21 K has been taken[70] to indicate that this is a pheno- menon involving the orientation of the hydrogen atoms. NMR line width measurements,[71] however, show no change in line width at this temperature. A sudden increase in rotational freedom is also eliminated as a possible cause, on the basis of no changes being observed in the IR spectrum of carbon monoxide in a methane matrix[72] in passing from 5–30 K. This evidence is, however, unsatisfactory since at the methane/carbon monoxide matrix ratio of 217 employed the band profile is predominantly that of carbon monoxide aggregates rather than isolated carbon monoxide mole- cules. Recent work[73] in our laboratories at $M/A = 5000$, unequivocally shows that carbon monoxide does not rotate in methane over the temperature range 20–40 K. Thus, whilst the explanation of this slight anomaly remains open, care should be taken in the interpretation of any spectral changes observed in the region of 21 K. A very recent X-ray study down to 4 K, by Hertczeg and Stoner,[73a] confirms that methane is f.c.c. at all temperatures below the m.p. Very small traces of impurities such as nitrogen, however, appear to cause the presence of h.c.p. methane crystallites in the same manner[67] which causes solid argon to crystallize in a h.c.p. structure. These authors suggest that the specific heat anomaly is probably due to nuclear spin conversion (see §3.6.1) rather than a change in phase.

The lattice constant of tetradeuteromethane is[73b] slightly smaller than that of methane and shows a change of slope near 21 K and perhaps a small anomaly near 27 K. Heat capacity measurements also exhibit anomalies at 22·2 K and 27·1 K. The X-ray measurements,[73b] however, show that the f.c.c. structure persists over the whole 4·2–80 K temperature range as is the methane lattice. No change in structure is observed after several hours between 17–20 K and after several days between 21–50 K.

Tetrafluoromethane. Tetrafluoromethane has a phase transition, $\alpha \to \beta$ at 76·2 K, but as its m.p. is 89·5 K, matrix studies will only be concerned (Table 2.6) with the low-temperature α-phase which has a monoclinic unit cell.

Neo-pentane. Neo-pentane provides a large spherical cavity and seems a matrix of potential interest for rotational (and other) studies. Experiments recently carried out in our laboratories have, however, shown that it forms extremely poor matrices at 20 K. They appear polycrystalline, are badly scattering and show poor isolating capability.

Non-spherically symmetrical molecules. Table 2.7 lists the dimensions of the cavities found in other matrix molecules together with mean substitutional site diameters.

Nitrogen. Nitrogen has an interesting phase change, $\alpha \to \beta$, occurring[74] at 35·6 K, accompanied by a considerable decrease in barrier to rotation of the nitrogen molecules.[75] The high-temperature β-phase, which is IR opaque, crystallizes[74] in a h.c.p. lattice in which the nitrogen molecules are apparently free to rotate. Thus above 35 K nitrogen does not act as a good matrix material, as diffusion would proceed rapidly. The low-temperature α-phase has[75] a f.c.c. structure, although a more recent study[75a] indicates that the nitrogen molecular centres are slightly moved from the f.c.c. sites and the symmetry thus lowered from Pa3 to P2$_1$3.

Oxygen. Oxygen has three solid phases $\alpha \to \beta \to \gamma$, the transition temperatures being 23·8 K and 43·8 K respectively. The m.p. is 54·4 K so that matrix studies will only be concerned with the $\alpha \to \beta$ transition. The α-phase is monoclinic and the β-phase is rhombohedral. This transition will clearly complicate any interpretations of diffusion studies in oxygen matrices.

Carbon monoxide. Carbon monoxide has two solid phases, α and β, with a transition temperature of 61·6 K, close to the m.p. of 68·1 K. The structures are isomorphous with the two phases of nitrogen but disordered with respect to the sense of the carbon monoxide molecules. The α-phase is f.c.c. and the β-phase hexagonal.

2.3.5 Infrared and Raman spectra

Finally, for a material to be of use as a support matrix it must have absorption-free regions where the solute absorptions occur. In the case of hydrogen-containing matrix molecules the accessible spectral range can be extended by the use of the deuterated compounds but these are usually too expensive for extensive matrix usage.

Noble gases

The noble gases have no absorptions in the vibrational region and are thus ideal spectroscopic solvents. However, they have low-frequency lattice vibrations, which may be greatly activated by the presence of impurities. (A discussion of these phonon bands is given in the introduction to Chapter 7.) These bands will complicate far IR and low frequency Raman studies of matrix-trapped species. Solid argon has[46] phonon band maxima at 64 cm^{-1} and 42 cm^{-1} and the presence of an impurity activates a band at 73 cm^{-1}.[76] Solid krypton has[77] a strong sharp phonon band at 43 cm^{-1} when activated by 1 per cent. impurity of xenon or argon; in the presence of methane this feature is weak but a second strong and broad band appears at 65 cm^{-1}.

Homonuclear diatomic molecules

Nitrogen, oxygen and halogen molecules have no IR-active vibration and are also ideal IR solvents. Their vibrations are R-active and this slightly restricts their use for Raman matrix experiments. Lattice imperfections can induce IR activity, for example, bands attributed to the oxygen stretching mode have been found[78] at 1671 cm^{-1}, 1591 cm^{-1} and 1549 cm^{-1} in the IR spectrum of solid oxygen. It has also been pointed out[75a] that the fact that the nitrogen molecules are slightly offset from the f.c.c. sites (see §2.3.4) may make the N–N frequency observable in the IR spectrum of α-N$_2$. The laser Raman spectrum of α-O$_2$ at 15 K shows[79] vibrational lattice modes at 44 cm^{-1} and 79 cm^{-1} in addition to the intense vibrational mode at 1552·5 cm^{-1}. The far IR of solid oxygen has also been reported.[80]

Lattice modes are also found, for example, at 69 cm^{-1} and 49 cm^{-1} in the far[81] IR spectrum of solid α-N$_2$. The laser Raman spectrum of α-N$_2$ over the temperature range 12–35·5 K shows[82] two bands at 31·5 cm^{-1} and 35·8 cm^{-1} (also reported[83] at 33·5 cm^{-1} and 37·5 cm^{-1} at 16 K).

The far IR of solid chlorine at 87 K shows[84] translational lattice modes at 88 cm^{-1} and 60 cm^{-1} whilst the laser Raman spectrum at 15 K shows[85] lattice modes at $83\,\text{cm}^{-1}$, $100\,\text{cm}^{-1}$ and $118\,\text{cm}^{-1}$ and intense stretching modes at $524.4\,\text{cm}^{-1}$, $532.2\,\text{cm}^{-1}$, $539.2\,\text{cm}^{-1}$ and $540.0\,\text{cm}^{-1}$. The laser Raman spectrum of solid Cl_2 at 15 K shows[85] lattice modes at $74\,\text{cm}^{-1}$, $86\,\text{cm}^{-1}$ and $101\,\text{cm}^{-1}$, and stretching modes at $296.1\,\text{cm}^{-1}$, $297.4\,\text{cm}^{-1}$, $299.5\,\text{cm}^{-1}$ and $303.2\,\text{cm}^{-1}$.

The near IR spectra of some other molecules which have IR-active vibrational frequencies but which are sometimes used for matrix supports are shown in Figures 2.14 to 2.19. The spectra purposely show intense blackouts in the regions of fundamental vibrational absorptions in order to depict the measure of background absorption that will obtain in a matrix experiment. When using high M/A ratios (i.e. dilute mixtures) it is necessary to deposit large amounts of mixture in order to obtain a reasonably intense spectrum of the solute. In so doing the thick layer of matrix will often obscure other regions, due to overtone and combination bands normally thought of as weak features. These reference spectra thus correspond to solvent spectra in liquid solution studies and use is made of the appropriate 'windows' in the background spectrum.

Carbon monoxide. The IR spectrum of solid α-CO at 20 K is shown in Figure 2.14. The region blacked-out corresponds to the overlapping fundamentals:[72,86] $v(^{12}C^{16}O)$, $2138\,\text{cm}^{-1}$; $v(^{12}C^{17}O)$, $2112\,\text{cm}^{-1}$; $v(^{13}C^{16}O)$, $2092\,\text{cm}^{-1}$; and $v(^{12}C^{18}O)$, $2088\,\text{cm}^{-1}$. The broad band structure at c. 2200 cm^{-1} is a combination of $v(CO)$ and lattice modes found[81] at $86\,\text{cm}^{-1}$ and $50.5\,\text{cm}^{-1}$. The laser Raman spectrum[83] of α-CO at 12 K exhibits a broad band $47.5\,\text{cm}^{-1}$.

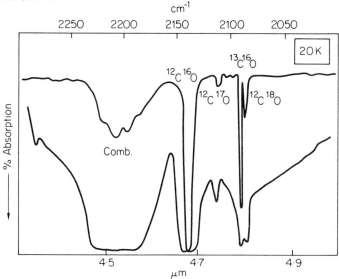

Figure 2.14 IR spectrum of solid carbon monoxide

Carbon dioxide. The spectrum of the low-temperature solid is depicted in Figure 2.15. Lattice modes in the far IR have been reported[81,87,88] at 114 cm^{-1} and 68 cm^{-1} and the laser Raman spectrum at 15 K shows[83] bands at 73·5 cm^{-1}, 91·5 cm^{-1} and 132 cm^{-1}.

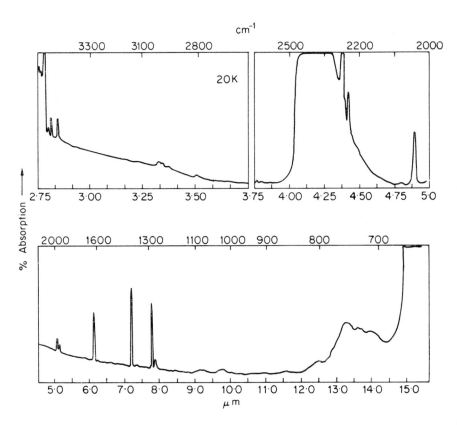

Figure 2.15 IR spectrum of solid carbon dioxide

Sulphur hexafluoride. The near IR spectrum of solid sulphur hexafluoride at 20 K is shown in Figure 2.16. The solid has not been studied in the far IR but the liquid at 243 K shows[89] bands at *c.* 150 cm^{-1}, 90 cm^{-1} and 55 cm^{-1}.

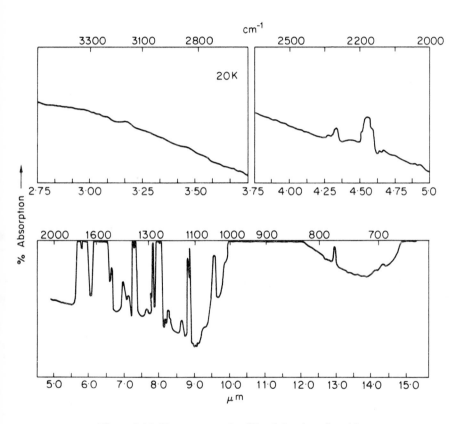

Figure 2.16 IR spectrum of solid sulphur hexafluoride

Methane. The near IR spectrum of solid methane (Figure 2.17) shows a complete blackout in the 3000 cm^{-1} region (v_3) and the 1300 cm^{-1} region (v_4). Solid deuteromethane shows corresponding unusable regions *c.* 2200 cm^{-1} and 900 cm^{-1}.

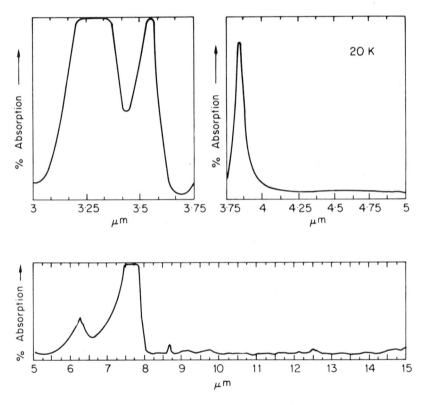

Figure 2.17 IR spectrum of solid methane

Ethylene. Figure 2.18 shows the near IR spectrum of solid ethylene at 20 K. Lattice modes are observed in the far IR[90] at 73 and 110 cm^{-1} for ethylene and at 69.5 cm^{-1} and 104 cm^{-1} for C_2D_4.

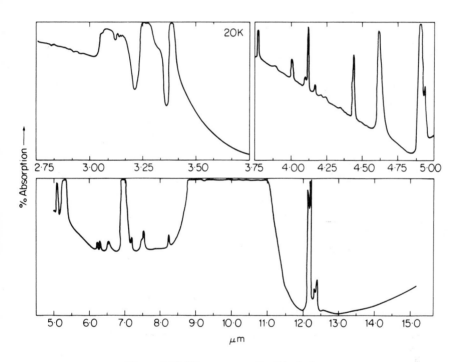

Figure 2.18 IR spectrum of solid ethylene

Tetrafluoromethane. Figure 2.19 depicts the near IR spectrum of solid carbon tetrafluoride at 20 K. The complete IR and R spectrum of α-CF$_4$ (and the β-form also) is given by Fournier *et al.*[91]

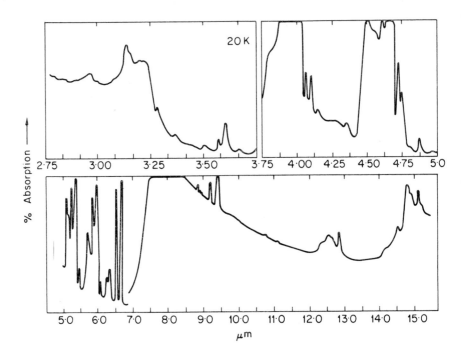

Figure 2.19 IR spectrum of solid tetrafluoromethane

Other molecules of possible matrix use but whose solid-state spectrum is not illustrated are listed below.

Nitrous oxide. The solid at 80 K shows fundamentals at 2238 cm^{-1} (ν_3), 1294 cm^{-1} (ν_1), and 591 cm^{-1} (ν_2) as well as combinations and overtones.[92] A lattice mode appears[87,88] in the far IR at 13 cm^{-1} and the laser-Raman spectrum at 15 K shows[83] bands at 68 cm^{-1}, 82 cm^{-1} and 124·5 cm^{-1}.

Sulphur dioxide. The solid shows fundamentals about 1360 cm^{-1} (ν_3), 1150 cm^{-1} (ν_1), and 520 cm^{-1} (ν_2) in the near IR. Bands are observed[87] in the far IR at 160 cm^{-1}, 141 cm^{-1}, 101 cm^{-1}, 79 cm^{-1} and 67 cm^{-1}.

Hydrogen sulphide. The solid at 78 K shows[93] fundamentals at 2540 cm^{-1} (ν_3), 2517 cm^{-1} (ν_1) and 1170 cm^{-1} (ν_2) with weaker features at 2722 cm^{-1}, 2622 cm^{-1} and 1184 cm^{-1}. The far IR spectrum is also available.[94]

Acetylene. The spectrum of a polycrystalline film at 63 K has been reported[95] in the range 4500–450 cm^{-1}. The far IR at 77 K shows[96] lattice modes at 122 cm^{-1} and 102 cm^{-1}.

Halocarbons. Only a limited amount of data is available for these compounds in the solid state. Since they are of potential use in matrix studies references are given in Table 2.8 to their IR spectra in other physical states.

Table 2.8 IR spectra of halocarbons

Halocarbon ( )^a	State	Range (μm)	References
CF$_4$ (14)	s (77K) Phase I; (70 K) Phase II.		91 (Figure 2.19)
CHF$_3$ (23)	g	2–24	97, 98
CHClF$_2$ (22)	g	2–30	97, 99
CHCl$_2$F (21)	g	2–30	97, 98
CCl$_2$F$_2$ (12)	g	2–25	97, 98
CClF$_3$ (13)	g	2–24	97, 98
CBrF$_3$ (13B1)	g		100, 101
C$_2$F$_6$ (116)	g		102, 103
C$_2$ClF (115)	g	2–38	101, 104
	l (228 K)	2–22	104
C$_2$Cl$_2$F$_4$ (114)	s (103 K)		105
C$_4$F$_8$ (C318)	g	2–250	106, 107
	l		108

a. Number given to Arcton, Freon, etc.

2.4 References

1. F. A. Mauer, Chapter 5 in *Formation and Trapping of Free Radicals*, eds. A. M. Bass and H. P. Broida, Academic Press, New York, 1960.
2. S. Dushman, *Scientific Foundations of Vacuum Technique*, J. Wiley and Sons, New York, 1949.
3. J. Yarwood, *High Vacuum Technique*, J. Wiley and Sons, New York, 1955.
4. G. W. Green, *The Design and Construction of Small Vacuum Systems*, Chapman and Hall Limited, London, 1968.
5. G. K. White, *Experimental Techniques in Low Temperature Physics*, Oxford University Press, London, 1959.
6. Science Research Council, London; *Cryogenic Equipment*, Issue No. 2, 1965.
7. W. Steckelmacher, *J. Sci. Instr.*, **28**, 10, (1951).
8. W. H. Keesom and C. J. Matthijs, *Physica*, **2**, 623, (1935).
9. National Bureau of Standards, Boulder, Colorado, U.S.A., Report No. 8750, 9249, 9712, 9719, 9721.
9a. D. G. Baddeley, *Handbook of Low Temperature Measurement*, Cryophysics SA, Geneva, Nov. (1971), compiled from material presented at the 5th Symposium on Temperature, Washington, June 1971.
9b. L. G. Rubin, *Cryogenics*, **10**, 14, (1970).

10. A. Wexler, *J. Appl. Phys.*, **22**, 1463, (1951).
11. S. G. Sydoriak and H. S. Sommers, jr., *Rev. Sci. Instr.*, **22**, 915, (1951).
12. L. J. Schoen, L. E. Kuentzel and H. P. Broida, *Rev. Sci. Instr.*, **29**, 663, (1958).
13. W. J. Duerig and I. L. Mador, *Rev. Sci. Instr.*, **23**, 421, (1952).
14. P. Rowland Davies, *Disc. Faraday Soc.*, **48**, 181, (1969).
15. A. K. Stober, *National Bureau of Standards Tech. News Bull.*, **43**, 146, (1959).
16. R. F. Broom and A. C. Rose-Innes, *J. Sci. Instr.*, **33**, 420, (1956).
17. H. P. Hernandez, J. W. Mark and R. D. Watt, *Rev. Sci. Instr.*, **28**, 528, (1957).
18. D. White and D. E. Mann, *Rev. Sci. Instr.*, **34**, 1370, (1963).
19. Air Products and Chemicals Inc. (Allentown, Pennsylvania, U.S.A.), *Cryo-Tip operation manual*.
20. Hymatic Engineering Co. Ltd., Redditch, Worcestershire, U.K.
21. J. M. Geist and P. K. Lashmet, *Advan. Cryog. Eng.*, **5**, 324, (1960).
22. J. M. Geist and P. K. Lashmet, *Advan. Cryog. Eng.*, **6**, 73, (1961).
23. S. N. Foner, F. A. Mauer and L. H. Bolz, *J. Chem. Phys.*, **31**, 546, (1959).
24. M. M. Rochkind, *Applied Spectroscopy*, **22**, 313, (1968).
25. J. W. L. Kohler, *Sci. Amer.*, **212**, 119, (1965).
26. North American Philips Co. Inc., Ashton, Rhode Island, U.S.A.
27. Cryogenic Technology Inc., Waltham, Mass., U.S.A.
28. A. Thomas, *Trans. Faraday Soc.*, **57**, 1679, (1961).
29. J. E. Bennett and A. Thomas, *Proc. Roy. Soc.*, **A280**, 123, (1964).
30. B. Mile, *Angew. Chem.*, **7**, 507, (1968).
31. J. E. Bennett, B. Mile and A. Thomas, *J. Chem. Soc.*, **A**, 1393, (1967).
32. M. J. Linevsky, *J. Chem. Phys.*, **34**, 587, (1961).
33. M. M. Rochkind, *Anal. Chem.*, **39**, 567, (1967); **40**, 762, (1968).
34. A. J. Barnes, H. E. Hallam and G. F. Scrimshaw, *Trans. Faraday Soc.*, **65**, 3172, (1969).
35. E. D. Becker and G. C. Pimentel, *J. Chem. Phys.*, **25**, 224, (1956).
36. G. C. Pimentel in Chapter 4, Ref. 1.
37. G. Boato, *Cryogenics*, **4**, 65, (1964).
38. G. K. White and S. B. Woods, *Phil. Mag.*, **3**, 785, (1958).
39. G. L. Pollack, *Rev. Mod. Phys.*, **36**, 748, (1964).
40. H. M. Roder, *Cryogenics*, **2**, 302, (1961–2).
41. A. N. Gerritsen and P. van der Star, *Physica*, **9**, 503, (1942).
42. J. T. Clarke and R. Gordon, jr., *J. Chem. Phys.*, **32**, 705, (1960).
43. D. R. Johnston, G. J. Oudemans and R. H. Cole, *J. Chem. Phys.*, **33**, 1310, (1960).
44. A. J. Eatwell and G. O. Jones, *Phil. Mag.*, **10**, 1059, (1964).
45. I. Lefkowitz, K. Kramer, M. A. Shields and G. L. Pollack, *J. Appl. Phys.*, **38**, 4867, (1967).
46. G. O. Jones and J. M. Woodfine, *Proc. Phys. Soc.*, **86**, 101, (1965).
47. R. L. Amey and R. H. Cole, *J. Chem. Phys.*, **40**, 146, (1964).
48. K. Kramer, I. Lefkowitz and G. L. Pollack, *Bull. Am. Phys. Soc.*, **10**, 31, (1965).
49. B. L. Smith and C. J. Pings, *J. Chem. Phys.*, **48**, 2387, (1968).
50. E. J. Stansbury, M. F. Crawford and H. L. Welsh, *Can. J. Phys.*, **31**, 954, (1953).
51. N. J. Bridge and A. D. Buckingham, *Proc. Roy. Soc.*, **A295**, 334, (1966).
52. D. E. Stogryn and A. P. Stogryn, *Mol. Phys.*, **11**, 371, (1966).
53. A. D. Buckingham, *Quart. Revs.*, **13**, 183, (1959).
54. S. Kielich, *Physica*, **31**, 444, (1965).
55. Krishnaji and V. Prakash, *Rev. Mod. Phys.*, **38**, 690, (1966).
56. R. J. W. LeFevre, *Dipole Moments*, Methuen and Co. Ltd., London, 1953.
57. J. W. Smith, *Electric Dipole Moments*, Butterworths, London, 1955.

58. P. Debye, *Polar Molecules*, Dover Publications Inc., N.Y., 1947.
59. C. P. Smyth, *Dielectric Behaviour of Structure*, McGraw-Hill Co. Inc., N.Y., 1955.
60. N. J. Bridge and A. D. Buckingham, *J. Chem. Phys.*, **40**, 2733, (1964).
61. A. D. Buckingham, *Chemistry in Britain*, **1**, 54, (1965).
62. A. D. Buckingham and R. L. Disch, *Proc. Roy. Soc.*, **A273**, 275, (1963).
63. W. E. Quinn, J. M. Baker, J. T. LaTourrette and N. F. Ramsey, *Phys. Rev.*, **112**, 1929, (1958).
64. G. Gräff and O. Rundfsson, *Z. Phys.*, **187**, 140, (1965).
65. R. H. Orcutt, *J. Chem. Phys.*, **39**, 605, (1963).
66. S. Kielich, *Mol. Phys.*, **9**, 549, (1965).
66a. A. E. Curson and A. T. Pawlowicz, *Proc. Phys. Soc.*, **85**, 375, (1965).
66b. E. M. Horl and J. A. Suddeth, *J. Appl. Phys.*, **32**, 2521, (1961).
67. C. S. Barrett and L. Meyer, *J. Chem. Phys.*, **42**, 107, (1965).
68. S. C. Greer and L. Meyer, *Z. Angew. Phys.*, **27**, 198, (1969).
69. K. Tomita, *Phys. Rev.*, **89**, 429 (1953).
70. H. M. James and T. A. Keenan, *J. Chem. Phys.*, **31**, 12, (1959).
71. J. T. Thomas, N. L. Alpert and H. C. Torrey, *J. Chem. Phys.*, **18**, 1511, (1950).
72. A. G. Maki, *J. Chem. Phys.*, **35**, 931, (1961).
73. J. B. Davies and H. E. Hallam, *J. Chem. Soc. Faraday II*, **68**, 509, (1972).
73a. J. Hertczeg and R. E. Stoner, *J. Chem. Phys.*, **54**, 2284, (1971).
73b. S. C. Greer and L. Meyer, *J. Chem. Phys.*, **52**, 468, (1970).
74. W. F. Giauque and J. O. Clayton, *J. Amer. Chem. Soc.*, **55**, 4875, (1933).
75. L. H. Bolz, M. E. Boyd, F. A. Mauer and H. S. Peiser, *Acta Cryst.*, **12**, 247, (1959).
75a. T. H. Jordan, H. W. Smith, W. E. Streib, and W. N. Lipscomb, *J. Chem. Phys.*, **41**, 756, (1964).
76. B. Katz, A. Ron and O. Schnepp, *J. Chem. Phys.*, **46**, 1926, (1967); A. J. Barnes, J. B. Davies, H. E. Hallam, G. F. Scrimshaw, G. C. Hayward and R. C. Milward, *Chem. Comm.*, 1089, (1969).
77. J. Obriot, P. Marteau, H. Vu and B. Vodar, *Spectrochim. Acta*, **26A**, 2051, (1970).
78. B. R. Cairns and G. C. Pimentel, *J. Chem. Phys.*, **43**, 3432, (1965).
79. J. E. Cahill and G. E. Leroi, *J. Chem. Phys.*, **51**, 97, (1969).
80. R. V. St. Louis and B. Crawford, jr., *J. Chem. Phys.*, **37**, 2156, (1962); *Canad. J. Chem.*, **40**, 1998, (1962).
81. A. Anderson and G. E. Leroi, *J. Chem. Phys.*, **45**, 4359, (1966); A. Ron and O. Schnepp, *J. Chem. Phys.*, **46**, 3991, (1967).
82. M. Britt, A. Ron and O. Schnepp, *J. Chem. Phys.*, **51**, 1318, (1969).
83. J. E. Cahill and G. E. Leroi, *J. Chem. Phys.*, **51**, 1324 (1969).
84. S. H. Walmsley and A. Anderson, *Mol. Phys.*, **7**, 411, (1964); J. G. David and W. B. Person, *J. Chem. Phys.*, **48**, 510, (1968); V. Wagner, *Phys. Letters*, **22**, 58, (1966).
85. J. E. Cahill and G. E. Leroi, *J. Chem. Phys.*, **51**, 4514, (1969).
86. G. E. Ewing and G. C. Pimentel, *J. Chem. Phys.*, **35**, 925, (1961).
87. A. Anderson and H. A. Gebbie, *Spectrochim. Acta*, **21**, 883, (1965).
88. A. Anderson and S. H. Walmsley, *Mol. Phys.*, **7**, 583, (1964).
89. A. Rosenberg and G. Birnbaum, *J. Chem. Phys.*, **52**, 683, (1970).
90. M. Brith and A. Ron, *J. Chem. Phys.*, **50**, 3053, (1969).
91. R. P. Fournier, R. Savoie, F. Bessette and A. Cabana, *J. Chem. Phys.*, **49**, 1159, (1968).
92. D. A. Dows, *J. Chem. Phys.*, **26**, 745, (1957), *Spectrochim. Acta*, **13**, 308, (1959); A. Le Roy and P. Jouve, *Compt. rend.*, **264B**, 1656, (1967).
93. J. B. Lohman and D. F. Hornig, *Phys. Rev.*, **79**, 235, (1950).

94. T. Osaka and S. Takahashi, *J. Phys. Soc. Japan*, **25**, 1654, (1968).
95. G. L. Bottger and D. F. Eggers, jr., *J. Chem. Phys.*, **40**, 2019, (1964).
96. A. Anderson and W. H. Smith, *J. Chem. Phys.*, **44**, 4216, (1966); Y. A. Schwartz, A. Ron and S. Kimel, *J. Chem. Phys.*, **51**, 1666, (1969).
97. E. K. Plyler and W. S. Benedict, *J. Res. Nat. Bur. Stand.*, **47**, 202, (1951).
98. H. W. Thompson and R. B. Temple, *J. Chem. Soc.*, 1422, (1948).
99. H. B. Weissman, A. G. Meister and F. F. Cleveland, *J. Chem. Phys.*, **29**, 72, (1958).
100. W. F. Edgell and C. E. May, *J. Chem. Phys.*, **22**, 1808, (1954).
101. M. Hauptschein, E. A. Nodiff and A. V. Grosse, *J. Amer. Chem. Soc.*, **74**, 1347, (1952).
102. J. R. Nielsen, C. M. Richards and H. L. McMurry, *J. Chem. Phys.*, **16**, 67, (1948).
103. D. G. Williams, W. B. Person and B. Crawford, jr., *J. Chem. Phys.*, **23**, 179, (1955).
104. J. R. Nielsen, C. Y. Liang, R. M. Smith and D. C. Smith, *J. Chem. Phys.*, **21**, 383, (1953).
105. R. E. Kagarise, *J. Chem. Phys.*, **26**, 381, (1957).
106. W. F. Edgell and D. G. Weiblen, *J. Chem. Phys.*, **18**, 571, (1950).
107. H. H. Claassen, *J. Chem. Phys.*, **18**, 543, (1950).
108. R. P. Bauman and B. J. Bulkin, *J. Chem. Phys.*, **45**, 496, (1966).

3 *Molecules trapped in low temperature molecular matrices*

H. E. HALLAM

Contents

3.1 Introduction

Although matrix-isolation spectroscopy is widely employed in the study of
unstable species, its usefulness as a tool in vibrational-rotational spectroscopy
generally is only just becoming recognized. The basis of its utility lies in the
fact that the IR spectrum of a matrix-isolated molecule at cryogenic temperatures usually consists of sharp, purely vibrational absorptions. Thus it
is possible to resolve features separately by as little as 1 cm^{-1} and to measure
the frequency of the absorption with a precision of the order of ± 0.1 cm^{-1};
hence an unambiguous assignment of a band can often be made from
isotopic shifts.

This chapter is devoted to IR studies of molecules condensed from the
vapour phase at about ambient temperatures and trapped in molecular
matrices at cryogenic temperatures. The spectroscopic parameters of diatomic
molecules are well-established and the prime aim of studying them under
isolated conditions is that of furthering our understanding of matrix effects.
In this sense the diatomic molecule acts as a probe of the intermolecular
force field in which it is situated and transmits information concerning its
environment by means of its spectrum. In the early matrix studies it was soon
evident that the environment, even of a noble gas, could have several effects
on the spectral features of the trapped guest species.

3.2 Matrix environmental effects

The spectral changes which are produced in solid solutions generally
resemble those that accompany the dissolution of molecules in liquid
solvents[1,2] but there are several additional effects which arise due to the
matrix environment. At present we can recognize six general effects all of
which may contribute to modify a vibrational band shape, intensity and
frequency and, in some instances, to cause a single vibrational mode to have a
multiplet structure.

3.2.1 Multiple trapping sites

It is assumed that a guest species is usually trapped in a substitutional
site, formed by the removal of one or more matrix molecules, but there are
interstitial holes which, for small solute molecules, might afford alternative
sites. The geometries of these sites are more fully discussed under matrix
properties in Chapter 2. The intermolecular forces between matrix and
absorber molecules will be different for each site and the resulting perturbations of the energy levels (see Chapter 4) may lead to two frequencies for a
diatomic absorber or two sets of frequencies for a polyatomic guest. Adjacent
vacancies may alter the existing site symmetry and crystal imperfections
may provide further sites—dislocation sites. It may be difficult to distinguish

multiple features due to multiple sites from those arising from aggregation. A convincing multiple site interpretation may require a detailed study, varying the matrix and/or deposition time.

3.2.2 Molecular rotation

Most entrapped species will be so tightly held as to prevent rotation and at low temperatures the pure vibrational modes will appear as very sharp absorptions. In a noble gas matrix possessing a sufficiently large cavity, it might be expected, however, that the rotational energy levels of a small guest species would be only slightly perturbed and thus rotation be relatively unhindered. At very low temperatures only the low rotational levels will be appreciably populated, thus for a diatomic molecule only the $R(1)(J = 2 \leftarrow 1)$, $R(0)(J = 1 \leftarrow 0)$, and $P(1)(J = 1 \rightarrow 0)$ transitions would be expected to be observed in the vibration-rotation band, and the $J = 1 \leftarrow 0$ and $J = 2 \leftarrow 1$ transitions in the pure rotation region. Rotational features are identified by reversible intensity changes on temperature cycling.

3.2.3 Medium effect; matrix shift

Under conditions of perfect isolation in any particular site, the guest molecule is subject only to solute-matrix interactions. These will perturb the solute's vibrational energy levels and be reflected in a frequency shift

$$\Delta v = (v_{matrix} - v_{gas})$$

which is analogous to a solvent shift and, for stretching modes, is usually to lower frequency. Like the solvent shift, the matrix-induced shift is an overall sum of bulk dielectric effects, dispersion forces, and specific solute-matrix interactions. In solid solutions repulsive forces can play an important role and matrix shifts thus provide an important source of information concerning the molecular interactions between absorber molecules and surrounding molecules. Theories of matrix shifts are discussed in Chapter 4.

3.2.4 Aggregation

True isolation is achieved only at a very high matrix/absorber ratio, usually greater than 1000; at low M/A ratios molecular aggregates may be formed and trapped in addition to monomers. Molecular association will be greatest for solutes capable of forming hydrogen bonds. In contrast to liquid solutions in which the absorption bands due to hydrogen-bonded multimeric species are usually very broad, considerable narrowing occurs on condensation in matrices. The multiple features due to self-association are usually readily identified from their concentration dependence and from warm-up experiments in which monomers diffuse to form dimers and higher multimers. Care, however, must be taken to exclude traces of impurity species which might hetero-associate.

3.2.5 Coupling with lattice vibrations

For a molecule in any physical state coupling may occur between certain internal vibrational modes. In the solid state additional features may arise due to coupling with external lattice modes. These interactions are of some consequence in the ionic matrices discussed in Chapter 7 but have yet to be reported in the molecular matrices discussed here.

3.2.6 Phonon bands

A trapped impurity species will disturb the host lattice symmetry and may activate otherwise inactive lattice modes which appear as the so-called phonon bands. The introduction to Chapter 7 provides a lucid description of these phenomena.

All these effects may complicate the analysis of matrix spectra and hence knowledge of the nature of environmental effects on the vibrational modes of trapped species is a necessary prerequisite for a complete interpretation of experimental results obtained by the matrix isolation method.

3.3 Diatomic molecules

3.3.1 Heteronuclear diatomic molecules

Hydrogen halides (HHal)

Of all stable molecules hydrogen chloride is the most thoroughly investigated[3-18] in solid matrices. The reasons are mainly due to (a) its ready availability, (b) its large $\partial\mu/\partial q$ and hence intense IR activity, (c) its hydrogen-bonding propensity and (d) the ease of isotopic substitution of D for H. The other hydrogen halides, hydrogen bromide and hydrogen iodide, have also been extensively examined and, to a lesser extent, so has hydrogen fluoride. These studies will be discussed at some length since all of the environmental effects described above are well illustrated in these systems.

The first aim was to study the rotational motions of these small molecules and it is thus of interest to compare the results with those of the same systems in other phases (Figure 3.1). When pressurized with argon or with nitrogen the discrete rotational fine structure of the P branch and R branch of the gas phase hydrogen chloride fundamental vibration is replaced by a broad band envelope together with an induced Q branch which greatly intensifies as the pressure of the foreign gas is increased. Similar features are observed[19] when hydrogen chloride is dissolved in liquid argon at 113 K and solid argon at 103 K but in each case with the Q branch even more dominant. These results by the CNRS group at Bellevue, Paris, are interpreted by them in terms of the hydrogen chloride molecules retaining some rotational freedom in all these phases, the Q branch arising due to hydrogen chloride/argon interactions.

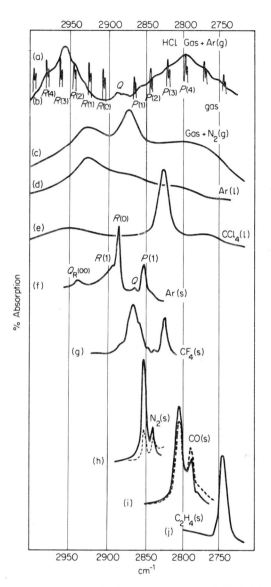

Figure 3.1 Monomer fundamental absorption region of HCl (*a*) gas (250 mmHg 1 = 10 cm), (*b*) gas (0·9 atm, 1 = 5 cm) pressurized with Ar (300 atm), (*c*) gas (*c*. 1 atm, 1 = 5 cm) pressurized with N_2 (372 atm); (*d*) solution in liquid Ar (C = 1:3000, 1 = 2·7 cm; 113 K); (*e*) solution in CCl$_4$ (C = *c*. 0·007 M, 1 = 15 mm), (*f*) Ar matrix (M/A = 1000, 20 K), (*g*) CF$_4$ matrix (M/A = 200, 20 K), (*h*) N_2 matrix (M/A = 200, 20 K), broken line refers to warm up experiment, after 10 min at 35 K and re-cooling to 20 K, (*i*) CO matrix (M/A = 200, 20 K), broken line, after 30 min at 40 K and re-cooling to 20 K, (*j*) C$_2$H$_4$ matrix (M/A = 500, 20 K). (Pressurized gas and liquid spectra re-drawn from data in ref. 19; matrix spectra re-drawn from data in ref. 14; CCl$_4$ spectrum re-drawn from L. Galatry, D. Robert, P. V. Huong, J. Lascombe, and M. Perrot, *Spectrochim. Acta* **25A**, 1693 (1969))

Broad wings are also observed[20] flanking the fundamental of hydrogen chloride when dissolved in non-polar solvents (e.g. carbon tetrachloride) at room temperature and these likewise have been given a perturbed rotational interpretation.

When, however, hydrogen chloride is isolated in a solid argon matrix at cryogenic temperatures, discrete rotational lines are observed.[3,4,6,7] Their reversible variation of intensity on temperature cycling (e.g. 20 K → 35 K → 20 K) are characteristic of low-lying energy states, and bands (Table 3.1) are readily assigned to $R(0)$, $R(1)$ and $P(1)$ transitions corresponding to quantized rotational motion of hydrogen chloride in the host lattice. A weak Q branch feature is also seen. The pure rotational transition $J = 1 \leftarrow 0$ has also been observed[15,17] in the far IR at $18 \cdot 6 \text{ cm}^{-1}$. The same absorption has been found for hydrogen chloride trapped in a quinol cage at 4 K in the form of a hydrogen chloride-quinol clathrate (see Chapter 8).

A translational mode of hydrogen chloride in an argon matrix is found[12] at 73 cm^{-1} and is able to couple with the vibrational mode to produce a weak band at 2944 cm^{-1}.

Barnes, Hallam and Scrimshaw[14] have recently examined hydrogen chloride isolated in a wider range of matrices at 20 K; some are depicted in Figure 3.1. It is seen that rotational features occur in carbon tetrafluoride and sulphur hexafluoride matrices. Multiple peaks are also observed in nitrogen, carbon dioxide, and carbon monoxide matrices but which cannot be accounted for as rotational fine structure since (a) the splitting of the main doublet is too small to correspond to an $R(0) - P(1)$ splitting, (b) on temperature cycling irreversible intensity changes occur, (c) in the case of hydrogen chloride/nitrogen no $J = 1 \leftarrow 0$ band is observed[15] in the far IR. The changes which occur on raising the temperature 20 K → 35 K show (Figure 3.1) that the main band, at the higher frequency, diminishes considerably in intensity compared with the minor band; on re-cooling to 20 K the changes are seen to be irreversible. Concomitantly, multimer bands grow in at lower frequencies (Figure 3.2). The only interpretation for this behaviour would seem[14] to be one of multiple trapping sites for a non-rotating monomer. The nature of the two sites is a matter of conjecture, the only information which the IR spectrum can provide indicates that the site giving rise to the lower band causes a slightly greater perturbation of the solute vibrational frequency and is a site less liable to allow diffusion therefrom.

The singlet feature observed in an ethylene matrix is shifted to lower frequency from the gas phase band centre by 138 cm^{-1}, indicative of a fairly strong specific interaction. Taking the mean frequency of $R(0)$ and $P(1)$ for the rotating monomers as a reference it is seen that the solute vibrational frequency shifts systematically downward in frequency in all of the matrices illustrated. This medium effect or matrix shift can be approximately accounted for by summing several interaction terms involving the electrical properties

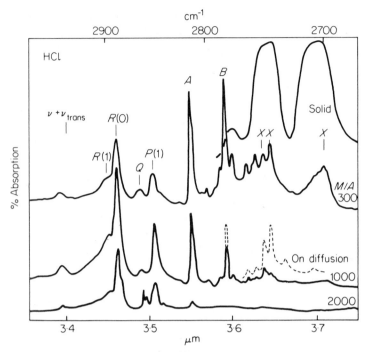

Figure 3.2 Spectra of fundamental region of HCl at various concentrations in Ar matrices at 20 K (reproduced by permission from A. J. Barnes, H. E. Hallam and G. F. Scrimshaw, *Trans. Faraday Soc.* **65**, 3150 (1969))

of solute and matrix molecules and the various treatments of the experimental data are discussed in paragraph 4.2.9.

Deuterium chloride has been examined[18] in the same range of matrices and additionally in methane; the main features are essentially similar to those of hydrogen chloride. In argon (Figure 4.12, Table 3.1), the $R(0)$ and $P(1)$ lines of deuterium chloride are resolved into isotopic doublets and also the $R(1)$ line is split by $c.$ 3 cm^{-1}; furthermore the $R(2)$ and $P(2)$ lines are observed (see Table 4.11). A detailed discussion of the interpretation and comparison of these results is more appropriately left until paragraph 4.3.4. In methane a doublet occurs (2062·6 cm^{-1} and 2052·3 cm^{-1}) in the 'monomer' region of deuterium chloride. The lower frequency and weaker feature is very concentration dependent and at $M/A = 860$ appears only as a weak shoulder, a behaviour indicative of a multimer species. The higher frequency band is thus taken to be that of a nonrotating monomer.

Hydrogen bromide,[14] deuterium bromide,[21] hydrogen iodide[14] and deuterium iodide[22] have also been examined in a similar range of matrices and hydrogen fluoride and deuterium fluoride have been studied[23] in all of the noble gases.

Table 3.1 Rotational features (in cm^{-1}) of hydrogen chloride in gas and argon matrix

Band	HCl		DCl	
	Gas	Ar matrix	Gas	Ar matrix
$J = 0 \rightarrow 1$	20·9	18·6	10·8	10·9
Translation				
$(J = 0; n = 0 \rightarrow 1)$		73		72
$v + v_{\text{trans}}Q_R(00)$		2944·0		2149·4
$R(2)$ $^{35/37}$Cl	2944·4	2913	2121·0	2108·0
$R(1)\begin{cases} T_{1u} \rightarrow E_g \\ T_{1u} \rightarrow T_{2g} \end{cases}$	2925·4	2897	2110·9	2098·7 / 2096·1
$R(0)\begin{cases} ^{35}\text{Cl} \\ ^{37}\text{Cl} \end{cases}$	2906·2 / 2904·1	2888·2 / 2885·9	2101·6 / 2098·6	2090·3 / 2087·8
$P(1)\begin{cases} ^{35}\text{Cl} \\ ^{37}\text{Cl} \end{cases}$	2865·1 / 2863·0	2854·2 / 2852·5	2080·3 / 2077·3	2070·1 / 2067·0
$P(2)$ $^{35/37}$Cl	2843·2	~2844	2068·3	2060·5

The results discussed above utilize the technique in the manner for which it was originally developed, i.e. for *isolating* monomeric species. The method has, however, great potentialities for the study of the aggregation behaviour of molecules, particularly those which possess a hydrogen atom capable of forming hydrogen bonds. A detailed study has been made of the association characteristics of hydrogen chloride, bromide and iodide.[14] These studies consist of examining the solute over a wider spectral region and over a large range of concentrations (Figure 3.2) and plotting growth curves of the multiple spectral features. These are carried out in conjunction with 'warm-up' experiments in which monomers are partly allowed to diffuse to form higher multimers.

For hydrogen chloride, at $M/A = 2000$, the spectrum obtained is that of the monomer, together with a weak band at 2818 cm^{-1}(A). As the matrix ratio is decreased, this band becomes very intense and a second strong band grows at 2787 cm^{-1}(B). A series of medium intensity bands also appears at 2781 cm^{-1}, 2768 cm^{-1}, 2761 cm^{-1}, 2754 cm^{-1} and 2748 cm^{-1}, of which the 2781 cm^{-1} band grows at a rate similar to band B, the 2768 cm^{-1} and 2761 cm^{-1} bands grow at comparable rates (faster than band B), while the 2754 cm^{-1} and 2748 cm^{-1} bands rapidly become very strong as the concentration of hydrogen chloride is increased. In the most concentrated matrix ($M/A = 300$) a broad absorption also appears at 2701 cm^{-1}. The 2754 cm^{-1}, 2748 cm^{-1} and 2701 cm^{-1} bands (X) are close to the frequencies found[25] for solid hydrogen chloride and are therefore probably due to large aggregates of hydrogen chloride molecules. The behaviour on diffusion is similar to that caused by increasing the concentration of the solute, but the high polymer bands (X) increase in intensity even more rapidly. From these growth rates, the other absorptions are assigned as $2818\ cm^{-1}$ dimer, $2787\ cm^{-1}$ and $2781\ cm^{-1}$ trimer, $2768\ cm^{-1}$ and $2761\ cm^{-1}$ tetramer. The multiplicity of bands have been interpreted by Keyser and Robinson[8] (KR) in terms of an intermolecular resonance interaction model. The principal assumptions underlying the KR treatment are: (a) that the multimers comprise hydrogen chloride molecules in adjacent substitutional sites, orientated in a similar manner to the pure solid halide, and (b) that intermolecular resonance interactions in the multimer will produce a number of components, split by amounts depending on the intermolecular force constant (f_{12}) of the pure halide.[24] There is some disagreement as to the sign to be taken for f_{12} but the most reasonable value seems to be $-6.5\ Nm^{-1}$, due to Savoie and Anderson,[25] which gives a fairly good prediction[14] of the bands due to dimer, trimer and tetramer species. The dimer is taken to have a cyclic structure since Katz, Ron and Schnepp[12] have shown in their far IR studies of hydrogen chloride, deuterium chloride and hydrogen chloride/deuterium chloride mixtures in noble gas matrices that the mixed dimer has only one low frequency hydrogen-bond stretching mode ($170\ cm^{-1}$) intermediate in frequency between the hydrogen chloride dimer ($185\ cm^{-1}$) and the deuterium chloride dimer ($147\ cm^{-1}$; all values in xenon). Furthermore no intense low frequency hydrogen-bond bending mode is observed as in the case of the linear hydrogen cyanide dimer.[26] Since the dimer is cyclic it is reasonably assumed[14] that the trimer is also cyclic. The assumption made by Keyser and Robinson of only open-chain multimers is thus discarded since in any case it leads to assignments incompatible with the experimental growth curves.

A near IR study[18] of the aggregation band pattern of hydrogen chloride/ deuterium chloride mixtures in both the $\nu(HCl)$ and the $\nu(DCl)$ regions allows a detailed picture of the trimer species to be built up and also provides a more rigorous test of the applicability of the KR model. The chloride to argon ratio was kept constant at 1 : 200 while the percentage deuterium chloride was varied from 25 to 85 per cent. Changes are seen in the spectral features in the deuterium chloride dimer and trimer regions compared with pure deuterium chloride/argon and corresponding changes are observed in the hydrogen chloride region as compared with the spectra[14] of pure hydrogen chloride/ argon. At high deuterium chloride concentrations the band at $2041.3\ cm^{-1}$ in the deuterium chloride dimer region is only a shoulder on the main $2040.0\ cm^{-1}$ band. At low deuterium chloride concentrations the $2041.3\ cm^{-1}$ feature becomes dominant and the $2040\ cm^{-1}$ band is lost beneath it. Similarly, in the hydrogen chloride dimer region at high hydrogen chloride concentrations, the main band is at $2819.3\ cm^{-1}$ with a shoulder at $2817.6\ cm^{-1}$ whereas at low hydrogen chloride concentrations the latter is the dominant feature. Clearly, therefore, the bands at $2041.3\ cm^{-1}$ and $2817.6\ cm^{-1}$ are due to dimeric HCl–DCl species. The KR model predicts a band at the *group centre frequency* (which is considered a more satisfactory description than the term *centre of gravity frequency* used by KR) of the dimer, λ_2^D for deuterium chloride (2033 cm^{-1}), and λ_2^H for hydrogen chloride (2808 cm^{-1}). The group centres are calculated from the observed value of $\lambda_2 + \lambda'$ for the *pure* dimer by using Savoie and Anderson[25] of f_{12} and $\lambda = 4\pi^2\nu^2$.

The trimer regions are depicted in Figure 3.3 from which it can be seen that the mixed trimer species give rise to bands at 2776·8 cm^{-1}, 2774·4 cm^{-1} and 2768·9 cm^{-1} in the hydrogen chloride region and at 2014·2 cm^{-1}, 2011·3 cm^{-1} and 2006·6 cm^{-1} in the deuterium chloride region (Table 3.2). These

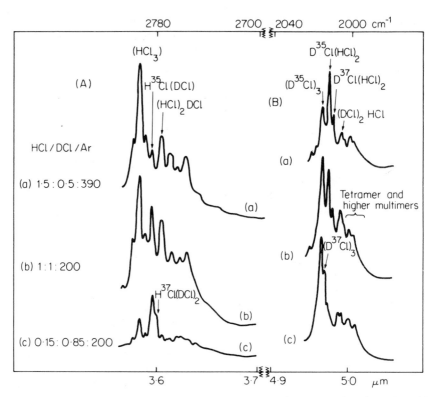

Figure 3.3 Spectra of mixed trimer species in (A) HCl fundamental region, (B) DCl fundamental region, at various HCl/DCl ratios in Ar at 20 K. (Re-drawn from data in ref. 18)

assignments provide a very sensitive test of the KR model since there can be no resonance coupling between bonded hydrogen chloride and deuterium chloride molecules, the splitting into components depending upon the number of isotopically identical molecules in nearest-neighbour contact. However, since the hydrogen-bonding capabilities of hydrogen chloride and

Table 3.2 Dimer and trimer assignments for HCl, DCl and their mixtures in argon matrices

Multimer	Band	DCl region Observed	DCl region Calculated	HCl region Observed	HCl region Calculated
(HCl)₂	$\lambda_2^H + \lambda'$			2818·0	
(DCl)₂	$\lambda_2^D + \lambda'$	{ 2040·0 (^{35}Cl), 2037·1 (^{37}Cl)			
HCl-DCl	$\lambda_2^{H,D}$	2041·3	2033	2817·6	2808
[cyclic trimer: Cl\ / D D / Cl······D—Cl]	$\lambda_3^D + \lambda'$	{ 2018·6 (^{35}Cl), 2017·0 (^{37}Cl)			
	$\lambda_3^D - \lambda'$		1997		
[cyclic trimer: Cl\ / D D / Cl······H—Cl]	$\lambda_3^D - \lambda'$	2006·6	2002		
	$\lambda_3^D + \lambda'$	—	2018		
	λ_3^H			{ 2776·8 (^{35}Cl), 2794 (^{37}Cl)	2776
[cyclic trimer: Cl\ / D H / Cl······H—Cl]	λ_3^D	{ 2014·2 (^{35}Cl) 2012, 2011·3 (^{37}Cl) 2010			
	$\lambda_3^H - \lambda'$			2768·9	2766
	$\lambda_3^H + \lambda'$			—	2786
[cyclic trimer: Cl\ / H H / Cl······H—Cl]	$\lambda_3^H + \lambda'$			2786·1	
	$\lambda_3^H - 2\lambda'$				2755

deuterium chloride are, to a first approximation, the same, the group centre of the components will depend on the total number of molecules in the aggregate whatever their isotopic identity. Thus, a cyclic trimer of type DCl(HCl)₂ would, with respect to hydrogen chloride, have the group centre of the trimer but the splitting of the dimer, while one of the type (DCl)₂HCl would have the group centre of the trimer and no splitting. With respect to deuterium chloride the latter would have the trimer group centre and the dimer splitting pattern.

In the trimer region of deuterium chloride the dominant features at 2018·6 cm^{-1} and 2017·0 cm^{-1} are assigned to the $\lambda_3^D + \lambda'$ band of (D^{35}Cl)₃ and (D$_3^{35}$Cl$_2^{37}$Cl) superimposed and D$_3^{35}$Cl^{37}Cl₂ respectively; these then give calculated values for the group centre frequency of these species of 2012 cm^{-1} and 2010 cm^{-1} respectively. The cyclic $(HCl)_2DCl$ trimer should have one band, λ_3^D, at the group centre frequency and two bands are observed at 2014·2 and 2011·3 cm^{-1} which are assigned to the λ_3^D band for this species for $D^{35}Cl$ and $D^{37}Cl$ respectively. It is seen that the agreement is much better than for the dimer. The variation in intensity of these bands with the deuterium to hydrogen ratio suggests that they are isotopic analogues and that they arise from a trimer with two hydrogen chloride molecules and one deuterium chloride molecule. The $HCl(DCl)_2$ cyclic trimer should have bands corresponding to $\lambda_3^D + \lambda'$ and $\lambda_3^D - \lambda'$ in the deuterium chloride region. The latter should fall around 2002 cm^{-1} and a band is observed at 2006·6 cm^{-1} which increases in intensity as the hydrogen to deuterium ratio goes from 15 per cent. to 50 per cent. but decreases when it reaches 67 per cent. This suggests that this band arises from a trimer with two deuterium chloride molecules and is assigned to the $\lambda_3^D - \lambda'$ band of the $HCl(DCl)_2$ cyclic trimer. The $\lambda_3^D + \lambda'$ band of this species would then fall about 6 cm^{-1} higher than the group centre at 2018 cm^{-1} and is presumably lost under the main trimer absorption at 2018·6 cm^{-1}.

In the hydrogen chloride trimer region the $\lambda_3^H + \lambda'$ band of the pure $(HCl)_3$ cyclic trimer is at 2786·1 cm^{-1} and this gives the calculated group centre band, λ_3^H, a value of 2776 cm^{-1}. The $(DCl)_2HCl$ cyclic trimer should thus have a band at this frequency and a band is observed at 2776·8 cm^{-1} with an isotopic shoulder at 2774 cm^{-1} which has its largest relative intensity at 15 per cent. At 50 and 67 per cent. hydrogen chloride it is weaker than the 2786·1 cm^{-1} band, so, clearly, it is due to a trimer containing only one hydrogen chloride molecule. Thus again there appears to be a good correlation between experiment and the KR theory. The $DCl(HCl)_2$ cyclic trimer should have two bands in the hydrogen chloride region, namely the $\lambda_3^H \pm \lambda'$ bands which should be approximately at 2786 cm^{-1} and 2766 cm^{-1}. The former band would fall under the most intense band in the region at 2786·1 cm^{-1} and so is lost. A band exhibiting the correct intensity-to-hydrogen-chloride-content relationship is observed at 2786·9 cm^{-1} and is thus assigned to the $\lambda_3^H - \lambda'$ band of the $DCl(HCl)_2$ trimer.

The inadequacy of the KR model to deal with the cyclic dimer is probably due to their configuration which involves two bent hydrogen bonds being far removed from that of the crystalline solid. In the case of the cyclic trimer the configuration involves essentially linear hydrogen bonds.

Hetero-association studies of binary mixtures of[14] hydrogen chloride, hydrogen bromide or hydrogen iodide with nitrogen, carbon monoxide, carbon dioxide or C_2H_4, (X) and of[18] DCl/X have been carried out in argon matrices at 20 K. Three regions of absorption are induced in the HHal or DHal spectrum due to the dopant molecule X. The most intense induced band (I) falls in the Q branch region of the monomer spectrum and is interpreted as arising from HHal—X interactions. The second induced band(s) (II) are observed at lower frequencies but still in the monomer region and are assigned to HHal—X—HHal interactions. The third band or group of bands (III) are alongside the dimer band (A) which is greatly reduced in intensity; feature III is thus assigned to (HHal)$_2$—X interactions. In the case of carbon monoxide as the dopant molecule, corresponding perturbations are observed in the carbon monoxide spectrum (Figure 3.4a). An intense pair of bands is induced[18] at 2156·2 and 2155 cm^{-1}, to *higher* frequency than the pure carbon monoxide monomer band (Figure 3.4d) and the intensity pattern of carbon monoxide aggregate bands (Figure 3.4c) is changed, with the two

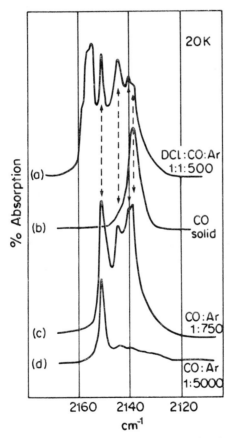

Figure 3.4 Spectra of CO fundamental region (*a*) CO/DCl in Ar Matrix ($M/A = 500$, 20 K), (*b*) solid (20 K), (*c*) Ar Matrix ($M/A = 750$, 20 K), (*d*) Ar matrix ($M/A = 5000$, 20 K). (Re-drawn from data in refs. 18 and 30)

low frequency bands being suppressed. An all-electron MO calculation performed[27] on carbon monoxide has shown that the molecular orbital associated with the oxygen sp lone pair has a slightly antibonding character. If, therefore, the deuterium chloride interacts with an electron in this orbital there would be a slight drain of electron density from the nodal region of the orbital and the C—O distance would contract slightly giving a small shift to higher frequency. The MO associated with the carbon lone pair has slight bonding character so that a shift to lower frequency would be expected if the deuterium chloride were hydrogen-bonded to the carbon. It is thus evident that deuterium chloride (and hydrogen chloride) is interacting with the oxygen atom of carbon monoxide.

Carbon monoxide (CO)

Next to the hydrogen halides carbon monoxide has been the subject of numerous matrix studies.[28,29,30] It appears to be notoriously difficult to isolate and, for such a simple molecule surprisingly there has been considerable controversy over its spectral interpretation. The earliest study, by Leroi, Ewing and Pimentel[28] showed two bands for carbon monoxide in argon at 20 K, one sharp one at 2148·8 cm^{-1} and one broad at 2138·0 cm^{-1}, whose relative intensities are very dependent upon experimental conditions, M/A ratio, deposition rate, etc. They interpreted the upper frequency band to nonrotating monomer and the lower one to carbon monoxide aggregates, being at the same frequency as that of solid carbon monoxide (Figure 3.4b). Charles and Lee[29] also observed doublets for carbon monoxide in krypton (2145·2 cm^{-1}, 2135·2 cm^{-1}) in xenon (2141·0 cm^{-1}, 2133·1 cm^{-1}) at 20 K, the two bands persisting at all mole ratios used down to $M/A = 3900$. In the krypton case the concentration behaviour is indicative of a monomer/aggregate phenomenon. However, their warm-up experiments dispelled this interpretation since they observed both bands to decrease and a new feature at 2147·5 cm^{-1} to grow in. They thus interpreted both bands of their doublets to monomeric species in substitutional and interstitial sites, but for some reason the two frequencies are reversed in the two matrices, i.e. krypton, 2145·2 cm^{-1}(*S*), 2135·3 cm^{-1}(*I*); xenon 2133·1 cm^{-1}(*S*), 2141·0 cm^{-1}(*I*).

Recent work by Davies and Hallam[30] has resolved the issue and provides the IR spectrum of truly isolated carbon monoxide (Figure 3.4d), an M/A ratio of 5000 in argon being necessary to achieve this. The singlet at 2149·4 cm^{-1} shows no variations other than broadening on temperature cycling between 20 K and 30 K and is clearly due to non-rotating monomer. At $M/A = 750$, three other, weaker absorptions appear to lower frequency, at 2143·3 cm^{-1}, 2139·6 cm^{-1} and 2138·0 cm^{-1}. These show no reversible changes in intensity between $20 \rightleftarrows 30$ K but at 40 K, diffusion occurs and the 2138·0 cm^{-1} band grows. As the concentration is increased to 500 : 1 the 2143·3 cm^{-1} band diminishes and at 50 : 1 it disappears leaving only the intense feature at $2138 \cdot 0\,\mathrm{cm}^{-1}$, corresponding to the frequency of solid carbon monoxide, and a shoulder at $2139 \cdot 6\,\mathrm{cm}^{-1}$. This shows that the $2138 \cdot 0\,\mathrm{cm}^{-1}$ band is due to aggregates similar to those in solid carbon monoxide. The $2139 \cdot 6\,\mathrm{cm}^{-1}$ absorption must also be due to large multimers whereas the $2143 \cdot 3\,\mathrm{cm}^{-1}$ feature is due to small multimers, probably dimers. Thus the original interpretation of Leroi et al.[28] of their two features is correct. Slightly better isolation is found[30] in methane matrices at 20 K; a similar aggregation pattern of frequencies is also obtained. It is also interesting to compare these spectra with those of carbon monoxide enclathrated in a quinol cage (Chapter 8).

Nitric oxide (NO)

Guillory and Hunter[31] have made a thorough investigation of nitric oxide trapped in nitrogen matrices, including studies of a 99 per cent. enriched sample of $^{15}\mathrm{NO}$ and a 35 per cent. enriched sample of $\mathrm{N}^{18}\mathrm{O}$. At an M/A ratio of 500 complete isolation of nitric oxide is achieved at 4 K yielding the following monomer frequencies: $1880\,\mathrm{cm}^{-1}$ ($^{14}\mathrm{N}\ ^{16}\mathrm{O}$), $1846\,\mathrm{cm}^{-1}$ ($^{15}\mathrm{N}\ ^{16}\mathrm{O}$) and $1830\,\mathrm{cm}^{-1}$ ($^{14}\mathrm{N}\ ^{18}\mathrm{O}$). Growth patterns of the absorptions occurring in the $1900-1700\,\mathrm{cm}^{-1}$ region when isolated nitric oxide at 4 K is warmed to, or deposited at, 15 K indicate the occurrence of three distinct multimer species. The most stable species is the $cis\mathrm{O}{=}\mathrm{N}{-}\mathrm{N}{=}\mathrm{O}$ dimer (v_{sym}, $1870\,\mathrm{cm}^{-1}$; v_{asym}, $1776\,\mathrm{cm}^{-1}$) having C_{2v} symmetry, while the other two unstable species appear to be the $trans\mathrm{O}{=}\mathrm{N}{-}\mathrm{N}{=}\mathrm{O}$ (v_{asym}, $1764\,\mathrm{cm}^{-1}$) and possibly another cis form (v_{sym}, $1870\,\mathrm{cm}^{-1}$, v_{asym}, $1785\,\mathrm{cm}^{-1}$). Attempts to corroborate these structures by observing their low frequencies proved unsuccessful. The force constants calculated using a valence-bond potential indicate that the dimers retain a strong double-bond character $\mathrm{N}{=}\mathrm{O}$ bond and a relatively weak $\mathrm{N}{-}\mathrm{N}$ bond. The production of these dimers by diffusion and by photolysis of nitrous oxide at 77 K indicates that these reactions occur with practically zero activation energies.

3.3.2 Homonuclear diatomic molecules

Free homonuclear diatomic molecules, because of their symmetry, possess no IR vibrational or rotational spectra. However, in the compressed state or in condensed phases such molecules acquire IR activity because of the dipole moment induced by intermolecular forces during collisions.

Hydrogen (H_2)

Kriegler and Welsh[32] have described a very elegant technique for growing transparent hydrogen-doped argon crystals near the normal freezing point of pure argon by very slow cooling (c. $0 \cdot 05\,^{\circ}\mathrm{C}$ per hour) of a 10 cm absorption cell containing liquid argon in which hydrogen had been dissolved at c. 25 atmospheres. They were able to obtain adequate transmission of an IR beam after twice passing through the resulting crystal, and achieve the feat of obtaining an IR spectrum in the 2·5 μm region of a guest species in a 20 cm thick solid matrix. They were thus able to record the induced IR fundamental band of hydrogen at an M/A ratio of 100 in argon at 82 K; this consisted (Figure 3.5) of three sets of maxima corresponding to the Q, $S(0)$, and $S(1)$ transitions expected at this temperature. The designations $Q(J)$ and $S(J)$

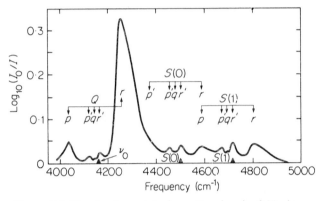

Figure 3.5 The fundamental absorption band of H_2 in a H_2-doped Ar crystal (reproduced by permission from R. J. Kriegler and H. L. Welsh, *Canad. J. Phys.* **46**, 1181 (1968)

refer to transitions in which the fundamental vibrational transition is accompanied by the rotational transitions, $\Delta J = 0$ and $\Delta J = 2$, respectively, J denoting the initial rotational quantum number. The positions of the $Q(0)$ ($\equiv \nu_0$, the band origin) $S(0)$, and $S(1)$ transitions, as calculated from the constants of the free gas molecule, are marked on the frequency axis in Figure 3.5. An analysis of the frequencies of the observed maxima shows that each of the three hydrogen transitions gives rise to the same symmetrical pattern of five lines, consisting of a central component, q, flanked by a higher-frequency pair of components, r' and r, and a lower-frequency pair, p' and p. The fact that structures associated with the $S(0)$ and $S(1)$ transitions are clearly identifiable in the spectrum shows that the hydrogen molecule has practically free rotation in solid argon, just as in solid hydrogen.[33] From the symmetry of the structures it seems that the central q components are to be associated with purely hydrogen transitions (zero-phonon lines), whereas the other components must be assigned to combinations of the hydrogen transitions with transitions in the lattice vibrational spectrum of the hydrogen-doped argon crystal.

The frequencies are shifted from the free-molecule gas values by the sum of a vibrational shift, $\Delta\nu_{vib} = 17\,\mathrm{cm}^{-1}$, and rotational shifts corresponding to

$\Delta B = -0.52 \text{ cm}^{-1}$. The three transitions show similar patterns of five maxima, each of which is analysed as a zero-phonon line (purely hydrogen transitions) and as sum-and-difference tones with lattice vibrational frequencies, 112 cm^{-1} and 22 cm^{-1}, of the hydrogen-doped argon crystal. Because of the inversion symmetry at the impurity sites, the model of single hydrogen molecules occupying substitutional sites in a f.c.c. argon crystal cannot account for the presence of zero-phonon lines. It is therefore considered probable that the structure of solid argon is h.c.p. in the neighbourhood of the hydrogen impurities, and is based on the reported observations[34] of such polymorphic changes for argon crystallized in the presence of impurities. The same interpretation has also been invoked by Von Holle and Robinson[17] for their observed 0.8 cm^{-1} splitting of the $J = 1 \leftarrow 0$ rotational transition of hydrogen fluoride in argon at 4 K. If argon crystallizes with h.c.p. in the presence of an impurity rather than c.c.p. for the pure element, the site symmetry of D_{3h} would split the $J = 1$ level[35] (see Figure 4.10).

The CNRS group at Bellevue, France, have also reported a series of similar studies[36] for H_2 and D_2. However, their use of high pressures (1700 atm for hydrogen) makes quantitative interpretation of their spectra difficult. In the case of deuterium the multiplet structure of the fundamental mode in the near IR can be readily interpreted as sum-and-difference modes of the observed phonon bands in the far IR.

In the light of the path lengths required in the above experiments it is not surprising that the induced IR fundamental of nitrogen or of oxygen in argon has not been observed,[14] even in the presence of interacting HX molecules, or that of the halogens, X_2, in the many photolytic studies.

3.4 Triatomic molecules

Triatomic molecules have been studied in low temperature matrices usually with the view to looking for rotational and aggregation features.

3.4.1 Linear triatomic molecules

Hydrogen cyanide (HCN)

Hydrogen cyanide was included as one of the first simple molecules studied by Becker and Pimentel[37] in establishing the MI technique. Monomeric and multimeric absorptions are reported in nitrogen at 20 K in a comparative study of the relative efficiencies of isolation of several molecules in various matrices. These preliminary observations have recently been greatly extended by King and Nixon[38] with the higher resolution and wider spectral range now available. They have examined hydrogen cyanide and deuterium cyanide in argon, nitrogen and carbon monoxide over a wide range of concentrations, at temperatures from 4.5 K to 20.5 K, with temperature cycling experiments over the range 4·5–45 K. The monomer frequencies in argon (Table 3.3) are not far removed from the gas-phase values but appreciable shifts are found in nitrogen and carbon monoxide matrices. With increasing solute concentrations a multiplicity of both sharp and broad absorption peaks arise, associated with each vibrational mode. Growth curves allow these additional features to be assigned to dimer and multimer species.

Table 3.3 Monomer frequencies (cm^{-1}) of HCN in the gas state at 293 K and in various matrices[38] at 20·5 K

Frequency	g	Ar	N_2	CO
v_3	3311	3303·3	3287·6	3261·2
v_1	2097	2093·4	2097·3	2104
v_2	712	720·2	$\begin{cases} 745·6 \\ 736·0 \end{cases}$	$\begin{cases} 761·3 \\ 739·0 \end{cases}$

There is no evidence for molecular rotation. No temperature dependence of monomer or dimer band intensities is observed in warm-up and cooldown experiments from 4·5 K to 20·5 K. Temperature cycling to 35–45 K resulted in changes in relative intensities due to monomer diffusion. If some form of restricted rotation of the monomer were occurring, the rotational lines should have a spacing somewhat smaller than that of the freely rotating gas molecule ($c.$ 3 cm^{-1}). In the v_2 region, however, where the spectral slit width was 0·8 cm^{-1}, the single peak of monomeric hydrogen cyanide in argon at 20·5 K has an apparent half-width of $c.$ 1·4 cm^{-1} and appears to be quite symmetrical. In this same spectral region, the bands assigned to the dimer, for which rotation in the matrix is hardly a reasonable possibility, have similar halfwidths as the monomer band.

Unequivocal evidence is provided for the structure of the dimer. In the 3 μm region the dimer has two peaks due to v(CH), 3301·3 cm^{-1} and 3202 cm^{-1}(Ar); the fact that one of these C—H stretches is almost coincident with that of the monomer while the other is shifted down by 101 cm^{-1} is strong evidence for a hydrogen-bonded linear or nearly-linear dimer. This is supported by the observation of an analogous pair of bands in the v(CN) stretching region and in the hydrogen cyanide bending region; a bent dimer should exhibit four hydrogen cyanide bending modes, the relative frequencies of which would depend upon the degree of non-linearity. Four modes are observed in both nitrogen and carbon monoxide but are attributed to site splitting of degenerate modes, since v_2 of the monomer is also split in both these matrices. The linear structure is confirmed by the observation of bands at 1440 cm^{-1} and 1555 cm^{-1} attributable to $2v_2$ (1415 cm^{-1} for hydrogen cyanide monomer) the appearance of such an overtone precluding a cyclic structure with a centre of symmetry. Finally, a linear dimer would have two degenerate low-frequency librational modes of the hydrogen cyanide units and these are observed in the far IR of a nitrogen matrix at 96 cm^{-1} and 137 cm^{-1} and 144 cm^{-1}, the latter being split by site effects. The Lippincott–Schroeder[39] one-dimensional model of the hydrogen-bond potential function is used to calculate frequencies for the stretching of the hydrogen bond of the dimer and for the torsion about the hydrogen bond. Fair agreement is obtained for the latter, 153 cm^{-1}, with the observed 137/144 cm^{-1} doublet.

The near IR spectra in nitrogen matrices exhibit many fewer multimer peaks under the same deposition conditions, indicating that, for hydrogen cyanide, nitrogen has a greater isolating efficiency. A similar pattern of results is found in carbon monoxide matrices, which is not surprising in view of the similarities in the crystal structures of carbon monoxide and nitrogen (see Chapter 2). The doublet spacings (of the monomer v_2 and the dimer bends) in carbon monoxide are about twice those in nitrogen, indicating that the field at the site in the matrix of the slightly polar carbon monoxide deviates further from cylindrical symmetry.

Hydrogen isocyanide (HNC)

Vacuum-UV photolysis[40] of hydrogen cyanide in argon and nitrogen matrices at 14 K leads to the production of HNC in concentration sufficient for IR observation of all three fundamentals: H ^{14}N ^{12}C, 3620 cm^{-1}, 2029·2 cm^{-1}, 477 cm^{-1} in argon. These results of Milligan and Jacox confirm their earlier identification[41] of HNC from photolysis of matrix-isolated methyl azide. Values are also given for the isotopically substituted species, H ^{15}N ^{12}C, H ^{14}N ^{13}C, and D ^{14}N ^{12}C and force constants and thermodynamic properties calculated. Force constants for HNC and DNC have also been derived from the above data by Ogilvie et al.[42] and compared with those of HNSi and HNGe in argon and nitrogen matrices. The latter compounds are assumed to be linear and bond orders for the NSi and NGe bonds are estimated to be 2·7 and 2·3 respectively.

Halogen cyanides (XCN)

Upon photolysis[40] of matrix-isolated FCN, two IR absorptions appear at 928 cm^{-1} and 2123 cm^{-1} which can be identified with the stretching fundamentals of FNC. In analogous experiments with ClCN and BrCN only the v(CN) of the ClNC and BrNC species can be tentatively assigned, e.g. Br^{14}N^{12}C, 2067 cm^{-1} compared with 2191 cm^{-1} for Br^{12}C^{14}N. A more thorough investigation of ClCN isolated in argon and neon matrices at temperatures between 4 K and 20 K has recently been reported by Murchison and Overend.[42a] Their main aim was to examine the effect of the matrix cage on the force constants of the intramolecular potential energy function of the ClCN molecule. Their results indicate that the quadratic and the anharmonic (cubic and quartic) intramolecular force constants are not significantly perturbed in going from the gas phase to argon and neon matrices. Freedman and Nixon[42b] have also made a detailed study of ClCN and BrCN in argon and krypton at 20 K in which evidence is presented for the monomeric species that the bending mode is split by the matrix environment. Their main concern however is the aggregation behaviour of XCN molecules, and in particular the structure of the dimers.

Previous MI studies of the structure of dimers of small molecules containing the $-$CN or the $-$NC groups have shown two distinct types of interaction. Hydrogen cyanide[38] on the one hand forms a linear dimer through a strong hydrogen-bonding interaction, while with methyl cyanide and methyl isocyanide the evidence suggests[42c] a 'CN dipole pair' interaction through an antiparallel arrangement of dipoles forming a dimer of C_{2h} symmetry. For ClCN and BrCN, from the pattern[42b] of two dimeric absorptions in the v_1, v_3, and $2v_2$ regions, the C_{2h} geometry can be eliminated; this configuration is centro-symmetric and for each normal mode of the monomer, only one phase of motion of the dimer units relative to one another would be infrared-active. Either a linear (as in the crystal) or bent chain configuration for the dimer is consistent with the observed spectra but the authors favour the linear

$$\text{XCN} \dots \text{XCN}$$
$$(B) \qquad (A)$$

arrangement and offer a consistent assignment of the bands in the v_1, v_2, and $2v_2$ regions relative to an arbitrary assignment of the two dimer peaks in the v_3 region to the two molecules, A and B, in the dimer.

Nitrous oxide (N_2O)

A detailed examination of nitrous oxide isolated in a nitrogen matrix at 15 K has recently been reported.[42d] The wavenumbers of the observed fundamentals, overtones and combinations for the species $^{14}N_2{}^{16}O$, $^{15}N_2{}^{16}O$, $^{14}N^{15}N^{16}O$, $^{15}N^{14}N^{16}O$ and $^{14}N_2{}^{18}O$ are utilized to calculate the force constants in an anharmonic force field which included cubic and quartic force constants. These are compared with gas-phase data and it is found that the small matrix shifts could be explained convincingly by adjusting only the quadratic force constants, taking the same set of anharmonic force constants for both gas-phase and matrix.

Other XY_2 species

The linear triatomic molecules carbon dioxide and carbon disulphide have been utilized as matrix supports but have not themselves been studied in detail as trapped solutes. Several linear dihalides of the noble gases have been

examined by MI but are more conveniently discussed later with other halogenated molecules.

3.4.2 Bent triatomic molecules

Water (H$_2$O)

The matrix spectrum of water in a variety of supports has been the subject
of many investigations.[43–53] It has been established by the work of several
schools[44,45,48,49] that water molecules undergo essentially free rotation in
noble-gas matrices of argon, krypton and xenon rotational assignments being
made of the prominent features in the v_3 (antisymmetric stretch) and v_2
(bend) bands. The time dependence of these features at liquid helium temperatures show[45,48,51] intensity changes over a period of a few hours which are
consistent with a nuclear-spin-species conversion process and fully confirms
the rotational assignment. This important aspect is discussed in detail for the
case of methane (paragraph 3.6.1).

Three studies have been made of water suspended in solid nitrogen. The
early study of Van Thiel, Becker and Pimentel[43] of water at 20 K has been
extended by Harvey and Shurvell[52] to D$_2$O at 5 K and more recently by
Tursi and Nixon[53] to H$_2$O, D$_2$O and HDO over the range 4–20 K. In the
first of these Van Thiel, *et al.* found, from a series of concentration studies at
20 K, a number of H$_2$O peaks which they assigned to the isolated monomer,
to the dimer, and to larger multimers. For the dimeric species, they suggested
a cyclic structure rather than an open or bifurcated one, their suggestion
being based upon a comparison of the number of observed dimer absorptions
with the numbers of infrared active fundamentals predicted for these several
plausible structures and upon the frequency pattern. They found just two
absorptions in the OH stretching region (3691 cm^{-1} and 3546 cm^{-1}) and one
in the bending region (1620 cm^{-1}) definitely attributable to the dimer. The
possibility that a second weaker band in the bending region (1615 cm^{-1})
might also be due to the dimer led them to speculate further that the cyclic
structure might be nonplanar and thus without a centre of symmetry. In the
second study, Harvey and Shurvell assign three broad bands (2725 cm^{-1},
2598 cm^{-1} and 1203 cm^{-1}) to the D$_2$O dimer but do not comment on the
structure of the dimer. They attribute several of sharper multiple peaks to
nearly freely rotating D$_2$O monomers in nitrogen as in the noble gases.

Tursi and Nixon, however, have conclusively shown that the frequencies
do not fit the pattern expected for a freely rotating molecule. Also, in contrast
to the case of H$_2$O in argon for which Hopkins *et al.*[51] detected changes of
10–20 per cent. in the intensities of rotational lines due to nuclear-spin
conversion (§3.6.1), Tursi and Nixon observed a slight sharpening of the
peaks and a concurrent slight decrease in intensities on lowering the temperature from 20 K to 4 K but no further change with time at 4 K. Tursi and Nixon

also reject a multiple-trapping site interpretation and assign with some confidence five fundamentals both to the H_2O and to the D_2O dimer. Two of these ($3714\cdot4$ cm^{-1} and $3625\cdot6$ cm^{-1} for H_2O) are only slightly displaced from the v_3 and v_1 frequencies of the monomer and the pair has essentially the same intensity ratio as the monomer v_3/v_1. This is suggestive that one unit of the dimer is an only slightly perturbed monomer. Of the remaining two frequencies in the stretching region, one ($3697\cdot7$ cm^{-1}) lies between v_3 and v_1 monomer, while the other ($3547\cdot5$ cm^{-1}) is displaced considerably to a lower frequency. This is not the distribution of stretching frequencies expected for a structure

with two equivalent 'free' OH groups and two equivalent hydrogen-bonded ones, which thus indicates an open structure

with a single hydrogen-bond for the dimer, rather than a bifurcated or a cyclic structure. The fifth dimer band at $1618\cdot1$ cm^{-1} is assigned to the bending mode of the hydrogen-bonded unit and a weaker, and more questionable, peak at $1600\cdot3$ cm^{-1} to the bending mode of the 'free' unit. This accounts for all of the dimer fundamentals, except for the lower frequency ones associated with the intermolecular bonding between the two water molecules. The absorption features of D_2O have an exact correspondence to those of H_2O (Table 3.4) similar features in the stretching region are assigned, but expectedly with less confidence, for HDO. The absence of the OH stretch of DOH . . . is interpreted as evidence for a stronger deuterium- than hydrogen-bond. Force constant calculations are made for the dimer which are reasonable for an open structure with a moderately weak hydrogen-bond.

The low frequency modes for water require detailed study, the only report to date being that of Miyazawa[46] for H_2O and D_2O in argon and nitrogen at 20 K in the region 400 cm^{-1} to 190 cm^{-1}. A monomer libration mode for H_2O is assigned to a band at 218 cm^{-1} in nitrogen and a feature at 243 cm^{-1} to a dimer hydrogen-bond stretching vibration. A band at 265 cm^{-1} is tentatively attributed to the trimer vibration. No bands were observed in the 490–190 cm^{-1} region in argon; the observations are thus in accord with the results of the near IR studies.

Hydrogen sulphide (H_2S)

Matrix-trapped hydrogen sulphide has recently been the subject of two simultaneous publications.[54,55] Despite the fact that the spectrum of this

Table 3.4 Monomer and dimer frequencies[53] (cm^{-1}) of H_2O and D_2O in nitrogen at 20 K

Assignment	Monomer		
	H_2O	D_2O	HDO
ν_3 antisymmetric stretch	3725·7	2764·6	3680·4
ν_1 symmetric stretch	3632·5	2655·0	2705·1
ν_2 bend	1596·9	1179·2	1405·4

	Dimer		
	H_2O		D_2O
OH stretch	$\begin{cases} 3714\cdot4 \\ 3625\cdot6 \end{cases}$	$\begin{matrix} \nu_3 \\ \nu_1 \end{matrix}$	$\left.\begin{matrix} 2756\cdot6 \\ 2650\cdot0 \end{matrix}\right\}$ OD stretch
OH...stretch	$\begin{cases} 3697\cdot7 \\ 3547\cdot5 \end{cases}$	$\begin{matrix} \nu_3 \\ \nu_1 \end{matrix}$	$\left.\begin{matrix} 2737\cdot6 \\ 2599\cdot1 \end{matrix}\right\}$ OD...stretch
HOH bend	(1600·3?)	ν_2	(1181·5?) DOD bend
HOH...bend	1618·1	ν_2	1193·3 DOD...bend

molecule has been extensively studied the antisymmetric stretching fundamental ν_3 has not been observed in any phase, even though this feature is usually the most intense one in the IR spectrum of triatomic molecules. The complexity of the spectral interpretation arises from the near coincidence of the two stretching frequencies, ν_1 and ν_3, and interactions between overtone-combination levels. In the gas phase[56] ν_1 and ν_2 are observed at 2614·6 cm^{-1} and 1182·7 cm^{-1} respectively and ν_3 is estimated from combination band frequencies to lie at 2627·5 cm^{-1}. Pacansky and Calder[54] have studied H_2S and D_2S in argon and krypton at c. 8 K and claim unambiguously to have identified ν_3, their assignments in argon being: ν_3, 2581·8 cm^{-1}; ν_1, 2568·8 cm^{-1}; ν_2, 1179·5 cm^{-1}, with ν_3 of comparable intensity to ν_1. Corresponding values are given for D_2S and general, quadratic, valence force constants are calculated from the isotopic shifts. No bands were observed due to HDS species, or to rotational features; very limited concentration studies were carried out to follow possible aggregation behaviour. Almost simultaneous with the above studies, Tursi and Nixon[55] reported IR spectra for H_2S in nitrogen at various M/A ratios at 20 K over the region 4000 cm^{-1} to 2500 cm^{-1}. They assign $\nu(SH)$ stretches of the isolated monomer to bands at 2632·6 cm^{-1} and 2619·5 cm^{-1} and no evidence is found for rotation. At low M/A ratios dimer bands are observed at 2631·1 cm^{-1}, 2625·3 cm^{-1}, 2617·8 cm^{-1} and 2580·3 cm^{-1} a frequency pattern for the SH stretching modes much the same as for the OH stretches in the H_2O dimer,[53] i.e. three frequencies falling nearly within the range covered by the two stretching frequencies of the monomer and one frequency somewhat lower. They correspondingly postulate the same open dimer structure with a single hydrogen-bond, the stretching frequencies at $2631\cdot1$ cm^{-1} and $2617\cdot8$ cm^{-1} are thus primarily associated with the proton-acceptor molecule of the dimer, and the ones at $2625\cdot3$ cm^{-1} and $2580\cdot3$ cm^{-1} with the proton donor. Spectra of D_2S were also recorded which yield a similar pattern but could not be interpreted in detail because of HDS contamination. An estimate, based on the very dubious procedure of using frequency shift data, indicates a hydrogen-bond strength roughly half that found in the matrix-isolated water dimer. At higher concentrations further bands are observed due to higher aggregates.

Recent studies in our laboratories[57] of H_2S in argon, krypton, nitrogen and carbon monoxide matrices confirms the nitrogen data but suggests that the argon data requires reassessment. We have carried out a detailed concentration study of H_2S in argon at 20 K, from M/A 2000 to 50, which shows conclusively that the band at $2568\cdot8$ cm^{-1} which Pacansky and Calder assign as v_1, is due to dimer. There are two monomer bands, at $2628\cdot3$ cm^{-1} and $2681\cdot8$ cm^{-1} with an intensity ratio of c. $1:9$, which are v_3 and v_1 respectively. A band at $2623\cdot8$ cm^{-1} shows a similar growth to the $2568\cdot8$ cm^{-1} feature, and we thus assign to v_3 of the dimer. The frequency separation and relative intensity of the two stretching modes exhibit large variations from the gas phase to matrix and from one matrix to another. This behaviour casts some doubt on the validity of the usual assumption that the vibrational spectrum is little perturbed in a noble-gas matrix environment.

Chlorine monoxide (Cl_2O)

IR studies[58] of chlorine monoxide (^{16}O and ^{18}O) have included MI spectra in argon and nitrogen at 20 K. The sharpness of the MI spectra were of value in revealing isotope shifts particularly for v_1, the symmetric stretch. In the solid-phase spectrum the breadth of v_1 interferes with the measurement of the chlorine isotope shift but the nitrogen matrix frequencies, accurate to $\pm0\cdot5$ cm^{-1}, allow the following values of v_1 to be assigned to OClO species: $639\cdot9$ cm^{-1} (35-16-35), $637\cdot0$ cm^{-1} (35-16-37), $634\cdot1$ cm^{-1} (37-16-37), $615\cdot3$ cm^{-1} (35-18-35), and $612\cdot3$ cm^{-1} (35-18-37).

Nitrogen dioxide (NO_2)

Nitrogen dioxide has been studied in oxygen, argon and other matrices at 4 K by Crawford *et al.*[59,60] Three bands are attributed to isolated nitrogen dioxide molecules and are assigned (in argon): 2900 cm^{-1} ($v_1 + v_2$), 1610 cm^{-1} (v_3), 749 cm^{-1} (v_2). In this instance the symmetric stretching frequency, which is very weak in the gas phase, is unobserved, but can be predicted from the combination band at 2900 cm^{-1}. Allavena *et al.*[61] also report a detailed examination of v_3NO_2 in krypton for comparison with SO_2 (see below). A doublet structure is observed (at $1607\cdot55$ cm^{-1} and $1606\cdot50$ cm^{-1} for

$^{14}N^{16}O_2$) which has an identical intensity variation with temperature as v_3SO_2 and NO_2 is thus inferred to have similar one-dimensional rotation about the pseudofigure axis.

A valuable comparison can be made with the ESR spectrum of nitrogen dioxide in an argon matrix at 4 K which has been interpreted[62] as indicating a rotation of the nitrogen atom about the O···O axis; and in neon at 4 K the spectrum is interpreted[63] in terms of three distinct sites: (a) a rigid (oriented) site, (b) a rotational site with motion of the nitrogen atom about the O···O axis, and (c) a hindered rotation about the C_{2v} symmetry axis, each site being of similar occupancy. The nitrogen dioxide in argon ESR spectra[63] are irreversibly changed by annealing at 30 K, populating site 3 at the expense of the other sites.

Sulphur dioxide (SO_2)

Allavena *et al.*[61] report the matrix IR spectrum of each isotopic combination of ^{32}S, ^{34}S, ^{16}O and ^{18}O, of SO_2 trapped in krypton at 20 K. With the exception of two of the vibrations of the species $^{34}S^{18}O_2$, all of the 18 absorptions of the six isotopic species were observed in a single scan. The matrix fundamentals $^{32}S^{16}O_2$ correspond closely to the gas-phase values. Analysis of the data on the basis of a valence force field which includes anharmonic corrections and in which the bond angle is treated as a variable parameter yields force constants very close to those derivable from gas-phase data. The sulphur dioxide bond angle which leads to the best fit of the experimental matrix data is 119° 37', in excellent agreement with the value of 119° 19' determined from microwave data. The agreement illustrates not only the negligible effect of the solid matrix on the bond angle, but also the fact that it can be deduced with considerable precision from the nearly pure vibrational spectra of the matrix-isolated molecule. The authors also show that in many cases gas-phase and matrix measured frequencies for only two isotopes can be used to establish upper and lower limits of the bond angle for triatomic molecules. Under high resolution v_3 is a doublet (at 1351·27 cm^{-1} and 1350·72 cm^{-1} for $^{32}S^{16}O_2$) whose lower-frequency component is temperature insensitive whereas the higher-frequency component shows a dramatic and completely reversible variation of intensity when the matrix temperature is cycled between 20 K and 38 K (Figure 3.6). The behaviour is similar to that observed for $R(0)$ in matrix-isolated hydrogen chloride discussed earlier and also in the case of sulphur dioxide can only be interpreted in terms of rotational-state population effects. For sulphur dioxide, the relationship of the rotational constants is $A \gg B \approx C$ so that the molecule approximates to a pseudosymmetric prolate top whose pseudofigure axis coincides with its least moment axis, and whose ordinary C_2 symmetry axis contains the intermediate-moment axis. It is thus proposed that one-dimensional rotation about the pseudofigure axis occurs which would

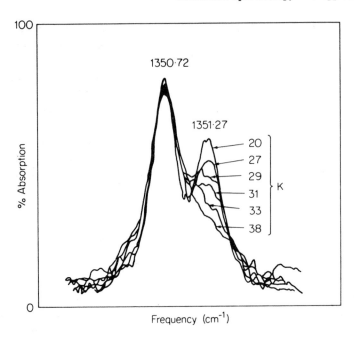

Figure 3.6 Variation of intensity with temperature of the two components of v_3 of $^{32}S^{16}O_2$ in Kr (reproduced by permission from M. Allavena *et al.*, *J. Chem. Phys.* **50**, 3399 (1969))

produce rotational features only in conjunction with v_3, as is observed. The separation of the first R_R ($K = 0 \rightarrow 1$) or P_R ($K = 1 \rightarrow 0$) sub-bands from the pure vibrational transition at the band centre would be A cm^{-1} with successive R_Q and P_Q sub-bands separated by $2A$ cm^{-1}, provided the one-dimensional rotation is free. If it is hindered, the spacings will be reduced. Thus the temperature-dependent feature in v_3 is attributable to a $^RR_0$ sub-band and the main feature to $^RQ_0$. It is interesting to note that band shape analysis of v_3 and v_1 of sulphur dioxide in a variety of liquid solvents has been interpreted[64] in terms of similar rotational motions.

In a simultaneous publication with Allavena *et al.*[61] Hastie *et al.*[65] report a matrix study of sulphur dioxide in argon or neon at 5 K by the vaporizing Knudsen method (Chapter 6). Their frequencies and bond angle ($117 \pm 1°$) are not as precise as Allavena *et al.* but the general agreement is most satisfactory. On controlled diffusion of sulphur dioxide in argon by warming the matrix to 37 K new features arise which are assigned to a new dimeric $(SO_2)_2$ species. Experiments performed on sulphur dioxide effusing from the Knudsen cell at temperatures ranging from 300 K to 1270 K gave essentially identical spectra indicating the absence of dimers or structural isomers in measurable concentrations in the vapours effusing under Knudsen cell conditions. These experiments were undertaken to assist in the understanding of the flash photolysis of sulphur dioxide which has been interpreted[66] in terms of the possible existence of a new isomeric form of sulphur dioxide at temperatures of around 770 K to 1170 K. It was suggested[65] that possibly photolysis of matrix-isolated sulphur dioxide may produce the SOO isomer in an analogous manner to the formation[67] of ClOO from OClO.

Selenium dioxide (SeO$_2$)

Hastie *et al.* have also studied the dioxides of selenium[65,68] and tellurium[68] although these, especially tellurium dioxide, are truly vaporizing molecule studies (Chapter 6). Cesaro *et al.*[68a] have extended the argon MI studies of Hastie *et al.*[65] and examined the IR spectrum of matrix-isolated $^{80}Se^{16}O_2$ over the range 5000–200 cm^{-1}. Three prominent bands were observed at 965·6 cm^{-1}, 922·0 cm^{-1} and 372·5 cm^{-1}, with a frequency accuracy of $\pm 0·5$ cm^{-1}; the two high frequency bands are in very close agreement with the v_3 and v_1 assignments of Hastie, whilst the band at 372·5 cm^{-1} is identified as v_2. The bending mode is the most important vibration to be considered in the calculation of thermodynamic functions for selenium dioxide and the authors calculate various thermodynamic functions of $^{80}Se^{16}O_2$, the most abundant isotopic species. In their study, Hastie *et al.*[65] performed diffusion studies which provided evidence for the formation of (SeO$_2$)$_2$ dimers. Evidence for the structure of these has very recently been obtained from Raman MI studies (§9.5.2).

Ozone (O$_3$)

Since both molecular and atomic oxygen (obtained from the electrical discharge of oxygen) are useful reagents for producing a variety of oxide species in low-temperature matrices (see Chapters 5 and 6) it is important to have the spectral characterization of ozone. This has recently been provided by Brewer and Wang[68b] who have synthesized ozone of the various isotopic ^{16}O and ^{18}O combinations and isolated them in krypton and xenon at 20 K. The results are tabulated in Table 3.5.

Table 3.5 IR absorption bands (in cm^{-1}) of ozone in a xenon matrix at 20 K

$^{16}O_3$	^{16}O – ^{16}O $\diagdown$ ^{18}O	^{16}O – ^{18}O $\diagdown$ ^{18}O	^{18}O – ^{16}O $\diagdown$ ^{16}O	$^{18}O_3$	Assignment
2110				1990	$v_1 + v_3$
1100				1038	v_1
1030 ⎫	1015	(1008)	995	978 ⎫	v_3
1025 ⎭				970 ⎭	
695				660	v_2

3.5 Tetra-atomic molecules

3.5.1 Planar tetra-atomic molecules

Borane (BH$_3$)

Attempts to isolate borane and identify it by optical spectroscopy have so far proved inconclusive although mass spectrometric studies of the pyrolytic decomposition of diborane and borane carbonyl have provided evidence of small concentrations of borane. Of the possible experimental methods for the spectroscopic examination of borane, isolation in an inert matrix clearly offers the best hope. This has recently been achieved by Kaldor and Porter[162a] using the reaction

$$BH_3CO \rightarrow BH_3 + CO$$

and combining the fast-flow pyrolysis technique with the MI technique. Bands observed at 2808 cm^{-1}, 1604 cm^{-1}, and 1125 cm^{-1} in argon at 5 K cannot be attributed to diborane, borane carbonyl, carbon monoxide or any known molecule containing boron and oxygen. Based on ^{10}B isotope shifts these are assigned to the simple borane, BH$_3$, with a planar D_{3h} structure: $v_2(A_2'')$ 1125 cm^{-1}, $v_3(E')$ 2808 cm^{-1}, and $v_4(E')$ 1604 cm^{-1}. The stretching force constant was calculated to be 4·08 mdyn/Å and the in-plane-bending force constant 0·46 mdyn/Å. Spectra obtained from the pyrolysis of BD$_3$CO were extremely weak and only v_2, at 845 cm^{-1}, was able to be observed for BD$_3$.

Boron trifluoride (BF$_3$)

Boron trifluoride has been studied[69] at 20 K in argon and krypton over a range of concentrations. In krypton at $M/A = 900$ the most intense peaks are essentially identical with the gas phase spectrum. Diffusion was followed by allowing the matrix to warm up above 30 K and 6 weaker bands grow in at the expense of the monomer bands. Only one series of new bands is found, either on diffusion or direct deposition of a concentrated mixture and it is suggested that the associated species is a simple dimer with a bridged structure of D_{2h} symmetry, similar to the AlX$_3$ dimers.

Boron trichloride (BCl$_3$)

Comeford *et al.*[70] have studied boron trichloride in argon and report detailed isotope shifts for all possible combinations of ^{10}B, ^{11}B, ^{35}Cl and ^{37}Cl. Bass *et al.*[71] have also included boron trichloride in their matrix study of HBCl$_2$.

Dichloroborane (HBCl$_2$)

Matrix spectra in argon at 10K are reported[71] for HBCl$_2$ and DBCl$_2$ and frequencies are listed for the ^{10}B and ^{11}B isotopes.

3.5.2 Pyramidal tetra-atomic molecules

Ammonia (NH_3)

An examination of ammonia in solid matrices is of special interest for two reasons. First, the matrix effect on the inversion as well as the rotational degree of freedom of the isolated monomer, inversion doubling being particularly sensitive to change in the potential function. Secondly, at high solute concentrations, the hydrogen bonding of the multimers of ammonia may be unusual, as suggested by the crystal structure of solid ammonia where apparently[72] each ammonia molecule acts as a base toward three adjacent ammonia molecules.

Becker and Pimentel[37] first examined ammonia suspended in argon, xenon and nitrogen in their early tests of matrix-isolating efficiencies. This was followed by Milligan *et al.*[73] trapping in argon and nitrogen and who present spectra in the v_2 region near 975 cm^{-1}. They found the v_2 (umbrella motion) band exhibited a sharp well-resolved fine structure which from their temperature dependence they conclude that rotation and inversion of ammonia occurs in both matrices. This interpretation has been contradicted for nitrogen matrices by Pimentel *et al.*[74] who feel that their concentration and diffusion studies preclude the assignment of the bands at 987 cm^{-1}, 1004·5 cm^{-1} and 1014·5 cm^{-1} to monomeric molecules, a premise used by Milligan *et al.*[73] in their interpretation. Instead, Pimentel *et al.* assign absorptions at 970 cm^{-1} and 1143 cm^{-1} to monomeric ammonia, implying that the nitrogen cage significantly restricts the motional freedom of the trapped molecules. This work does not negate the conclusions of Milligan *et al.* for rotation and inversion in argon. However, Hopkins *et al.*[51] have made a study of the time evolution of the fine structure in argon similar to that discussed in paragraph 3.6.1 on nuclear spin conversion. The ammonia/argon matrix was allowed to warm to *c.* 30 K for several minutes and then rapidly quenched to 7 K. The spectrum was then repeatedly scanned as a function of time. Only two lines show a clear time dependence. These are the strongest line at 974 cm^{-1} which grows in intensity and is assigned to $R(0)$, and a weak line at 956 cm^{-1} which decreases in intensity and is probably $Q(1)$. Because the lines are not time dependent it appears likely that most of the fine structure in ammonia does not arise from rotation. The line $R(1)$ which should be more intense than $Q(1)$ is missing. The lines behave similarly to the 1593 cm^{-1} band of water which was assigned to non-rotating H_2O.

The bands which Pimentel *et al.*[74] reject as monomer absorptions, they assign to dimeric and higher aggregates. The dimer absorbs at 3404 cm^{-1}, 1004·5 cm^{-1}, 987 cm^{-1} and possibly also at 3313 cm^{-1}, 3246 cm^{-1} and 3237 cm^{-1}; tentative assignments of these are proposed and it is suggested that the structure of the dimer is one in which a single hydrogen-bond links the two nitrogen atoms.

Monochlorodifluoroammonia $(ClNF_2)$

Comeford[75] has reported the argon matrix-isolated spectrum of $ClNF_2$ at 20 K and is able to resolve v_4 366 cm^{-1} and v_5 382 cm^{-1} which appear overlapping in the gas phase at 377 cm^{-1}.

3.5.3 Other tetra-atomic molecules

Hydrogen peroxide (H_2O_2)

The IR spectra between 4000 cm^{-1} and 200 cm^{-1} of H_2O_2 and D_2O_2 matrix isolated in argon and nitrogen at 8 K has recently been reported by Lannon *et al.*[76] which extend earlier work of Catalano and Sanborn[77] on hydrogen peroxide in nitrogen. The assignments of the monomer fundamentals are straightforward, for H_2O_2 in argon the two O—H stretching modes being v_1 at 3589·2 cm^{-1} and v_5 at 3597·4 cm^{-1}, and the O—H bending motions at $v_2 = 1276·9$ cm^{-1} and $v_6 = 1270·6$ cm^{-1}. The corresponding values for D_2O_2 are $v_1 = 2645·8$ cm^{-1}, $v_5 = 2652·0$ cm^{-1}, v_2 shoulder on $v_6 = 980·9$ cm^{-1}, but the spectrum is a little complicated due to the presence of HOOD and traces of D_2O. Aggregation in argon is apparent in their spectra even at $M/A = 1000$, but isolation is significantly better in nitrogen. Aggregation bands are assigned on the basis of dilution studies which confirm their disappearance at 3000 : 1; the authors do not discuss the possible multimer species but simply refer to all the features as 'cluster' peaks. In the hydrogen peroxide/argon matrix these appear at 3580 cm^{-1} (*w*), 3470 cm^{-1} (*s*), 3460 cm^{-1} (*w*), 1293 cm^{-1} (*s*) and 1273 cm^{-1}, the latter being ill-defined as it appears in the 6·3 cm^{-1} gap between the monomer v_2 and v_6. Without the availability of the original spectra at the several concentrations it is not possible to propose a structure for the multimer(s), but it is tempting to suggest that, like the case of water,[53] it is an open-chain dimer. This would allow the two intense and strongly displaced peaks, 3470 cm^{-1} and 1293 cm^{-1}, to be assigned to the singly hydrogenbonded OH group and the other less displaced features to that unit of the dimer which is an only slightly perturbed monomer. The most interesting aspects of these matrix spectra are, however, the extraordinarily broad bands in the torsional region, v_4. In argon, H_2O_2 exhibits three peaks in the 370 cm^{-1} region whose intensities change reversibly on temperature cycling between 9 K and 20 K, the strong central peak at 373 cm^{-1} decreasing in relative intensity with increasing temperature. It is difficult to determine if any change occurs in the two subsidiary peaks at 395 cm^{-1} and 357 cm^{-1}. At lower frequencies a second band occurs with two maxima at 286·6 cm^{-1} and 264·0 cm^{-1} which show no intensity change with temperature. A similar triplet at 270 cm^{-1} (*m*), 251 cm^{-1} (*s*) and 237 cm^{-1} (*m*) is observed for D_2O_2 in argon but the second band probably lies outside the far IR limit (200 cm^{-1}) of the spectrometer used. These torsional bands are unusually broad and so far seem to be unique in matrix spectra of *isolated* small molecules. The multiple peaks and overall breadth is interpreted by Lannon *et al.* in terms of a coupling of rotational and translational motion similar to that invoked by Friedmann and Kimel to explain the spectrum of HCl and DCl in argon (see Chapter 4). It is thus believed that H_2O_2 and D_2O_2 perform internal rotational motion in argon similar to the gas phase but that there is stronger coupling between internal rotation and hindered translation (rattling) and/or hindered overall rotation (libration) in the matrix than in the gas. It is also proposed that the torsional frequency, v_4, forms combination sum-and-difference bands with these other motions. The v_4 band in nitrogen is broad and apparently devoid of structure which indicates that the internal rotation is much more hindered. The band breadth is interpreted in the same manner, i.e. a coupling between the hindered internal rotation and translation and/or librational motion and an unresolved pattern of sum-and-difference bands.

Hydrazoic acid (HN_3)

Hydrazoic acid has been examined in various matrices at 20 K by Pimentel's group.[37,78,79,80] All six fundamentals of the monomer are observed (Table 3.6) although some doubt exists over v_6 whose weakness

Table 3.6 Monomer and dimer frequencies[79] (cm^{-1}) of HN_3 in nitrogen at 20 K

Assignment	Monomer	Dimer	Δv
v_1 N—H stretch	3324	3314	−10
		3174	−150
v_2 N—N—N asym stretch	2150	2162	+12
		2143	−7
v_3 N—N—N bend-sym stretch	1273	1279	+6
v_4 H—N—N sym stretch-bend	1168	1177	+9
v_6 N—N—N bend	588 (?)	605 (?)	
v_5 N—N—N bend	527	545 (?)	
v_7 H bond tors		387	

in the nitrogen matrix is a mystifying feature. Corresponding data are also given[80] for the species $H^{15}NNN$ and $HNN^{15}N$ and their deuterated analogues. A detailed study is given[79] of the association of HN_3 in a nitrogen matrix and a complete assignment offered for the dimer species (Table 3.6). There are two dimeric N—H stretching frequencies, one at $3314 \, cm^{-1}$ is shifted only $10 \, cm^{-1}$ from the monomer, and the other, at $3174 \, cm^{-1}$, is shifted considerably to lower frequency ($-150 \, cm^{-1}$). This is indicative of an open dimeric structure, either I or II,

$$\bar{N}=\overset{+}{N}=N$$

(structures I and II showing open dimeric forms of hydrazoic acid with H bonds)

(I)   (II)

and is inconsistent with the two cyclic structures, III and IV:

(structures III and IV showing cyclic forms)

(III)   (IV)

Furthermore, these frequency shifts, and their intensities, are similar to those of diols which[81] have an intramolecular hydrogen bond,

(structure showing O—H···H—O diol with free hydrogen)

in which one hydrogen atom is free. The absorption most informative of the dimer structure is the band at 387 cm^{-1} which shifts by a factor of 1·28 to 302 cm^{-1} in DN_3. This frequency has no monomer counterpart and is thus assigned to the torsional mode of a hydrogen bond. The fact that only two bands are observed, 387 cm^{-1} and 302 cm^{-1}, irrespective of the H/D ratio, shows that the $HN_3 \cdots DN_3$ dimer absorbs at the same frequencies as $HN_3 \cdots HN_3$ and $DN_3 \cdots DN_3$. This behaviour is consistent only with an open dimeric structure. The authors suggest structure I as the most likely because of the charge distribution implied by the bonds. They point out, however, that a triple-bonded resonance structure that places negative charge on the first nitrogen atom, makes structure II also plausible. This writer favours II which better explains the -10 cm^{-1} shift for νNH_{free}. A comparison of the frequency shift of 150 cm^{-1} with those of other N—H···N bonds of known strength, $NH_3 \cdots NH_3$ and pyrrole/pyridine, suggests a dimer hydrogen-bond energy of 9·6 kJ mol^{-1}.

Diimide (N_2H_2)

The above-described aggregation studies were undertaken as a necessary prerequisite for an understanding of photolysis studies[82] of hydrazoic acid.

These latter studies have identified, in addition to the free radical NH, the reactive molecule diimide, both the *trans* and the *cis* isomers. The ^{15}N and ^2H substitution studies yield a triplet near 1286 cm^{-1}, a quartet c. 1058 cm^{-1} and a triplet c. 946 cm^{-1} which represents the multiplet splitting expected from a N_2H_2 molecule containing two equivalent hydrogens and two equivalent nitrogens. The formaldehyde structure for diimide, $H_2\overset{+}{N}{=}\overset{-}{N}$, is rejected because the non-equivalent nitrogen atoms would produce a system of three quartets. Diimide is thus of interest as the nitrogen counterpart of ethylene and comparisons are made between their spectra. The matrix spectra provide strong evidence that the *trans*-N_2H_2 has expected planar structure with a centre of symmetry (C_{2h}) and frequencies are assigned on this basis. Of particular significance is the identification of v_2, the N=N stretch, at 1481 cm^{-1} in *trans*-N_2HD (infrared-inactive in *trans*-N_2H_2 and *trans*-N_2D_2) which is lower than that of ethylene, 1623 cm^{-1}, by an amount largely attributable to the mass change. There has been considerable controversy about the location of the N=N stretching frequency in azo compounds (summarized well by Bellamy[83]) so that the diimide study is of importance in that it establishes the frequency of the prototype N=N double bond.

Nitrous acid (HONO)

Nitrous and deuteronitrous acid have been studied[84] in nitrogen matrices at 20 K. Nitrous acid is available in the gas phase only at low partial pressure in an equilibrium mixture containing nitric oxide, nitrogen dioxide, and water. Hence the advantage of the MI technique was utilized and HONO prepared *in situ* by the photolysis of ammonia and oxygen in solid nitrogen. Assignments are given for both *cis* and *trans* isomers and evidence is provided for the infrared-induced isomerization. This is considered to be a new type of photochemical reaction, the study of which is possible only because of the unique potentialities of the MI technique. At normal temperatures, any reaction that could be induced by a quantum of infrared radiation would be extremely rapid through thermal excitation. Filter and kinetic experiments were performed based on the intensity behaviour of the v_4 (N—O stretch) band pair: 865 cm^{-1} (*cis*) and 815 cm^{-1} (*trans*). Other pairs of bands were found to behave similarly, but the v_4 bands, being intense and in a region free from interfering bands, were best suited for accurate photometry. A narrow range of frequencies is responsible for *cis* to *trans* isomerization, 3650–3200 cm^{-1} for HONO and 4100–3500 cm^{-1} for DONO. Since isomerization appears to cease after long exposure, it seems that both *cis* → *trans* and *trans* → *cis* isomerizations occur. An isomerization mechanism is proposed, that involves a highly efficient intramolecular transfer of energy between vibrational modes, and is postulated to be the same for

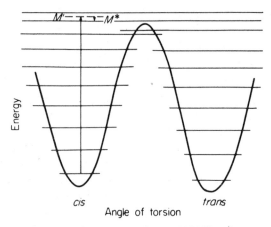

Figure 3.7 The proposed *cis* HONO $\rightleftharpoons$ *trans*
HONO isomerization mechanism (reproduced by
permission from R. T. Hall and G. C. Pimentel,
J. Chem. Phys. **38**, 1889 (1963))

isomerizations in both directions (Figure 3.7). The molecule first absorbs
the effective radiation in a non-torsional mode, e.g. v (OH), and is excited
to a non-torsional excited state, M' (dashed level). Then, through intramolecular energy transfer, the energy is transferred from level M' to M^*, an
odd overtone of the torsional mode (of same symmetry as the in-plane
fundamentals, A'). The molecule is now in a 'two-well' level and has about
a 50 per cent. chance of decaying into the other well. Once so trapped in this
well, the molecule ultimately will decay to the ground state of the other
isomer. The height of the potential barrier to isomerization is estimated to
be 41 kJmol^{-1}.

Tautomeric species HNCO *and* HOCN

Jacox and Milligan[85] in their photolytic free radical studies (Chapter 5)
report the spectra of HNCO in argon and nitrogen. One of the dominant
processes involves the detachment of NH, which recombines with CO to
produce, in addition to HNCO, the unstable tautomer HOCN, for which
the following assignments in nitrogen are proposed: O—H stretch, 3506
cm^{-1}; C$\equiv$N stretch, 2294 cm^{-1}; O—H bend, 1241 cm^{-1}; O—C stretch,
1098 cm^{-1}; OCN deformation, 460 cm^{-1}.

Isothiocyanic acid (HNCS)

Durig and Wertz[86] in their study of isothiocyanic acid have included an
MI study in argon at 20 K in order to better evaluate the actual band centres.
The HNCS molecule is planar and the six fundamentals divide up into five
A' type and one A''. All five A' species are readily assigned: v_1, N—H stretch,

3505 cm^{-1}; v_2, C=N stretch, 1979 cm^{-1}; v_3, C=S stretch, 988 cm^{-1}; v_4, N=C=S in-plane bend, 577 cm^{-1}; v_5, N—H bend, 461 cm^{-1}. Values are also given for DNCS and the observed product

$$\prod_{i=1}^{5} \frac{v_i(\text{HNCS})}{v_i(\text{DNCS})} = 1 \cdot 88$$

which compares well with the theoretical value of $1 \cdot 91$ for A' vibrations. The study resolves the ambiguities[87] of the complex spectrum of the gas in the low-frequency region and reassigns v_4 and v_5. The N=C=S out-of-plane bending mode is not observed in the matrix spectrum but is found in the solid at 780 cm^{-1}. The same feature is also absent[79] in the MI spectrum of HN_3 and was accounted for[80] due to Coriolis coupling leading to intensity sharing in the gas phase, a feature which would be absent in the MI phase. Durig and Wertz infer a similar explanation in the case of HNCS.

M/A ratios of 20, 100, and 1000 were employed, but as extensive hydrogen bonding was apparent in all but the lowest dilution only the $1000:1$ spectra are illustrated. Bands at 3540 cm^{-1} and 3485 cm^{-1}, on either side of the monomer v (N—H), are assigned to a dimer species and one at 3250 cm^{-1} tentatively to a multimer. A weaker absorption at 2009 cm^{-1}, on the high frequency side of the monomer v (C=N) is also assigned to dimer. Similar features are found in the spectrum of DNCS. A detailed study of the aggregation behaviour is clearly necessary before the structure of the dimer can be discussed.

Formaldehyde (HCHO)

Formaldehyde has been examined by Harvey and Ogilvie[88] in argon and nitrogen at M/A ratios of 300, or less in some cases. Most bands exhibit fine structure which is not rotational but is accounted for in terms of multiple trapping sites. The triplets observed in argon are assigned to substitutional sites in which 1, 2 or 3 argon atoms are displaced. For the nitrogen matrix the fine structure was less complex and only two types are proposed, either 1 or 2 nitrogen molecules being displaced. However, in the light of the high concentrations employed and the known better isolating capability of nitrogen it would appear that the band multiplicity is almost certainly due to aggregation, a view which is confirmed by some recent[89] matrix studies of fluorescence spectra. Smith and Meyer[89] find at 20 K and ratios of $M/A = 1000$, that isolation of HCHO is better than 99 per cent. in a sulphur hexafluoride matrix, 90 per cent. in xenon, and only 50 per cent. in krypton.

Carbon trioxide (CO_3)

Reactions of oxygen atoms with carbon dioxide molecules to give an unstable species identified as carbon trioxide have been observed[90] in three systems: (a) the UV-photolysis of solid CO_2 at 77 K; (b) the UV-photolysis of O_3 in a CO_2 matrix at 50–60 K ; and (c) the trapping at 50–70 K of products of the RF-discharge of CO_2 gas. Isotopic studies using ^{18}O or ^{13}C enriched CO_2 show that the molecular formula is CO_3; frequency assignments and isotopic product-rule calculations favour a planar C_{2v} molecule in which the carbon atom is bonded to

$$\begin{matrix} O \\ | \\ O \end{matrix} \Big\rangle C = O$$

the unique oxygen atom by a strong carbonyl bond [v (C=O) for $^{12}C\ ^{16}O_3$, 2045 cm^{-1}] and to two equivalent oxygen atoms by weaker bonds. There is presumably covalent bonding between the two equivalent oxygen atoms. These conclusions have since been confirmed in detail by Jacox and Milligan.[90a] They fit the vibrational frequency pattern for the planar modes of CO_3 to a seven-constant valence-force potential for both an open structure, with an O$-\hat{C}-$O angle of 80°, and the three-membered ring structure, with an O$-\hat{C}-$O angle of 65°. However, the O$-$C$-$O bending force constant and the interaction constants for the open-structure are exceptionally large and on this basis the cyclic structure is favoured for CO_3 in its ground state.

3.6 Polyatomic molecules

It is more convenient to discuss the larger molecules which have been studied by MI, either individually or collectively under various headings.

3.6.1 Methane; nuclear-spin-species conversion

Preliminary work[91,92] on methane in argon matrices failed to reveal rotational features but a detailed study of Cabana et al.[93] provides conclusive evidence for quantized rotation of methane in argon, krypton, and xenon crystals. They observed rotational fine structure on both the v_3 (stretching) and v_4 (bending) bands, a four line pattern being observed in each case. The pattern is that expected if nuclear spins are equilibrated at the low temperature. In xenon the most intense line at 3013·4 cm^{-1} is R(0) ($J = 1 \leftarrow 0$); on the high frequency side R(1) ($J = 2 \leftarrow 1$) appears at 3018·3 cm^{-1} and on the low frequency side Q(1) ($J = 1 \leftarrow 1$) and P(1) ($J = 1 \rightarrow 0$) appear at 3005·8 cm^{-1} and 2997·8 cm^{-1} respectively (similar in appearance to Figure 3.8 for methane in argon). The corresponding features for v_4 appear at 1309·2 cm^{-1}, 1304·5 cm^{-1}, 1299·9 cm^{-1} and 1294·9 cm^{-1}. Rotation is considered to be only slightly hindered with barriers of the order of 50 cm^{-1} high. King's[94] theory of the hindered rotational levels of a rigid tetrahedron moving in an octahedral field is shown to give a satisfactory semi-quantitative account of the observations. The effect of rotation-translation coupling (see Chapter 4) on the energy levels of a tetrahedral molecule in an octahedral cell has been considered by Friedmann et al.[94a]

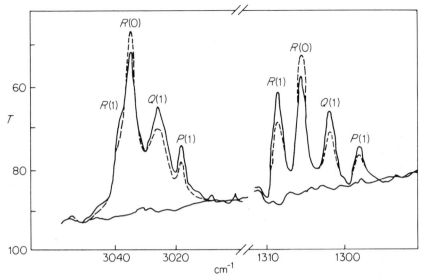

Figure 3.8 Variation of intensity with time of the components of v_3 and v_4 of CH_4 in Ar (M/A 500) at 4 K. For the 3000 cm^{-1} region the solid curve is after 3 min and the dashed curve is after 73 min. For the 1300 cm^{-1} region the solid curve is after 25 min and the broken curve is after 228 min. (Reproduced by permission from F. H. Frayer and G. E. Ewing, *J. Chem. Phys.* **48**, 781 (1968))

In the only Raman study of a matrix-isolated molecule reported before Chapter 9 of this book was written, Cabana *et al.*[95] have confirmed the IR observations. Even though the Raman v_3 band was not resolved the fact that its breadth covers a range wide enough to accommodate the predicted structure of $P(1)$, Q, $R(0)$, $R(1)$, $S(0)$, and $S(1)$ and is about twice the range of the IR v_3 where only P, Q and R branches exist, supports the view of free rotation. v_1 should consist of a relatively sharp band and it is found to be so.

Since the discovery of *ortho*- and *para*-hydrogen the phenomenon of nuclear-spin-species conversion has aroused considerable interest. It has been possible to prepare separated isomers in the gas phase only for hydrogen, deuterium, and tritium and there are extensive experimental and theoretical studies of these molecules. However, it is of interest to investigate other molecules and the observation of rotational fine structure in matrices at cryogenic temperatures has afforded such an opportunity. The method involves a study of the time dependence of the rotational fine structure of a species which has been rapidly quenched to a very low temperature. The first attempts to do this were[45,48] for water but the most conclusive case is for methane which is described in a very elegant experiment by Frayer and Ewing.[96] In the case of hydrogen only singlet ($I = 0$) spin states are possible for even J levels (*para*-H_2) while triplet ($I = 1$) states are associated with odd J levels (*ortho*-H_2). In methane, however, the arrangement of nuclear spin with rotational levels is more complicated since the four protons can be coupled into quintet ($I = 2$) (*meta*-CH_4), triplet ($I = 1$) (*ortho*-CH_4), or singlet ($I = 0$) (*para*-CH_4) nuclear spin states. Because the Pauli principle requires the total molecular wave function to be antisymmetric to interchange of identical protons, only certain combinations of spin states are allowed. A nuclear spin quintet is thus associated with the $J = 0$ rotational level while the $J = 1$ state is a triplet; both triplet and singlet states are possible with $J = 2$. The distribution of spin-states is thus temperature dependent and as the temperature is changed it is necessary for forbidden spin-state changes to occur before thermal equilibrium is attained. Except for hydrogen and its homonuclear isotopes, the effects of these spin conversions are too small to be detected at ambient temperatures. Frayer and Ewing[96] prepared an equilibrium matrix of $Ar/CH_4 = 500$ at 77 K which they rapidly quenched (*c.* 5 min) to 4 K and maintained at this temperature for several hours, whilst recording the IR spectrum in the v_3 and v_4 regions (Figure 3.8). $R(0)$ is seen to increase in intensity with time, while $R(1)$, $Q(1)$ and $P(1)$ decrease, the effect being completely reversible on temperature cycling between 4 K and 35 K. Nuclear spin conversion is very rapid (minutes) for singlet $\rightarrow$ triplet transitions at 4 K because both spin isomers reside in the $J = 2$ level and inter-system crossing is efficient. Triplet $\rightarrow$ quintet relaxation is however slow (hours) at 4 K ($t_{1/2}$ *c.* 60 min) but rapid at 40 K because the $J = 3$ level is populated and can contain both spin isomers to make intersystem crossing possible. The perturbation which mixes the spin wave functions of the $J = 0$ and $J = 1$ rotational levels is probably spin-spin interaction within the molecule and thus relaxation is slow. When a paramagnetic impurity (e.g. as little as 0·2 per cent. oxygen) is present the magnetic dipole provides the perturbation and acts as a catalyst causing a rapid (minutes) conversion.

The above work on methane has been confirmed by Hopkins *et al.*[51] and extended to CH_3D, H_2O and NH_3 in argon, krypton and xenon matrices. The work indicates how the study of the time dependence of fine-structure features of MI spectra can be a powerful aid in discriminating rotational structure from structure arising from multiple-site trapping or aggregation. The oxygen catalysis provides a valuable check on an interpretation of time dependence in terms of rotational structure. The directions of the intensity changes often provide an aid to the assignment of rotational quantum numbers.

Silane (SiH₄)

The IR spectra of SiH_4 and its deuterated species isolated in solid noble gases, carbon monoxide and methane at 6 K have been reported.[96a] There is no evidence for free or nearly free rotation in any of these matrices but at least two types of trapping sites are indicated. Since nuclear-spin conversion has been observed with methane,[96] it seems reasonable to assume that silane also exhibits this phenomenon. The authors therefore carried out oxygen-doping experiments in order to catalyse the nuclear-spin conversion but were unable to detect any spectral changes. They were also unable to observe any products of a silane/oxygen reaction, such as disiloxane.

3.6.2 Hydrocarbons

Reference 91 provides argon MI spectra at low M/A ratios of 10–50 for methane, ethane, propane, butane, pentane, ethylene, 1-butene, and 1-pentene, but with little discussion of the data. Spectra of mixtures of hydrocarbons and their deuterated analogues have been systematically examined by Rochkind in his pseudo-matrix isolation technique and these are discussed in detail in §3.7. Cyclopentane and benzene have also been briefly reported[97] in argon at 10 K.

The dihedral angle in cyclobutane has been determined[98] as $37 \pm 6°$ from a matrix study in argon at 21 K. In the IR spectrum several bands are observed which are assigned to motions which would be inactive (B_{2g} or E_g symmetry) for a planar structure (D_{4h}) of cyclobutane, but active (B_2 or E) in a bent (D_{2d}) structure. Assuming the transition moments are localized in the CH_2 groups, then the relative intensity of a B_{2g} or E_g band compared to the analogous allowed A_{2u} or E_u band is related to the dihedral angle of the molecule. Dows and Rich[98] compared the 747 cm^{-1} (E_g) and 625 cm^{-1} (A_{2u}) bands which are analogues (CH_2 rocking motion) and obtained the angle from the relationship $I_{747}/I_{625} \tan^2 \frac{1}{2}\alpha$; the value obtained is in agreement with the electron diffraction value[99] of 35°. Comparison of analogous bands at ambient temperatures is not feasible due to gross overlap of bands and the presence of hot bands.

The IR spectra between 3200 cm^{-1} and 350 cm^{-1} for 1,3-cyclohexadiene, cis- and trans-1,3,5-hexatriene have been reported[99a] in a photolysis MI study in argon at 20 K.

The IR spectrum of benzene in a hydrogen chloride 'matrix' with a hydrogen chloride to benzene mole ratio of c. 10 : 1 has recently been reported by Szczepaniak and Person.[99b] A number of absorption bands, forbidden in the spectrum of the pure benzene crystal, appear weakly in the spectrum of benzene in the matrix. Also the intensity of the gas-phase allowed fundamentals v_{18} and v_{20} show a surprising decrease in the hydrogen chloride matrix. The quantitative comparison of these relatively small changes suggests that benzene in the hydrogen chloride matrix is in a site of approximately D_{6h} symmetry, but is strongly perturbed.

3.6.3 Halogen-containing molecules

Halogen chemistry provides a fruitful field both for free-radical matrix studies (Chapter 5) and for studies of matrix-isolated molecules. Much work centres around the synthesis of fluorine compounds utilizing the ease of photolytic dissociation of the fluorine molecule.

Hypo-halous acids (HOX)

Until recently the IR spectrum was known only of hypochlorous acid, but MI studies have succeeded in preparing hypobromous[100] and hypofluorous acids.[101] Hypobromous acid and hypochlorous acid were synthesized by Schwager and Arkell[100] by photolysis of HX, O_3/Ar mixtures at 4 K; their frequencies are listed in Table 3.7. The use of isotopic starting materials, HX, DX and $^{16,18}O_3$ provided data to enable force constants to be calculated. Noble and Pimentel[101] stabilized HOF by photolysing F_2, H_2O/N_2 mixtures at 14 K and 20 K. Growth and diffusion studies, isotopic labelling, and a normal co-ordinate analysis allows an unequivocal assignment of the fundamental frequencies which are compared with those of hypobromous acid and hypochlorous acid in Table 3.7. The force field

Table 3.7 Vibrational frequencies (cm^{-1}) of hypohalous acids

Frequency	HOF[a]	HOCl[b]	HOBr[b]
ν_3 antisymmetric stretch	3537·1	3581	3590
ν_2 bend	1359·0	1239	1164
ν_1 symmetric stretch	886·0	729	626

a. In nitrogen matrices at *c.* 8 K, from ref. 101c.
b. In argon matrices at 4 K, from ref. 98.

calculated for HOF implies that the bonding is similar to the bonding in the two prototype molecules, H_2O and F_2O. The OH bond is slightly weaker than in H_2O and the OF bond somewhat stronger than for F_2O but these differences are small compared with the dramatic differences observed for O_2F. The vibrational potential function provides no evidence for the instability of HOF which the authors suggest must be explained in terms of some rapid chemical reaction which does not permit the compound to exist unless it is trapped in an unreactive medium. The reaction proposed is the highly exothermic decomposition to hydrogen fluoride and oxygen

$$2HOF = 2HF + O_2$$

of ΔH *c.* -315 kJ and ΔG more than 335 kJ negative, very much favouring products. An additional band at 3777 cm^{-1} is attributed to HF hydrogen-bonded in the same matrix cage to HOF.

Very recently HOF has been prepared[101a] and found to be stable for long enough to undertake a microwave study.[101b] The structural parameters obtained [$r(O—H) = 0.964$ Å, $r(O—F) = 1.442$ Å and $H\hat{O}F = 97.2 \pm 0.6°$]

are consistent with the conclusion of the IR work. The same preparative method has also been used to obtain a purer matrix-isolated specimen,[101c] and its IR spectrum confirms the earlier[101] observations and interpretations. From the small discrepancies in the fundamental frequencies it is suggested that the earlier values were all perturbed by HOF hydrogen-bonded to a HF molecule.

The values in Table 3.7 clearly allow a prediction of the frequencies for hypoiodous acid, provided account is taken of the matrix shift of nitrogen in the case of HOF compared with the other values in argon.

Oxygen fluorides

Photolysis of matrix-trapped fluorine in the presence of oxygen has produced[102] O_2F and O_2F_2 by fluorine-atom addition. In the presence of nitrous oxide, instead of the usual fluorine-atom addition reaction, abstraction of an oxygen atom leads[103] to the formation of OF_2.

Noble gas compounds

Pimentel *et al.*[104,105] have elegantly demonstrated the usefulness of the matrix technique as a preparative method for noble gas halides. On photolysing argon/fluorine mixtures they found[104] no evidence for the production of argon fluorides but in the presence of krypton two new absorptions appeared which are assigned to the antisymmetric stretching mode and bending mode of a linear symmetrical KrF_2 molecule (Table 3.8). Photolysis

Table 3.8 Vibrational frequencies (cm^{-1}) of noble-gas halides in argon matrices (except $XeCl_2$, see text)

Compound	v_2	v_3
KrF_2	236	580
XeF_2	(213 gas)	547
$XeCl_2$	< 200	313 (with satellites)
XeF_4	v_2 290	v_6 568

of argon/xenon/fluorine mixtures gave absorptions characteristic of XeF_2 (also linear, $D_{\infty h}$) and XeF_4 (square planar, D_{4h}). The bending mode of XeF_2 (213 cm^{-1} gas) is below the range of the spectrometer and is unobserved, as is the lowest (v_7, 123 cm^{-1} in gas) of the three infrared-active modes of XeF_4. Xenon dichloride was prepared[105] by passing mixtures of xenon and chlorine through a microwave discharge and condensed on a matrix-support window at 20 K. A complex multiplet band was observed *c.* 313 cm^{-1} which is assigned to v_3 of a symmetric linear molecule. The multiplicity is expected because of the isotopic complexity—xenon has 7 isotopes above 1 per cent. natural abundance.

Claassen et al.[106] have prepared the oxyfluoride XeO_2F_2 and trapped it in an argon matrix at 4 K and observed all of the nine infrared-active fundamentals due to a molecule of pseudo-bipyramidal structure, where the two fluorine atoms are axial to the xenon, and the two oxygen atoms with a lone pair orbital are equatorial. The same group of workers have also recently reported[106a] the IR and Raman spectra of XeO_3F_2 matrix-isolated in neon and argon. The same trigonal-bipyramidal structure is indicated, the lone pair having become a bonding pair to an oxygen atom. All eight fundamentals consistent with D_{3h} symmetry are observed: there are four bands observed only in the Raman effect and two observed only in the IR spectrum. This work is noteworthy as it is one of the first to report the Raman spectrum of a molecule matrix-isolated at cryogenic temperatures (§9.5.4).

Polyhalogen compounds

The spectra of matrix-isolated ClF_3 and BrF_3 have been studied[107] in the region 800–33 cm^{-1}. The gaseous molecules are known from microwave spectroscopy to be T-shaped (C_{2v} symmetry) and X-ray studies show that this monomer structure is substantially preserved in the solid state. IR and Raman spectra of ClF_3 and BrF_3 vapours confirm[108] the structures and provide an assignment of all six fundamentals although the two lowest frequencies in each case could not be separated and positively identified. The matrix-isolation data,[107] in argon and nitrogen, confirm all the assignments, add isotopic frequency shifts and separate the two low-frequency modes:

	ClF_3		BrF_3	
	gas	Ar	gas	N_2
$\nu_6(B_2)$		332		249
	328		242	
$\nu_3(A_1)$		328		240

For BrF_3 in argon the data have been simultaneously and independently[108a] confirmed and some thermodynamic properties computed.

Several absorption bands found[107] at low M/A ratios are attributed to dimers whose structure seems best represented with two fluorine bridges formed using the long-bond fluorine atoms. Considered as transient liquid-state species, these dimers suggest the likely fluorine atom exchange mechanism which can account for the exchange of non-equivalent fluorine atoms which is observed in the liquid state for these systems.

The pentafluoride, BrF_5, has also been examined[107] in argon. Its structure is C_{4v} with four long XF bonds and one short XF bond. The four long-bonded fluorine atoms are tipped slightly towards the short-bonded fluorine atom, which plays the role of the equatorial fluorine atom in ClF_3 and BrF_3. The assignments are: $v_1(A_1)$, 681 cm^{-1}, symmetric short-bond stretch (X—F'); $v_2(A_1)$, 582 cm^{-1}, symmetric long-bond stretch (X—F); $v_3(A_1)$, 366 cm^{-1}, symmetric long-bond bend; $v_7(E)$, 636 cm^{-1}, antisymmetric long-bond stretch; $v_8(E)$, 415 cm^{-1}, F—X—F' antisymmetric to yz plane; $v_9(E)$ 240 cm^{-1}, F—X—F bend antisymmetric to xz plane; values which are in close agreement with gas phase values.[109] Dimer bands are observed at 648 cm^{-1} and 623 cm^{-1} and, with very low intensity, at 578 cm^{-1} and 522 cm^{-1}. The two higher-frequency dimer bands are interpreted in terms of two monomer molecules associated through the long Br—F bonds in a similar way to ClF_3.

A detailed study has recently been made[110] of ClF_5 in argon and nitrogen at 4 K with warm-up studies up to 37 K and at M/A ratios 10 000–400. The assignments to a square-pyramidal C_{4v} structure, in particular the unusual triple coincidence of two deformation (v_3 and v_8) and one stretching (v_4) mode at 480 cm^{-1}, previously suggested by Begun et al.[109] for the gas phase, is confirmed. In the matrix experiments the observed ^{35}Cl-^{37}Cl isotope splittings permit assignment of the bands to the individual modes:

$$493 \text{ cm}^{-1} \text{ mw } v_3(A_1)\ ^{35}\text{Cl} \qquad 482 \text{ cm}^{-1} \text{ m } v_8(E)\ ^{35}\text{Cl}$$

$$483 \text{ cm}^{-1} \text{ sh w } v_3(A_1)\ ^{37}\text{Cl} \qquad 479 \text{ cm}^{-1} \text{ mw } v_8(E)\ ^{37}\text{Cl}$$

$$v_4 \text{ is only Raman-active}$$

At $M/A = 400$ additional bands appeared between 740 cm^{-1} and 700 cm^{-1} which diffusion experiments confirm are due to associated species. Matrix site splitting is also seen for v_7.

Several new chlorine-bromine polyhalogen molecules have been produced[111] by passing a mixture of noble gas (argon, krypton, or xenon) chlorine and bromine through a microwave discharge and condensing at 20 K. The IR spectra indicate that several T-shaped molecules, analogous to ClF_3, possessing either the Cl—Br—Cl or the Cl—Br—Br linear unit are
                | |

formed, the most definitely identified species being Cl—Br—Br.
                                             |
                                          Cl

Other fluorides

The pentafluorides of antimony and arsenic suspended in argon and neon matrices have been recorded[112] between 4000 cm^{-1} and 100 cm^{-1}. For antimony pentafluoride there are six intense absorptions which persist at the highest M/A ratios and are attributed to monomer. These frequencies are assigned on the basis of C_{4v} symmetry, since there are only five infrared-active fundamentals for D_{3h} symmetry, and only four of these have appreciable intensity in the case of arsenic pentafluoride and phosphorus pentafluoride.[113] The vibrational assignment for monomeric antimony pentafluoride parallels the assignment for other C_{4v} molecules.[109] At low M/A ratios multimer bands appear which are attributed to dimers and higher species, associated through Sb—F—Sb bridge bonds—a connection favoured by the C_{4v} monomer structure.

The matrix spectrum of arsenic pentafluoride is similar to that of the vapour[113] but quite dissimilar from that of antimony pentafluoride and it appears unambiguous that antimony pentafluoride possesses a different symmetry, namely D_{3h}. At low M/A ratios or on diffusing, the four observed monomer fundamentals are replaced by a single polymeric absorption, and it is evident that arsenic pentafluoride does not have the fluorine-bridge-forming capability of antimony pentafluoride.

The IR MI spectrum[114] of sulphur tetrafluoride in argon at 4 K is in line with the gas-phase spectrum and shows that sulphur tetrafluoride has a very similar structure to ClF_3, even to the extent of having its axial fluorine atoms tipped slightly in towards the equatorial fluorine atoms:

Redington et al.[114] make a detailed comparison with their data[107] on ClF_3 and propose a reversal of the early assignments for the antisymmetric axial and equatorial SF stretching vibrations. Aggregation is observed at high concentrations and the data suggest a discrete sulphur tetrafluoride dimer which has the probable structure:

The MI spectrum of SOF_2 is observed as an impurity with SF_4, and that of SO_2F_2 in argon has been reported[115] in a separate study.

Tetrahedral halogenated molecules

Abramowitz and Comeford[116] have studied the v_3, $2v_4$ and v_3, $v_1 + v_4$ Fermi resonance doublets of CF_4 and CCl_4 in argon, krypton and nitrogen matrices at 4 K and 20 K. They find that the resonance interaction differs considerably from that found in the gas phase and increases as the matrix becomes more concentrated, and was greatest in the pure solid phase. Their resolution was insufficient, however, to observe all the isotopic lines for CCl_4.

King[117] has examined the spectra over the range 3800–200 cm^{-1} of H_3CCl, $C_2H_5CH_2Cl$, H_2CCl_2, $XCCl_3$ (X = H, F, Cl, Br, or CH_3) and $SiCl_4$ in an argon matrix at 20 K. The vibrational frequency shift due to different Cl isotopes of a chlorinated molecule is small (c. 3 cm^{-1}) and usually cannot be observed in the IR spectrum at room temperature but can be detected in a low-temperature matrix. This detailed study identifies most of the isotopic frequencies and establishes the isotopic splitting pattern of the vibrational absorption bands for chlorinated tetrahedral molecules. The patterns should prove useful in the analysis of the vibrational spectra of chlorinated molecules and free radicals. For carbon tetrachloride the intensities and frequencies of the bands at 789 cm^{-1} and 768 cm^{-1} show no drastic change in going from room temperature to 20 K and their assignment is confirmed as arising from v_3 and $v_1 + v_4$ in Fermi resonance and not to a pair of hot bands. Verderame and Nixon[118] have examined the v_5 band of CH_3I and CD_3I in argon and in nitrogen at 10 K, and attribute the observed splitting to Fermi resonance with the combination band $v_3 + v_6$.

There has been considerable interest in the rotation of methyl halides in condensed phases. For all the methyl halides in solution the shapes of the perpendicular bands approximately overlap the contours of the gas phase bands which is interpreted that in solution hindered rotation occurs about the C_3 axis. In the case of CH_3F in solution[119] the parallel bands are also broad indicating rotation also about an axis at right angles to C_3. Hallam et al.[120] have examined all of the methyl halides in argon matrices at 20 K and find no evidence for rotation of the trapped CH_3X monomer. This is in accord with the quinol/CH_3Hal clathrate data (Chapter 8) and with the work of Dows[121] who studied the vibrational spectra of crystalline CH_3Cl, CH_3Br and CH_3I at 80 K, and established that the molecules are not free to rotate in the solid. Concentration studies[120] of CH_3Hal in argon show the growth of peaks which are assigned to multimeric species.

Other halogenated molecules

IR matrix studies have been reported for trifluoromethylacetylene,[122] trifluoroethanol,[123] and hexafluoroacetone.[124] The four-membered ring chlorides C_4H_7Cl, C_4H_6DCl and $C_4H_3D_4Cl$ have been studied[125] in argon at 20 K. *Cis-* and *trans*-1,2-difluoroethylene and their deutero-analogues have been studied[126] in argon at 20 K and xenon at 60 K and 77 K and a complete fundamental assignment given; the matrix work was necessitated by the near-degeneracy of v_7 and v_{12} in the *trans* isomer.

The IR spectra of diboron tetrachloride and tetrafluoride have been investigated[127] in argon. X-ray diffraction studies[128] of B_2Cl_4 and B_2F_4 solids have shown both molecules to be planar (V_h molecular symmetry) while electron diffraction measurements[129] on B_2F_4 gas have been interpreted in terms of a staggered configuration (V_d). The IR and Raman spectra of the gas, liquid, and solid states also provide evidence to support the staggered V_d symmetry for the gas and liquid, and the planar one for the solid. There is remarkable agreement between the principal features of the matrix spectra and those of the gaseous spectra which together with detailed [11]B—[10]B isotopic shift data[127] support a staggered configuration for both of the matrix-isolated molecules.

Two studies, one of bromine monoxide,[130] Br_2O, and one[131] of nitryl bromide, NO_2Br, and nitryl chloride, utilize solid bromine as a matrix support but the solute concentrations are comparable to those of the bromine.

3.6.4 Alkanols, thiols; molecular association and conformational isomerism

Methanol

The superiority of matrix-isolation spectroscopy for vibrational analysis is well demonstrated in the case of methanol and ethanol. Barnes and Hallam[132] have studied CH_3OH and CH_3OD between 4000 cm^{-1} and 40 cm^{-1} suspended in argon matrices at 20 K and CD_3OD, $^{13}CH_3OH$, and $CH_3^{18}OD$, between 4000 cm^{-1} and 650 cm^{-1}. The small bandwidths obtained under cryogenic matrix conditions have allowed a revised assignment of the methanol fundamentals to be made. In particular, the positions of the methyl rocking modes and the methyl antisymmetric A' and A'' stretching and deformation frequencies could be distinguished, which are badly overlapping in the vapour and in condensed phases. For example for CH_3OH, a band is observed at 1477 cm^{-1} in the vapour phase, which has been assigned both to the A' CH_3 deformation mode[133] and to the A'' mode.[134] This band splits into two in the argon matrix spectrum: 1474·1 cm^{-1} (A') and 1465·8 cm^{-1} (A''). Although it is debatable that the assignments might be reversed, it does demonstrate the value of the MI technique for the splitting of near-coincident vibrational frequencies. The same study[132] also provides a thorough investigation of the association of methanol in argon, in extension of the earlier work of Van Thiel *et al.*[135] for nitrogen matrices. In the $vO-H(D)$ stretching region a series of bands are observed as the M/A ratio is lowered, which, from their concentration dependence (Figure 3.9) may be assigned to monomer, dimer, trimer, tetramer and high multimer species. The improvement over the spectra in all other phases is clearly shown in Figure 3.9. In the vapour phase, difficulties arise from the rotational structure of the vibrational band, although Inskeep *et al.*[136] have assigned bands due to dimer and to tetramer, within the rotational envelope, and even attempted quantitative measurements on them. In the pure liquid or solid phases, the $OH(D)$ modes give bands of enormous breadth, due to the presence of large multimeric species. In solution in non-polar solvents, some control of the species present is obtained by varying the concentration; at extreme dilution solute-solute interactions are minimized and the monomer spectrum obtained, whereas at high concentrations spectra approximating to the

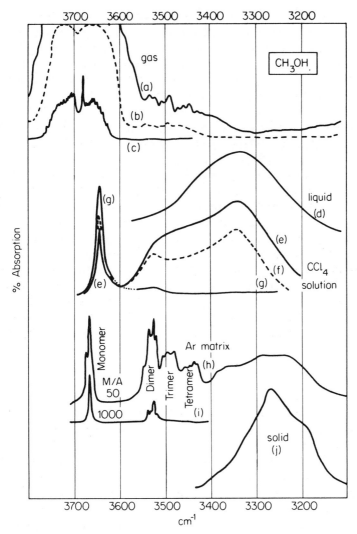

Figure 3.9 IR spectra of $v(OH)$ stretching region of methanol in the gas phase and in CCl_4 solution at ambient temperatures and in an Ar matrix and the pure solid phase at 20 K

pure liquid are obtained. On progressive dilution the broad multimer absorption gradually shows indications of a shoulder on the high frequency side, which gives rise to a poorly defined dimer band at $c.$ 3500 cm^{-1}, before the monomer species becomes dominant. The breadth of the bands, however, makes it impossible to locate bands due to multimeric species intermediate between dimer and high polymer.

Van Thiel et al.[135] observed a similar series of bands in a nitrogen matrix and assigned the bands with low-frequency shifts to cyclic dimer and trimer, while an absorption showing a frequency shift close to that of the pure solid was assigned to open chain tetramer. Barnes and Hallam,[132] however, from their far-infrared data have shown that the argon-trapped dimer has an open chain structure. A cyclic dimer should show an intense hydrogen-bond stretching mode, but only a very weak deformation mode, as for hydrogen chloride,[12] whereas an open chain dimer should show stretching and deformation modes of comparable intensity, as for hydrogen cyanide.[26] Two strong bands, at 222 cm^{-1} and 116 cm^{-1}, are in fact observed[132] for methanol and the low deuteration shift (to 213 cm^{-1}) of the stretching mode lends further support for an open-chain structure for the dimer. The trimer and tetramer which absorb at slightly lower frequencies than the dimer are also assumed to have open-chain structures. The multiplet structure of each of these absorptions is not fully understood and may be due to slightly different angular orientations of the hydrogen bonds and/or multiple trapping sites. In addition to these absorptions, a band is observed on the high frequency side of the monomer band whose concentration dependence is not clear but it appears to grow as the multimers increase. It is interpreted as being due to the free terminal OH group of an open-chain multimer although an assignment to a monomer in another site is not entirely eliminated.

Ethanol

Barnes and Hallam[137] have also studied EtOH and EtOD in solid solution in argon at 20 K. In liquid solutions in 'inert' solvents at ambient temperatures the monomer OH stretching frequency has been reported by some authors[138,139] as displaying asymmetry while others[140] report the band as symmetrical. The small bandwidths in the low-temperature matrix reveal[137] two monomer OH absorptions at $3657 \cdot 6 \text{ cm}^{-1}$ and $3662 \cdot 2 \text{ cm}^{-1}$. A possible interpretation of the doublet is multiple trapping sites but the strength of the second component makes this an unlikely explanation. There are two distinguishable conformers of ethanol, with the OH group *trans* or *gauche* with respect to the methyl group; the two *gauche* structures being spectroscopically equivalent:

trans (C_s)   *gauche* (C_1)

The monomer OH stretching absorption at $3657 \cdot 6 \text{ cm}^{-1}$ is thus assigned to the *trans* conformer and that at $3662 \cdot 2 \text{ cm}^{-1}$ to the *gauche*. Comparison of the intensities of the two absorptions indicates that the *trans* : *gauche* ratio is *c.* 2 : 1. Since the matrix sample is trapped and quenched to 20 K very rapidly, the relative intensities in low concentration matrices presumably reflect the proportion of the two conformers in the vapour phase at ambient temperature. The small splitting of 4·6 cm^{-1} and the similar intensity of the two OH stretching modes accounts for the controversy over the asymmetry of $v(O-H)$ in solution. Bands due to the subsidiary *gauche* component are observed for several modes, and a complete vibrational assignment is given for the *trans* conformer. The most significant splitting is the methyl out-of-plane rocking mode; in EtOH this is resolved into two components at 1083·2 cm^{-1} and 1076·6 cm^{-1}, while in EtOD it is a single band at 1079·1 cm^{-1}. In EtOH interaction may occur with the COH bending mode (1240 cm^{-1}) only in the *gauche* form (in the *trans* form it is symmetry forbidden), whereas the corresponding bending mode in EtOD is at much lower frequency. Thus the 1076·6 cm^{-1} band must be due to the *gauche* conformer and the 1083·2 cm^{-1} band due to the *trans* species. The latter band is the more intense absorption of the doublet, and thus the stronger band of each of the doublets may be assigned to the *trans* rotamer. This again illustrates the great advantage of a matrix study compared with the detailed study by Perchard and Josien[139] of 12 isotopic species of ethanol in vapour, liquid, solution and solid phases which was inconclusive, the authors tentatively suggesting that the molecule is essentially in the *gauche* form. Concentration studies[137] yield an association pattern of $v(O-H)$ frequencies similar to that of methanol. Two types of open-chain dimer are identified from their $O-H(D)$ stretching modes, and from their strong hydrogen-bond stretching modes at *c.* 215 cm^{-1} (EtOH) and 214 cm^{-1} (EtOD) with weaker bands at 228 cm^{-1} (EtOH) and *c.* 220 cm^{-1} (EtOD), and comparatively weak hydrogen-bond deformation modes at about 125 cm^{-1}. One of the dimers is presumed to have a linear and the other a non-linear hydrogen-bond. Other $v(O-H)$ bands are assigned to open-chain trimer and tetramer and cyclic tetramer species.

Thiols

Similar studies have been made[141] of methanethiol and ethanethiol in argon at 20 K. In the case of EtSH a monomer doublet at 2597 cm^{-1} and 2600 cm^{-1}, analogous to that of EtOH, is assigned to *trans* and *gauche* conformers. These investigations suggest that matrix isolation spectroscopy is a powerful technique for the detection and structural investigation of conformational isomerism.

Cyclobutanol

Durig and Green[142] have partly utilized MI spectroscopy in their extensive studies of *cyclo*butanol, and *cyclo*butanol-d_1, -d_4, and -d_5. The splitting of the O—H in-plane bending vibrations, for example 1262 cm^{-1} and 1307 cm^{-1} for cyclobutanol-d_5 in argon, leads to the conclusion that more than one molecular conformation exists. The molecule has a puckered 4-membered ring

which could lead to equatorial and axial conformers as found[143] for cyclo-butyl halides. Durig and Green feel that the existence of both equatorial and axial conformations at matrix temperatures is very unlikely because only the more stable form should be present; consequently the splitting of (OH) is believed to result from different orientations of the hydroxyl proton with respect to the other atoms. The different proton-orientation interpretation is a reasonable one although the proposal that both equatorial and axial conformers are unlikely to co-exist at matrix temperatures is questionable. Most matrix studies presume that any equilibrium mixture at ambient temperature is frozen out in a low temperature matrix. This important issue is one which requires detailed experimental corroboration.

3.6.5 Nitrogen-containing molecules

Higher oxides of nitrogen

In their studies of nitrogen dioxide, Crawford *et al.*[59,60] performed concentration and warm-up studies and were able thus to investigate some higher oxides. A complicated pattern of bands is observed but they are able to assign five bands to stable N_2O_4 of V_h symmetry. They also identify an unstable molecule, possibly a 'twisted' N_2O_4 having V_d symmetry. Eight further absorptions are assigned to an unstable isomer of dinitrogen tetroxide having the structure $ONO-NO_2(D)$, believed to be planar, and seven more bands to D' a rotational isomer of D. The transformation $D \rightarrow D'$ is considered a *cis-trans* isomerization analogous to alkyl nitrites and nitrous acid.[84] Dinitrogen trioxide and pentoxide molecules are also reported.[59]

In a more recent study Varetti and Pimentel[143a] have examined dinitrogen trioxide in a nitrogen matrix at 20 K and find that it displays a vibrational spectrum very similar to that of the gas-phase asymmetrical N_2O_3 molecule. On near-infrared irradiation at 20 K this is converted into a symmetrical form with absorptions at 1689.7 cm^{-1}, 1661.0 cm^{-1}, 969.4 cm^{-1}, 704.3 cm^{-1}, 387.4 cm^{-1} and 365.4 cm^{-1}. Both ^{15}N and ^{18}O labelling show that sym-N_2O_3 has conventional, nitrite-like bonding and an extended, symmetric configuration, $O=N-O-N=O$. The authors also identify molecule A in the earlier study[59] as N_2O_3 in the 'normal' gas-phase form and B as the extended $O=N-O-N=O$.

Imines

Matrix photolysis of methyl azide[144] and diazomethane[145] dimers produces methyleneimine, $H_2C=NH$ which is thus stabilized for spectroscopic study. Due to the presence of other species and of matrix splittings, however, some of the vibrational assignments are controversial.

Le Brumant[146] has studied ethyleneimine in argon and nitrogen at 20 K but the matrices were too scattering to allow an examination of the $v(N-H)$ stretching frequency. Below 1500 cm^{-1} the spectra are satisfactory but no assignments are given and the system clearly requires more than this cursory study.

Diazomethane

Moore and Pimentel[147] have studied CH_2N_2, $CHDN_2$, CD_2N_2, $CH_2^{15}N^{14}N$ and $CH_2^{14}N^{15}N$ either in argon or nitrogen matrices at 20 K and compared the frequencies with gas phase values. A number of bands are also assigned to second sites and to dimers. Ogilvie[148] has studied diazomethane in argon at 4 K and interpreted the spectra in terms of the structure H_2NNC.

Methylamine and substituted methylamines

As part of a comprehensive study of the methylamine molecule, Durig et al.[149] have reported the spectra of the four isotopes CH_3NH_2, CH_3ND_2, CD_3NH_2 and CD_3ND_2, matrix-isolated in argon at 20 K. The observed frequencies are extremely close to those of the gas-phase but the presence of band splitting due to multiple-site trapping, multimeric absorptions, and isotopic exchange during deposition precluded a detailed analysis of the matrix spectra.

The spectra of N-methylene methylamine, CH_3NCH_2 and CD_3NCH_2 have also been examined[150] in argon at 20 K and detailed assignments given.

The molecules $(CH_3)_3N$, $(CD_3)_3N$ and $(SiH_3)_3N$ have also been reported[151] in argon and complete assignments of the fundamental vibrational modes given. Barriers to internal rotation are calculated for the trimethylamines from their torsional frequencies. No torsional frequencies have been observed for trisilylamine down to 70 cm^{-1} which suggests that the silyl group undergoes almost free internal rotation. If this is so and it persists in the solid matrix it may account for the greater breadth of the IR bands of matrix-isolated trisilylamine relative to those of trimethylamine, in terms of multiple small splittings resulting from the coupling of internal rotation and vibration.

A partial assignment of the fundamental vibrational modes of methyl-disilylamine and dimethylsilylamine has been proposed[152] based on the IR spectra of the matrix-isolated molecules, with a model of a pyramidal skeleton for the monosilyl molecule and a planar skeleton for the disilyl molecule. The work also provides direct evidence for the presence of intermolecular association in $(CH_3)_2SiH_3N$ which had been suggested[153] to account for its unusual properties, e.g. high m.p. and ΔH of sublimation, compared with $CH_3(SiH_3)_2N$ and $(SiH_3)_3N$. As the M/A ratio is lowered from 800 to 50 changes are observed in the spectrum of dimethylsilylamine which are very similar to those observed for various hydrogen-bonded substances (see, for example, refs. 43, 53 and 132 and discussions in earlier paragraphs of this chapter). The intermolecular association is interpreted in terms of an interaction involving the nonbonding electrons on the nitrogen atom of one dimethylsilylamine molecule and an empty silicon d orbital on an adjacent molecule. This is indicated by contrasting the relatively small changes in the C—H stretching and CH_3 deformation regions to the more pronounced spectral changes below 1300 cm^{-1}. The absence of this type of aggregation behaviour in methyldisilylamine and trisilylamine is suggested to be the result of the more extensive delocalization of the nitrogen nonbonding electrons by means of $p \rightarrow d \pi$ bonding with the silicon atoms of the same molecule.

Amides

The IR spectrum of formamide in argon at 20 K has been studied by King.[154] This simplest primary amide is of interest because the $C-NH_2$ group is slightly non-planar and although the 'inversion' frequency has been calculated from the satellite intensities in the microwave spectrum,[155] direct observation of the inversion mode in the IR has not been reported. King finds all the frequencies observed in the vapour with only small matrix shifts; his spectra are, however, unfortunately contaminated with the decomposition products, carbon monoxide, ammonia and carbon dioxide from his hot cell. The splitting of $\nu(C-H)$ at 2881 cm^{-1} and 2871 cm^{-1} is assumed to be due to different sites but the multiple structures observed for all the other bands are mainly due to imperfect isolation. The NH_2 wagging or 'inversion' mode is assigned to a sharp feature at 303 cm^{-1} in the matrix. The complicated vapour spectrum from 450 cm^{-1} to 250 cm^{-1} is attributed to the presence of hot bands and the band centred at $288 \cdot 7 \text{ cm}^{-1}$ is considered the most likely to be the inversion fundamental.

More recently, King[155a] has extended these studies to acetamide, urea and urea-d_4 isolated in argon. The results indicate that the structure of the amide group for the isolated urea molecule is very similar to that of acetamide and other amides, namely a non-planar structure with the $C-NH_2$ group forming a very shallow pyramid; the NH_2 inversion transitions were observed for acetamide and urea at 268 cm^{-1} and 227 cm^{-1} respectively.

Pyridine

Pyridine and pyridine-d_5 have been studied[156] in nitrogen at 5 K. Three fundamentals show a splitting: $\nu_{10}(A_1)$ at $606 \cdot 0 \text{ cm}^{-1}$ and $607 \cdot 5 \text{ cm}^{-1}$,

$\nu_{26}(B_2)$ at 706 cm^{-1} and 709.0 cm^{-1}, and $\nu_{14}(B_1)$ at 1442.0 cm^{-1} and 1443.0 cm^{-1}, at frequencies higher than the monomer absorptions at 602.8 cm^{-1}, 702.2 cm^{-1} and 1441.4 cm^{-1} respectively. These doublets are assigned to pyridine-pyridine intermolecular coupling arising from pair association of pyridine molecules facing each other with the nitrogen atom opposite to the hydrogen atom of the partner 3.4 Å apart.

3.6.6 Carboxylic acids

In their study of dimeric and crystalline formic acid, Millikan and Pitzer[157] briefly report the 3500–2500 cm^{-1} region of a sample trapped in a nitrogen matrix at 20 K. They note that the matrix bands are considerably sharper than in the vapour and that the most intense bands for the dimer lie near the 3100 cm^{-1} peak observed in the vapour. Miyazawa and Pitzer[158] have made a detailed comparison of the four isotopic species of formic acid, HCOOH, DCOOH, HCOOD and DCOOD, in the vapour phase and in a nitrogen matrix at 20 K, in the region 800–400 cm^{-1}. Their work was undertaken to investigate the internal rotation of formic acid monomer and M/A ratios of 1000 or 3000 were used. Assignments are given for the $\delta(C-H)$ bending, $\delta(OCO)$ bending and the OH(D) torsions for the matrix-isolated monomer. The torsional vibrations of the terminal OH groups of short chain multimers of HCOOH were observed in the matrix at 685 cm^{-1} and 694 cm^{-1}.

Like formic acid the IR spectrum of acetic acid has been extensively investigated. However, a new assignment of the fundamental frequencies is proposed as a result of IR studies[159] between 4000 cm^{-1} and 400 cm^{-1} of the four species $CH_3COOH(D)$ and $CD_3COOH(D)$ isolated as monomers in argon and nitrogen matrices at 4 K. The interference caused by over-lapping absorptions in vapour-phase studies was eliminated and the matrix spectra yield clear and unambiguous monomer absorptions in all spectral regions. At $M/A = 1000$, weak bands due to acetic acid dimers were observed. These have been the subject of a separate study by Redington and Lin[159a] who have made a detailed examination of the hydroxyl stretching region of $CF_3COOH(D)$ and CH_3COOH matrix-isolated in neon and argon at 5 K. Fundamental frequencies for the cyclic dimer ring vibrations are suggested and the well-known band broadening phenomenon in the $\nu(OH)$ region is accounted for as being due to combination bands of the low-frequency ring vibrations and $\nu(OH)$ and excited binary combinations of the COH(D) bending and the C—O stretching vibrations.

Grenie et al.[159b] have reported the spectra of CD_3COOH, CH_3COOH and $CH_3C^{18}O^{18}OH$ dimers in an argon matrix at 20 K. Their results do not agree with Redington and Lin's assignment but partially confirm those of Haurie and Novak.[159c] They point out, however, that it is difficult to correlate clearly the gas and matrix frequencies in the 3200–2700 cm^{-1}

region as the whole absorption is lowered in intensity and frequency relative to the bands observed between 2500 and 2700 cm^{-1} when going from gas to matrix phase. It is anticipated that many further MI studies of carboxylic acids will be forthcoming in the near future.

The matrix spectrum of acetyl fluoride is also reported[159] for comparison purposes.

3.6.7 Transition metal carbonyls

In combination with UV visible absorption spectroscopy, IR matrix-isolation spectroscopy is playing a significant rôle in structural and photolytic studies of transition metal carbonyls under cryogenic conditions. The IR spectra of the trimetal carbonyls, $Os_3(CO)_{12}$, $Ru_3(CO)_{10}(NO)_2$, and $Fe_3(CO)_{12}$, in argon or nitrogen at 20 K, have been published by Turner et al.[160] from a selection of a large number of trimetal dodecacarbonyls and derivatives studied in matrices. In the $v(CO)$ stretching region the spectrum of $Os_3(CO)_{12}$ shows a pattern of four bands, 2075 cm^{-1}, 2045 cm^{-2}, 2018·5 cm^{-1}, and 2004 cm^{-1}, identical to the solution spectrum; this indicates the same structure for $Os_3(CO)_{12}$ in solution and matrix, presumably the unbridged D_{3h} structure found in the solid. The solution and matrix spectra of $Ru_3(CO)_{10}(NO)_2$ are also almost identical and the molecule is considered to have the doubly-bridged C_{2v} structure analogous to $Fe_3(CO)_{12}$ in the solid, with bridging NO groups in place of the CO bridges. On the other hand, the spectrum of $Fe_3(CO)_{12}$ in argon is very different from that in solution, but is very similar to both the argon matrix and solution spectra of $Ru_3(CO)_{10}(NO)_2$. These spectra strongly suggest that in argon (but not necessarily in the gas phase) $Fe_3(CO)_{12}$ has the C_{2v} structure found in the solid; the matrix spectrum thus provides, for the first time, the vibrational frequencies of 'free' $Fe_3(CO)_{12}$ in the C_{2v} form.

Photolysis of trapped metal carbonyls leads to the isolation of unstable carbonyl fragments, for example[161] $HMn(CO)_4$ from $HMn(CO)_5$ and $Ni(CO)_3$ from $Ni(CO)_4$. These active fragments are able to react with other species. For example,[162] if $Ni(CO)_4$ is photolysed in an argon matrix containing a little nitrogen there is strong evidence for the formation of $Ni(CO)_3N_2$; similarly, photolysis of $M(CO)_6$ (where M = Cr, Mo, W) gives $M(CO)_5N_2$. Such studies clearly offer interesting possibilities for the synthesis of novel inorganic and organometallic compounds.

All four carbonyls, $Ni(CO)_{1-4}$ and possibly six carbonyls of tantalum, $Ta(CO)_{1-6}$ have been identified[162b] in argon matrices at 4 K. The carbonyls were prepared by vaporization of the metal atoms and condensation into a carbon monoxide/argon mixture. Careful warming of the matrix results in the growth and disappearance of $v(CO)$ bands in the 2000 cm^{-1} region. For the nickel experiments these bands appear at 2052 cm^{-1}, 2017 cm^{-1}, 1967 cm^{-1} and 1996 cm^{-1} and are assigned to $Ni(CO)_4$, $Ni(CO)_3$, $Ni(CO)_2$

and NiCO, respectively. Specific assignments for tantalum carbonyls are less certain, but at least five molecular species are definitely formed during the diffusion experiments. For both the nickel and the tantalum series the general trend is that v(CO) increases with increasing co-ordination number, which is predicted on the basis of simple bonding theory. A similar study[162c] for uranium also provides evidence of the successive formation of carbonyl species; at least five groups of bands are involved during diffusion which show a close correspondence to the tantalum data[162b] and are thus tentatively assigned to the species $U(CO)_{1-6}$.

3.6.8 Boron compounds

The MI spectra[69,70] of BF_3, BCl_3, and $HBCl_2$ have been discussed earlier. The report[162a] of the MI spectrum of BH_3 also includes the spectra of diborane and borane carbonyl, BH_3CO, matrix-isolated in argon. A MI study in argon has also been reported[162d] for borazine and boroxine. For borazine the spectra of the isotopic species $H_3B_3N_3H_3$, $H_3^{10}B_3N_3H_3$ and $D_3B_3N_3H_3$ are presented. One of the E' modes, v_{16}, for the planar D_{3h} structure, is reassigned to 1068 cm^{-1} and two of the inactive A_2' modes, v_6 and v_7, are assigned to 1195 cm^{-1} and 782 cm^{-1} respectively. The literature assignment of the other modes is confirmed. A complete assignment of boroxine, $H_3B_3O_3$, is also presented.

3.6.9 Miscellaneous compounds

The alkali-metal matrix reaction technique (see Chapter 5) is also suited for the synthesis of alkali-metal inorganic compounds. For example, in their methyl radical studies Tan and Pimentel[163] have produced evidence of a new molecular type, a methyl alkali-metal halide complex, CH_3MX (M = Li, Na, or K; X = Br or I).

Argon and nitrogen matrix spectra have been used[164] in a vibrational study of disiloxane, $(SiH_3)_2O$. The IR spectrum of matrix-isolated vinylene carbonate is reported[165] and a complete vibrational assignment given; where observed in both states, the frequencies agree well with gas-phase studies.

King[166] has described a simple heated injection system whereby strongly hydrogen-bonded, high-boiling compounds can be effectively isolated. He reports with partial assignments the IR spectra of monomeric pyrazole, imidazole and dimethylphosphinic acid, isolated in argon at 20 K.

3.7 Analysis of gas mixtures

A novel method for the qualitative and quantitative analysis of multi-component gas and volatile mixtures of stable molecules known as pseudo-matrix isolation (PMI) has recently been developed by Rochkind.[167]

overlapping rotational fine structure effectively prevents gas phase IR spectroscopic analysis of complex mixtures, while intermolecular interactions lead to band broadening and spectral perturbations in the solid phase. Matrix isolation should yield narrow bands at approximately the gas phase frequencies by elimination of rotational effects and solute-solute intermolecular interactions. This is illustrated in Figure 3.10, which shows the gas phase, solid phase and PMI spectra of a five-component mixture.

Rochkind's experimental technique of PMI retains the salient features of conventional matrix isolation, but replaces the slow continuous condensation by a controlled-pulse deposition. The mixture of gases, or the vapours of volatile liquids (a vapour pressure of about 1 torr or greater is required in practice), is diluted 100 : 1 with nitrogen and the diluted mixture then deposited at 20 K in a series of discrete pulses, each of up to 0·5 mmol. Although the substrate temperature rises during deposition of the pulse, the temperature attained is apparently not sufficiently high to cause diffusion and hence aggregation of the solute species. The annealing effect reduces the tendency for the solutes to be isolated in alternative trapping sites, and provides matrices which do not scatter the IR radiation significantly. The time required for deposition of the sample is reported to be much less than that conventionally used for continuous condensation. However there is no obvious reason why pulsed deposition should lead to greater isolation efficiency than continuous deposition at the same overall rate. A mixture of, for example, hydrogen chloride in nitrogen may be deposited continuously at a rate of up to 20 mmolh^{-1} without loss of isolation compared with slower rates of deposition. This gives a similar overall rate to one pulse per minute on the pulsed deposition technique. The use of nitrogen as the support matrix avoids the possible appearance of rotational fine structure for small trapped molecules since no molecule has yet been found to rotate in a nitrogen matrix. The resultant spectra are characterized by extremely sharp, needle-like ($\Delta v_{1/2}$ $c.$ 2 cm^{-1}) absorptions which contrast markedly with gas-phase spectra (Figure 3.10). One striking example (Figure 3.11) is the PMI spectrum of a mixture of deuterated ethylenes: 2·3 μmol ethylene, 3·5 μmol [^{2}H]ethylene, 1·5 μmol [1,1-^{2}H$_2$]ethylene, 2·3 μmol [$trans$-1,2-^{2}H$_2$]-ethylene, and 4·9 μmol [^{2}H$_4$]ethylene deposited as a total of only 14·7 μmol in nitrogen at 20 K. The three ^{2}H isomers are readily identified by bands at 1295 cm^{-1}, 847 cm^{-1} and 751 cm^{-1} belonging, respectively, to the [$trans$-1,2-], [cis-1,2-] and the [1,1-] molecules. Each of these bands is free from overlap with that of any other ethylene, methane or deuterated methane. This contrasts with a recent-chromatographic study[168] of isotopically labelled ethylenes which failed to separate the cis and $trans$-1,2-isomers even using a 900 foot silver nitrate-ethylene glycol packed column and a retention time of 17 hours. The absorption bands of this and other mixtures described are sharp enough to allow peak absorbances to be used for band intensities for quantitative assay.

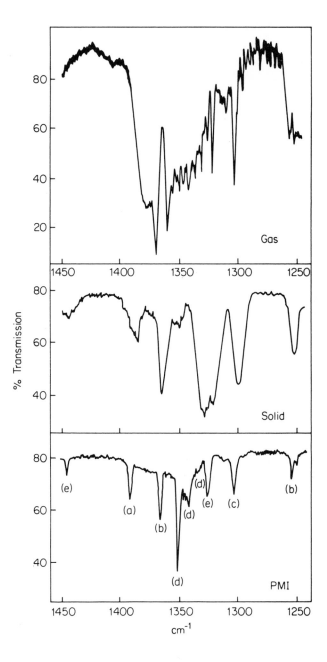

Figure 3.10 IR spectra between 1450–1250 cm^{-1} of a five-component mixture: 25 per cent. allene (a), 31 per cent. dimethyl propane (neopentane) (b), 20 per cent. methane (c) 9 per cent. sulphur dioxide (d), and 15 per cent. vinyl methyl ether (e). Gas spectrum: 45 torr, 1 = 10 cm; solid spectrum: 85 μmol, 23 K; PMI spectrum: N$_2$ matrix, M/A = 100, 57 μmol, 23 K. (Redrawn from ref. 167)

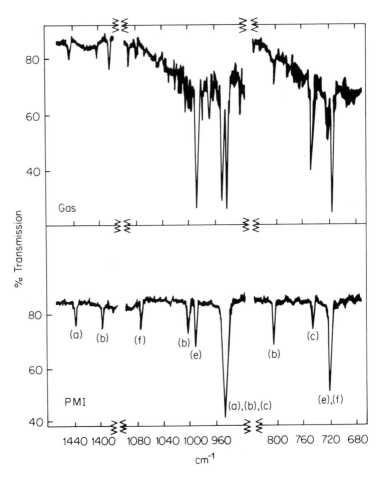

Figure 3.11 IR spectra between 1450–700 cm^{-1} of a mixture of deuterated ethylenes: 2·3 μmol ethylene (*a*), 3·5 μmol [^{2}H] ethylene (*b*), 1·5 μmol [1,1 $-\,^2$H$_2$] ethylene (*c*), 2·3 μmol [*trans*-1,2 $-\,^2$H$_2$] ethylene (*e*), and 4·9 μmol [^{2}H$_4$] ethylene (*f*). Gas spectrum: 30 torr, $l = 10$ cm with mole ratios slightly different from above; PMI spectrum: N$_2$ matrix, $M/A = 100$, total of 14·7 μmoles deposited in three *c*. 0·5 mmol pulses. (Redrawn from ref. 167)

Application of the technique to quantitative analysis depends upon the applicability of the Lambert-Beer law to matrix conditions. The law states that the *absorbance*, $\log(I_0/I) = \alpha cl$, where α is the absorption coefficient for the transition, c the concentration of the absorbing species and l the path length, thus cl for a matrix experiment equals the number of absorbing molecules trapped in the matrix which lie in the beam of the spectrometer. Rochkind has carried out quantitative tests of Beer's law for a variety of structurally dissimilar molecules under pseudomatrix isolation conditions. For each molecule a single, sample mixture was prepared, and successive pulses deposited while monitoring the reservoir pressure and recording spectra between each pulse. A plot of absorbance against micromoles deposited then yielded straight lines with a general accuracy of better than 5 per cent., but, as Rochkind points out, this can be greatly improved, for individual systems.

Snelson[169] has also tested the Lambert–Beer law under the severe experimental conditions for the trapping of vaporizing molecules (Chapter 6), using monoisotopic ^{7}LiF in an argon matrix. The lithium fluoride was superheated to 1700–1800° and matrix ratios in excess of 80 000 were used to ensure that, within the limits of detection, only monomer was trapped. The mass of lithium fluoride vaporized, determined from the loss of weight of the effusion cell (corrected for the loss of weight of the cell itself), is proportional to cl, and thus a plot of mass vaporized against absorbance should yield a straight line. The monomer absorption of lithium fluoride in an argon matrix is a doublet, due to alternative trapping sites, at 835 cm^{-1} and 840 cm^{-1}. Figure 3.12 shows the Beer's law plots for each of the peaks,

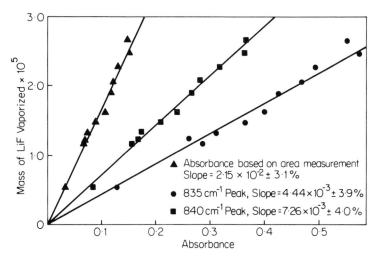

Figure 3.12 Beer's law plot for LiF isolated in argon matrices (reproduced by permission of *J. Phys. Chem.* **73**, 1922 (1969))

and for the overall area of the absorption. The error limit of the slope was found to be ±3·1 per cent., compared with a minimum expected precision of ±2·6 per cent. indicating that the Lambert-Beer law is obeyed to within ±0·5 per cent.

The number of micromoles deposited can be used only as a relative measure of the quantity of material observed, since the proportion actually condensed on the window will be determined by the geometry of the apparatus and by the sticking coefficient of the substance under the conditions used (see Chapter 2). The latter will depend on the rate of deposition, or the size and timing of the pulses in the PMI technique, the temperature of the support window, etc. However, this does not invalidate the quantitative correlation since, provided that the experimental conditions are kept constant, results should be reproducible on a particular apparatus. The problem of transferability of absorption coefficients between different laboratories are no different from those pertaining for liquid solutions and may be overcome by the use of an internal standard as a scaling parameter.

The technique is amenable to larger molecules and more polar species, provided that good (pseudo) isolation is achieved and intermolecular interactions minimized. Qualitative analysis of a gas mixture may be carried out by a comparison of the PMI spectrum with a computer-generated catalogue of frequencies and absorbances compiled by Rochkind[170] from data of 89 molecules. A frequency-sequenced list of absorptions determines the possible components of the mixture, and comparison of relative peak intensities in the standard spectra with those observed enables a definite identification to be made. Comparison of the observed intensities of the strongest peaks in each spectrum with the listed sensitivities enables estimates to be made of the *relative* concentrations of the identified components of a mixture.

The use of pulsed deposition appears to overcome problems which arise from multiple trapping sites by annealing out the less favoured sites. However, in general PMI involves a lower degree of molecular isolation than is acceptable for the normal MI technique, thus in many instances difficulties will arise from solute-solute intermolecular interactions. At a matrix ratio of 100, strongly hydrogen-bonding molecules, such as hydrogen halides and alkanols, are strongly associated (§3.3.1 and §3.6.4) and even relatively non-polar substances such as carbon monoxide are extensively aggregated at this concentration.[30] Interactions between molecules of different substances may also lead to new spectral features; for example (§3.3.1) the hydrogen halides interact with a wide range of molecules giving strong absorptions due to mixed dimeric and multimeric species.[14,18] Although solute-solute interactions do not cause serious interference within the range of molecules so far examined,[167] under PMI conditions it is a possibility which must be considered for every new mixture of solutes.

The PMI technique has obvious applications as a supplement to gas chromatography, analysis of chromatographic fractions by matrix isolation allowing the separation of species with similar retention times or the identification of unknown fractions. The technique has definite advantages compared with gas chromatography in the analysis of isotopic mixtures. Rochkind has shown that the technique is eminently suited for hydrocarbon and halogenocarbon mixtures and applications seem most appealing in the area of gas-phase and cryogenic photochemistry, where kinetic and mechanistic studies often require the use of isotopic tracers.

3.8 Conclusions

It is evident from the large number and variety of molecules now studied in inert matrices that the matrix technique has now become fully established in most spectroscopic research laboratories as a standard technique for vibrational assignment. These applications are increasing rapidly and will continue to do so as our understanding of matrix interactions increases and as Raman matrix-isolation spectroscopy develops. The use of the technique for the synthesis of reactive and stable new molecules is only just beginning to be exploited and considerable developments are anticipated in this area.

3.9 References

1. H. E. Hallam, in *Infrared Spectroscopy and Molecular Structure*, ed. Mansel Davies, p. 405, Chapter XII, Elsevier, 1963.
2. H. E. Hallam, p. 245, 'Proc. 3rd Inst. Petroleum Hydrocarbon Research Group Conf.', in *Spectroscopy*, ed. M. J. Wells, Inst. Petroleum, London, 1962.
3. L. J. Schoen, D. E. Mann, C. M. Knobler and D. White, *J. Chem. Phys.*, **37**, 1146, (1962).
4. M. T. Bowers and W. H. Flygare, *J. Chem. Phys.*, **44**, 1389, (1966).
5. M. T. Bowers and W. H. Flygare, *J. Mol. Spectr.*, **19**, 325, (1966).
6. J. M. P. J. Verstegen, H. Goldring, S. Kimel and B. Katz, *J. Chem. Phys.*, **44**, 3216, (1966).
7. D. E. Mann, M. Acquista and D. White, *J. Chem. Phys.*, **44**, 3453, (1966).
8. L. F. Keyser and G. W. Robinson, *J. Chem. Phys.*, **44**, 3225, (1966); *J. Chem. Phys.*, **45**, 1694, (1966).
9. L. C. Brunel and M. Peyron, *Compt. rend.*, **262**, 1297, (1966).
10. H. Vu, M. R. Atwood and M. Jean-Louis, *Compt. rend.*, **262C**, 311, (1966); *J. Physique*, **28**, 31, (1967).
11. R. Ranganath, T. E. Whyte, jr., T. Theophanides and G. C. Turrell, *Spectrochim. Acta.*, **23A**, 807, (1967).
12. B. Katz, A. Ron and O. Schnepp, *J. Chem. Phys.*, **46**, 1926, (1967); **47**, 5303, (1967).
13. K. B. Harvey and H. F. Shurvell, *Chem. Comm.*, 490, (1967); *Canad. J. Chem.*, **45**, 2689, (1967); *Canad. Spectroscopy*, **14**, 1, 32, (1969).
14. A. J. Barnes, H. E. Hallam and G. F. Scrimshaw, *Trans. Faraday Soc.*, **65**, 3150, 3159, 3172, (1969).
15. A. J. Barnes, J. B. Davies, H. E. Hallam, G. F. Scrimshaw, G. C. Hayward and R. C. Milward, *Chem. Comm.*, 1089, (1969).
16. B. Katz and A. Ron, *Chem. Phys. Letters*, **7**, 357, (1970).
17. W. G. Von Holle and D. W. Robinson, *J. Chem. Phys.*, **53**, 3768, (1970).
18. J. B. Davies and H. E. Hallam, *Trans. Faraday Soc.*, **67**, 3176, (1971).
19. R. Coulon, L. Galatry, B. Oksengorn, S. Robin and B. Vodar, *J. Phys. Radium*, **15**, 641, (1954); H. Vu. and B. Vodar, *Compt. rend.*, **248**, 2082, (1959); M. R. Atwood, H. Vu and B. Vodar, *Spectrochim. Acta.*, **23A**, 553, (1967); M. R. Atwood, M. Jean-Louis and H. Vu, *J. Physique*, **28**, 31, (1967).

20. J. Lascombe, P. V. Huong and M. L. Josien, *Bull. Soc. Chim.*, 1175, (1959); W. J. Jones and N. Sheppard, *Trans. Faraday Soc.*, **56,** 625, (1960).
21. P. Washbrook and H. E. Hallam, unpublished results.
22. A. J. Barnes, J. B. Davies, H. E. Hallam and J. D. R. Howells, *J.C.S. Faraday II,* 69, 246, (1973).
23. M. T. Bowers, G. I. Kerley and W. H. Flygare, *J. Chem. Phys.*, **45,** 3399, (1966); M. G. Mason, W. G. Von Holle, and D. W. Robinson, *J. Chem. Phys.,* **54,** 3491, (1971).
24. D. F. Hornig and G. L. Hiebert, *J. Chem. Phys.*, **27,** 752, (1957).
25. R. Savoie and A. Anderson, *J. Chem. Phys.*, **44,** 548, (1966).
26. C. M. King and E. R. Nixon, *J. Chem. Phys.*, **48,** 1685, (1968).
27. A. D. McLean and M. Yoshimine, *IBM J. Res. Dev.*, **12,** 206, (1968).
28. G. E. Leroi, G. E. Ewing and G. C. Pimentel, *J. Chem. Phys.*, **40,** 2298, (1964).
29. S. W. Charles and K. O. Lee, *Trans. Faraday Soc.*, **61,** 614, (1965).
30. J. B. Davies and H. E. Hallam, *J.C.S. Trans. Faraday II,* **68,** 509, (1972).
31. W. A. Guillory and C. E. Hunter, *J. Chem. Phys.*, **50,** 3516, (1969).
32. R. J. Kriegler and H. L. Welsh, *Canad. J. Phys.*, **46,** 1181, (1968).
33. H. P. Gush, W. F. J. Hare, E. J. Allin and H. L. Welsh, *Canad. J. Phys.*, **38,** 176, (1960).
34. L. Meyer, C. S. Barrett and P. Haasen, *J. Chem. Phys.*, **40,** 2744, (1964); C. S. Barrett and L. Meyer, *J. Chem. Phys.*, **42,** 107, (1965); **43,** 3502, 1965); C. S. Barrett, L. Meyer and J. Wasserman, *J. Chem. Phys.*, **44,** 998, (1966).
35. A. F. Devonshire, *Proc. Roy. Soc.*, **A153,** 601, (1936).
36. J. Obriot, P. Marteau, H. Vu and B. Vodar, *Spectrochim. Acta*, **26A,** 2051, (1970); H. Vu, M. R. Attwood and E. Staude, *Compt. rend.*, **257,** 1771, (1963); M. Jean-Louis, M. Bahreini and H. Vu, *Compt. rend.*, **268B,** 41, (1969).
37. E. D. Becker and G. C. Pimentel, *J. Chem. Phys.*, **25,** 224, (1956).
38. C. M. King and E. R. Nixon, *J. Chem. Phys.*, **48,** 1685, (1968).
39. E. R. Lippincott and R. Schroeder, *J. Chem. Phys.*, **23,** 1099, (1955).
40. D. E. Milligan and M. E. Jacox, *J. Chem. Phys.*, **47,** 278, (1967).
41. D. E. Milligan and M. E. Jacox, *J. Chem. Phys.*, **39,** 712, (1963).
42. J. F. Ogilvie and M. J. Newlands, *Trans. Faraday Soc.*, **65,** 2602, (1969); S. Cradock and J. F. Ogilvie, *Chem. Comm.*, **1966,** 364.
42a. C. B. Murchison and J. Overend, *Spectrochim. Acta*, **27A,** 1509, (1971).
42b. T. B. Freedman and E. R. Nixon, *J. Chem. Phys.*, **56,** 698, (1972).
42c. T. B. Freedman and E. R. Nixon, unpublished data, quoted in ref. 42b.
42d. D. Foss Smith, Jr., J. Overend, R. C. Spiker and Lester Andrews, *Spectrochim. Acta,* **28A,** 87, (1972).
43. M. Van Thiel, E. D. Becker and G. C. Pimentel, *J. Chem. Phys.*, **27,** 486, (1957).
44. E. Catalano and D. E. Milligan, *J. Chem. Phys.*, **30,** 45, (1959).
45. J. A. Glasel, *J. Chem. Phys.*, **33,** 252, (1960).
46. T. Miyazawa, *Bull. Chem. Soc. Japan*, **34,** 202, (1961).
47. M. E. Jacox and D. E. Milligan, *Spectrochim. Acta*, **17,** 1196, (1961).
48. R. L. Redington and D. E. Milligan, *J. Chem. Phys.*, **37,** 2162, (1962); **39,** 1276, (1963).
49. D. W. Robinson, *J. Chem. Phys.*, **39,** 3430, (1963).
50. B. R. Cairns and G. C. Pimentel, *J. Chem. Phys.*, **43,** 3432, (1965).
51. H. P. Hopkins, R. F. Curl and K. S. Pitzer, *J. Chem. Phys.*, **48,** 2959, (1968).
52. K. B. Harvey and H. F. Shurvell, *J. Mol. Spectry.*, **25,** 120, (1968).
53. A. J. Tursi and E. R. Nixon, *J. Chem. Phys.*, **52,** 1521, (1970).
54. J. Pacansky and V. Calder, *J. Chem. Phys.*, **53,** 4519, (1970).

55. A. J. Tursi and E. R. Nixon, *J. Chem. Phys.*, **53**, 518 (1970).
56. H. C. Allen and E. K. Plyler, *J. Chem. Phys.*, **25**, 1132, (1956).
57. A. J. Barnes and J. D. R. Howells, *J.C.S. Trans. Faraday II*, **68**, 729, (1972).
58. M. M. Rochkind and G. C. Pimentel, *J. Chem. Phys.*, **42**, 1361, (1965).
59. W. G. Fateley, H. A. Bent and B. Crawford, *J. Chem. Phys.*, **31**, 204, (1959).
60, R. V. St. Louis and B. Crawford, *J. Chem. Phys.*, **42**, 857, (1965).
61. M. Allavena, R. Rysnik, D. White, V. Calder and D. E. Mann, *J. Chem. Phys.*, **50**, 3399, (1969).
62. F. J. Adrian, *J. Chem. Phys.*, **36**, 1692, (1962).
63. G. H. Myers, W. C. Easley and B. A. Zilles, *J. Chem. Phys.*, **53**, 1181, (1970).
64. J. G. David and H. E. Hallam, *J. Mol. Struc.*, **5**, 31, (1970); *Trans. Faraday Soc.*, **65**, 2838, (1970).
65. J. W. Hastie, R. H. Hauge and J. L. Margrave, *M. Inorg. Nucl. Chem.*, **31**, 281, (1969).
66. R. G. W. Norrish and G. A. Oldershaw, *Proc. Roy. Soc.*, **A249**, 498, (1959).
67. A. Arkell and I. Schwager, *J. Amer. Chem. Soc.*, **89**, 5999, (1967).
68. D. W. Muenow, J. W. Hastie, R. H. Hauge, R. Bautista and J. L. Margrave, *Trans. Faraday Soc.*, **65**, 3210, (1969).
68a. S. N. Cesaro, M. Spoliti, A. J. Hinchcliffe and J. S. Ogden, *J. Chem. Phys.*, **55**, 5834, (1971).
68b. L. Brewer and J. Ling-Fai Wang, *J. Chem. Phys.*, **56**, 759, (1972).
69. J. M. Bassler, P. L. Timms and J. L. Margrave, *J. Chem. Phys.*, **45**, 2704, (1966).
70. J. J. Comeford, S. Abramowitz and I. R. Levin, *J. Chem. Phys.*, **43**, 4536, (1965).
71. C. D. Bass, L. Lynds, T. Wolfram and R. E. DeWames, *Inorg. Chem.*, **3**, 1063, (1964).
72. I. Olovsson and D. H. Templeton, *Acta Cryst.*, **12**, 832, (1959).
73. D. E. Milligan, R. M. Hexter and K. Dressler, *J. Chem. Phys.*, **34**, 1009, (1961).
74. G. C. Pimentel, M. O. Bulanin and M. Van Thiel, *J. Chem. Phys.*, **36**, 500, (1962).
75. J. J. Comeford, *J. Chem. Phys.*, **45**, 3463, (1966).
76. J. A. Lannon, F. D. Verderame and R. W. Anderson, *J. Chem. Phys.*, **54**, 2212, (1971).
77. E. Catalano and R. H. Sanborn, *J. Chem. Phys.*, **38**, 2273, (1963).
78. G. C. Pimentel and S. W. Charles, *Pure Appl. Chem.*, **7**, 111, (1963).
79. G. C. Pimentel, S. W. Charles and Kj. Rosengren, *J. Chem. Phys.*, **44**, 3029, (1966).
80. C. B. Moore and Kj. Rosengren, *J. Chem. Phys.*, **44**, 4108, (1966).
81. L. P. Kuhn, *J. Amer. Chem. Soc.*, **74**, 2492, (1952).
82. Kj. Rosengren and G. C. Pimentel, *J. Chem. Phys.*, **43**, 507, (1965).
83. L. J. Bellamy, *The Infra-Red Spectra of Complex Molecules*, 2nd ed. p. 272, J. Wiley, New York, 1958.
84. G. C. Pimentel, *J. Amer. Chem. Soc.*, **80**, 62, (1958); J. D. Baldeschwieler and G. C. Pimentel, *J. Chem. Phys.*, **33**, 1008, (1960); R. T. Hall and G. C. Pimentel, *J. Chem. Phys.*, **38**, 1889, (1963).
85. M. E. Jacox and D. E. Milligan, *J. Chem. Phys.*, **40**, 2457, (1964); **47**, 5157, (1967).
86. J. R. Durig and D. W. Wertz, *J. Chem. Phys.*, **46**, 3069, (1967).
87. C. Reid, *J. Chem. Phys.*, **18**, 1512, (1950).
88. K. B. Harvey and J. F. Ogilvie, *Canad. J. Chem.*, **40**, 85, (1962).
89. J. J. Smith and B. Meyer, *J. Chem. Phys.*, **50**, 456, (1969).
90. N. G. Moll, D. R. Clutter and W. E. Thompson, *J. Chem. Phys.*, **45**, 4469, (1966).
90a. M. E. Jacox and D. E. Milligan, *J. Chem. Phys.*, **54**, 919, (1971).
91. J. J. Comeford and J. H. Gould, *J. Mol. Spectroscopy*, **5**, 474, (1960).
92. S. Abramowitz and H. P. Broida, *J. Chem. Phys.*, **39**, 2383, (1963).

93. A. Cabana, G. B. Savitsky and D. F. Hornig, *J. Chem. Phys.*, **39**, 2942, (1963).
94. H. F. King, Ph.D. thesis, Princeton University, New Jersey, (1960); H. F. King and D. F. Hornig, *J. Chem. Phys.*, **44**, 4520, (1966).
94a. H. Friedmann, A. Shalom and S. Kimel, *J. Chem. Phys.*, **50**, 2496, (1969).
95. A. Cabana, A. Anderson and R. Savoie, *J. Chem. Phys.*, **42**, 1122, (1965).
96. F. H. Frayer and G. E. Ewing, *J. Chem. Phys.*, **46**, 1994, (1967); **48**, 781, (1968).
96a. R. E. Wilde, T. K. K. Srinivasa, R. W. Hassel and S. G. Sanker, *J. Chem. Phys.*, **55**, 5681, (1971).
97. A. Le Roy and E. Dayan, *Compt. rend.*, **268B**, 48, (1969).
98. D. A. Dows and N. Rich, *J. Chem. Phys.*, **47**, 333, (1967).
99. L. J. Stief, *J. Chem. Phys.*, **44**, 277, (1966).
99a. P. Datta, T. D. Goldfarb and R. S. Boikess, *J. Amer. Chem. Soc.*, **93**, 5189, (1971).
99b. K. Szczepaniak and W. B. Person, *Spectrochim. Acta*, **28A**, 15, (1972).
100. I. Schwager and A. Arkell, *J. Amer. Chem. Soc.*, **89**, 6006, (1967).
101. P. N. Noble and G. C. Pimentel, *Spectrochim. Acta*, **24A**, 797, (1968).
101a. M. H. Studier and E. H. Appelman, *J. Amer. Chem. Soc.*, **93**, 2349, (1971).
101b. H. Kim, E. F. Pearson and E. H. Appelman, *J. Chem. Phys.*, **56**, 1, (1972).
101c. J. A. Groleb, H. H. Claassen, M. H. Studier and E. A. Appelman, *Spectrochim. Acta*, **28A**, 65, (1972).
102. A. Arkell, *J. Amer. Chem. Soc.*, **87**, 4057, (1965); R. D. Spratley, J. J. Turner-and G. C. Pimentel, *J. Chem. Phys.*, **44**, 2063, (1966).
103. J. S. Ogden and J. J. Turner, *J. Chem. Soc.*, **A**, 1483, (1967).
104. J. J. Turner and G. C. Pimentel, in *Noble Gas Compounds*, ed. H. H. Hyman, p. 101, University of Chicago Press, 1963; J. J. Turner and G. C. Pimental, *Science*, **140**, 974, (1963).
105. L. Y. Nelson and H. C. Pimentel, *Inorg. Chem.*, **6**, 1758, (1967).
106. H. H. Claassen, E. L. Gasner, H. Kim and J. L. Huston, *J. Chem. Phys.*, **49**, 253, (1968).
106a. H. H. Claassen and J. L. Huston, *J. Chem. Phys.*, **55**, 1505, (1971).
107. R. A. Frey, R. L. Redington and A. L. K. Aljibury, *J. Chem. Phys.*, **54**, 344, (1971).
108. H. H. Claassen, B. Weinstock and J. G. Malm, *J. Chem. Phys.*, **28**, 285, (1958); H. Selig, H. H. Claassen and J. H. Holloway, *J. Chem. Phys.*, **52**, 3517, (1970).
108a. K. O. Christe, E. C. Curtis and D. Pitipovich, *Spectrochim. Acta*, **27A**, 931, (1971).
109. G. M. Begun, W. H. Fletcher and D. F. Smith, *J. Chem. Phys.*, **42**, 2236, (1965).
110. K. O. Christie, *Spectrochim. Acta*, **27A**, 631, (1971).
111. L. Y. Nelson and G. C. Pimentel, *Inorg. Chem.*, **7**, 1695, (1968).
112. A. L. K. Aljibury and R. L. Redington, *J. Chem. Phys.*, **52**, 453, (1970).
113. L. C. Hoskins and R. C. Lord, *J. Chem. Phys.*, **46**, 2402, (1967); H. Selig, J. H. Holloway, J. Tyson and H. H. Claassen, *J. Chem. Phys.*, **53**, 2559, (1970).
114. R. L. Redington and C. V. Berney, *J. Chem. Phys.*, **43**, 2020, (1965).
115. D. R. Lide, jr., D. E. Mann and J. J. Comeford, *Spectrochim. Acta*, **21**, 497, (1965).
116. S. Abramowitz and J. J. Comeford, *Spectrochim. Acta*, **21**, 1479, (1965).
117. S. T. King, *J. Chem. Phys.*, **49**, 1321, (1968).
118. F. D. Verderame and E. R. Nixon, *J. Chem. Phys.*, **45**, 3476, (1966).
119. H. E. Hallam and T. C. Ray, *Trans. Faraday Soc.*, **59**, 1983, (1963).
120. A. J. Barnes, H. E. Hallam, J. D. R. Howells and G. F. Schrimshaw, *J.C.S. Faraday II*, **69**, in press, (1973).
121. D. A. Dows, *J. Chem. Phys.*, **29**, 484, (1958).
122. R. H. Sanborn, *Spectrochim. Acta*, **23A**, 1999, (1967).

123. A. J. Barnes and H. E. Hallam, (unpublished).
124. C. V. Berney, *Spectrochim. Acta*, **21**, 1809, (1965).
125. J. R. Durig and A. C. Morissey, *J. Chem. Phys.*, **46**, 4854, (1967).
126. N. C. Craig and J. Overend, *J. Chem. Phys.*, **51**, 1127, (1969).
127. L. A. Nimon, K. S. Seshadri, R. C. Taylor and D. White, *J. Chem. Phys.*, **53**, 2416, (1970).
128. A. Atoji, P. J. Wheatley and W. Lipscomb, *J. Chem. Phys.*, **27**, 196, (1957). L. Trefonas and W. N. Lipscomb, *J. Chem. Phys.*, **28**, 54, (1958).
129. K. Hedberg and R. Ryan, *J. Chem. Phys.*, **41**, 2214, (1964); **50**, 4986, (1969).
130. C. Campbell, J. P. M. Jones and J. J. Turner, *Chem. Comm.*, 888, (1968).
131. H. D. von Zerssen and H. Martin, Paper 80 in Abstracts of Bunsen meeting, Hanover, May 1971.
132. A. J. Barnes and H. E. Hallam, *Trans. Faraday Soc.*, **66**, 1920, (1970).
133. M. Falk and E. Whalley, *J. Chem. Phys.*, **34**, 1554, (1961).
134. M. Margottin-Maclou, *J. Phys. Radium*, **21**, 634, (1960).
135. M. Van Thiel, E. D. Becker and G. C. Pimentel, *J. Chem. Phys.*, **27**, 95, (1957).
136. R. G. Inskeep, J. M. Kelliher, P. E. McMahon and B. G. Somers, *J. Chem. Phys.*, **28**, 1033, (1958).
137. A. J. Barnes and H. E. Hallam, *Trans. Faraday Soc.*, **66**, 1932, (1970).
138. M. Oki and H. Iwamura, *Bull. Chem. Soc. Japan*, **32**, 567, 950, (1959); F. Dalton, G. D. Meakins, J. H. Robinson and W. Zaharia, *J. Chem. Soc.*, 1566, (1962).
139. J. P. Perchard and M.-L. Josien, *J. Chim. Phys.*, **65**, 1834, 1856, (1968).
140. E. L. Saier, L. R. Cousins and M. R. Basila, *J. Chem. Phys.*, **41**, 40, (1964).
141. A. J. Barnes, H. E. Hallam and J. D. R. Howells, *J.C.S. Faraday II*, **68**, 737, (1972).
142. J. R. Durig and W. H. Green, *Spectrochim. Acta*, **25A**, 849, (1969).
143. I. O. C. Ekejiuba and H. E. Hallam, *J. Mol. Structure*, **6**, 341, (1970).
143a. E. L. Varetti and G. C. Pimentel, *J. Chem. Phys.*, **55**, 3813, (1971).
144. D. E. Milligan, *J. Chem. Phys.*, **35**, 1491, (1961).
145. C. B. Moore, G. C. Pimentel, and T. D. Goldfarb, *J. Chem. Phys.*, **43**, 63, (1965).
146. J. Le Brumant, *Compt. rend.*, **270B**, 898, (1970).
147. C. B. Moore and G. C. Pimentel, *J. Chem. Phys.*, **40**, 342, (1964).
148. J. F. Ogilvie, *J. Mol. Structure*, **3**, 513, (1969); *Canad. J. Chem.*, **46**, 2472, (1968).
149. J. D. Durig, S. F. Bush and F. G. Baglin, *J. Chem. Phys.*, **49**, 2106, (1968).
150. J. Hinge and R. F. Curl, *J. Amer. Chem. Soc.*, **86**, 5068, (1964).
151. T. D. Goldfarb and B. H. Khare, *J. Chem. Phys.*, **46**, 3379, (1967).
152. T. D. Goldfarb and B. H. Khare, *J. Chem. Phys.*, **46**, 3384, (1967).
153. S. Sujishi and S. Witz, *J. Amer. Chem. Soc.*, **76**, 4631, (1954).
154. S. T. King, *J. Phys. Chem.*, **75**, 405, (1971).
155. C. C. Costain and J. M. Dowling, *J. Chem. Phys.*, **32**, 158, (1960).
155a. S. T. King, *Spectrochim. Acta*, **28A**, 165, (1972).
156. G. Taddei, E. Castellucci and F. D. Verderame, *J. Chem. Phys.*, **53**, 2407, (1970); E. Castellucci, G. Sbrana and F. D. Verderame, ibid, **51**, 3762, (1969).
157. R. C. Millikan and K. S. Pitzer, *J. Amer. Chem. Soc.*, **80**, 3515, (1958).
158. T. Miyazawa and K. S. Pitzer, *J. Chem. Phys.*, **30**, 1076, (1959).
159. C. L. Berney, R. L. Redington and K. C. Lin, *J. Chem. Phys.*, **53**, 1713, (1970).
159a. R. L. Redington and K. C. Lin, *J. Chem. Phys.*, **54**, 4111, (1971).
159b. Y. Grenie, J.-C. Cornut and J.-C. Lassegues, *J. Chem. Phys.*, **55**, 5844, (1971).
159c. M. Haurie and A. Novak, *J. Chim. Phys.*, **62**, 146, (1965).
160. A. J. Rest and J. J. Turner, *Proc. 4th Internat. Conf. Organometallic Chem.*, Bristol, 1969; M. A. Graham, A. J. Rest and J. J. Turner, *J. Organometallic Chem.*, **24**, C54, (1970); M. Poliakoff and J. J. Turner, *Chem. Comm.*, 1008, (1970); *J. Chem. Soc.*, (A), 654, (1971).

161. A. J. Rest and J. J. Turner, *Chem. Comm.*, 375, 1026, (1969).
162. A. J. Rest, quoted in J. S. Ogden and J. J. Turner, *Chem. in Britain*, 7, 186, (1971).
162a. A. Kaldor and R. F. Porter, *J. Amer. Chem. Soc.*, 93, 2140, (1971).
162b. R. L. DeKock, *Inorg. Chem.*, 10, 1205, (1971).
162c. J. L. Slater, R. K. Sheline, K. C. Lin and W. Weltner, *J. Chem. Phys.*, 55, 5129,·(1971).
162d. A. Kaldor and R. F. Porter, *Inorg. Chem.*, 10, 775, (1971).
163. L. Y. Tan and G. C. Pimentel, *J. Chem. Phys.*, 48, 5202, (1968).
164. R. F. Curl, jr., and K. S. Pitzer, *J. Amer. Chem. Soc.*, 80, 2371, (1958).
165. J. R. Durig, J. W. Clark and J. M. Casper, *J. Mol. Structure*, 5, 67, (1970).
166. S. T. King, *J. Phys. Chem.*, 74, 2133, (1970).
167. M. M. Rochkind, *Analyt. Chem.*, 39, 567, (1967); 40, 762, (1968); *Science*, 160, 196, (1968); *Appl. Spectroscopy*, 22, 313, (1968); *Spectrochim. Acta*, 27A, 547, (1971).
168. J. G. Atkinson, A. A. Russell and R. S. Smart, *Canad. J. Chem.*, 45, 1963, (1967).
169. A. Snelson and K. S. Pitzer, *J. Phys. Chem.*, 67, 882, (1963); A. Snelson, *J. Phys. Chem.*, 73, 1919, (1969).
170. M. M. Rochkind, in *Spectrometry of Fuels*, ed. R. A. Friedel, p. 280, Plenum Press, 1970.

4 *Theoretical treatment of matrix effects*

A. J. BARNES

Contents

4.1 Introduction

While the technique of matrix isolation affords a number of advantages in the spectroscopic study of stable molecules, its greatest value lies in the possibility of stabilizing radicals and other 'active' species, such as monomers of hydrogen-bonding substances, in the solid phase for a sufficient time to allow leisurely examination by spectroscopy. Since the spectra of such species may previously have been unknown, it is particularly important to understand the various possible effects on the spectrum.

The matrix environmental effects which result in shifts or splitting of infrared absorption bands of trapped species are:

(i) Vibrational shift—in matrices, as in any condensed phase, the band centre is shifted from its gas phase value, normally to lower frequency. This is the analogue of the solvent shift in solution.

(ii) Multiple trapping sites—in a matrix a molecule may be trapped in two or more distinct sites, each of which will give rise to a different vibrational shift, and thus several bands will appear in the spectrum.

(iii) Rotational, or other, motion of the solute molecule within its cage—at the low temperatures used, few rotational levels will be populated but nevertheless several absorptions could result from this effect. Rotation has been observed for a number of small molecules and radicals. Inversion (ammonia) and nuclear spin conversion (water and methane) complicate the rotational spectra of certain molecules in matrices.

(iv) Aggregation—even where the interaction between two solute molecules is small, a group of two or more solute molecules will give rise to a frequency slightly different from that of the isolated monomer. In the case of hydrogen-bonding materials, the shift may be very large.

(v) Splitting of degenerate frequencies, or the appearance of infrared 'in-active' bands—the matrix cage may perturb the solute molecule sufficiently to lift the degeneracy of two levels, or to induce IR inactive bands (for example the hydrogen stretching frequency). Also the low temperature, and the isolation of monomeric solute molecules, often causes near-degenerate bands to be resolved.

Considerable progress has been made towards a theoretical understanding of the vibrational and rotational shifts in matrices, the established theoretical treatment of solvent shifts and hindered rotors being extended to solutes trapped in matrices, the extra factor involved—the rigidity of the cage—being taken into account.

4.2 Vibrational shift

4.2.1 Solvent effects

The effect of the environment on IR absorption frequencies of substances in (liquid) solution has been extensively studied,[1] and is interpreted in terms of specific solute-solvent dipolar interactions[2] (particularly $\overset{\delta+}{X}-\overset{\delta-}{H}\cdots S$) superimposed on non-specific—inductive and dispersive—contributions. Various theories have been put forward to account for the non-specific solvent shifts, based on a model[3] of the solute as a point dipole in a spherical cavity within the solvent medium of uniform dielectric constant ε' and refractive index n, for example: Kirkwood–Bauer–Magat (KBM)[4]

$$\frac{\Delta v}{v} = C\frac{(\varepsilon' - 1)}{(2\varepsilon' + 1)} \qquad (4.1)$$

Buckingham[5]

$$\frac{\Delta v}{v} = C_1 + \frac{1}{2}(C_2 + C_3)\frac{(\varepsilon' - 1)}{(2\varepsilon' + 1)} \quad \text{non-polar solvents} \tag{4.2}$$

$$\frac{\Delta v}{v} = C_1 + C_2\frac{(\varepsilon' - 1)}{(2\varepsilon' + 1)} + C_3\frac{(n^2 - 1)}{(2n^2 + 1)} \quad \text{polar solvents} \tag{4.3}$$

David and Hallam[6]

$$\frac{\Delta v}{v} = C_1' + C_2'\frac{(\varepsilon' - 1)}{(\varepsilon' + 2)} + C_3'\frac{(n^2 - 1)}{(n^2 + 2)} \tag{4.4}$$

Caldow and Thompson[7] attempted to take account of specific interactions by adding a fourth term to equation (4.3), namely $C_4\sigma^*$—where σ^* is the Taft inductive factor for the residue R of a solvent RH.

In all these equations, the solvent shift Δv is defined as

$$\Delta v = (v_{gas} - v_{solution}) \tag{4.5}$$

thus making it normally positive, at least for stretching modes, since such bands always shift to lower frequency in solution.

In principle, it might be expected that these equations could also be applied to the vibrational shifts of solutes in matrices, since the only difference between solutions and matrices lies in the fact that in a matrix the solvent molecules surrounding the trapped species are fixed in position, forming a rigid cage, whereas in solution the solute occupies a flexible cavity. However the rigidity of the cage introduces an important interaction—repulsive forces—which will have to be taken into account if these equations are to be extended to cover matrix shifts.

4.2.2 Intermolecular forces

If a pair of molecules are sufficiently far apart that electron exchange may be neglected, the interaction Hamiltonian may be treated as a perturbation of the Hamiltonian of the free molecule. The first order perturbed energy results from the electrostatic interaction—the interaction between the permanent charge distributions of the two molecules. The dependence of the potential energy on the separation and mutual orientation of the two molecules is the same in classical as in quantum mechanics, the only difference lying in the determination of the multipole moments involved. Thus the electrostatic interaction is conveniently obtained classically, using experimental values of the dipole and quadrupole moments.

The second order perturbed energy includes both the induction and dispersion energies. Induction forces are caused by the interaction between the permanent charge distribution of one molecule and the charge distribution induced in the other molecule, the first term being the dipole-induced dipole interaction (r^{-6} dependent), and higher terms the quadrupole-induced quadrupole interaction (r^{-8} dependent), etc. Again the interaction energy may be obtained classically, using experimental values of polarizabilities, but the classical formula appears to differ from that derived from quantum mechanics.[8]

The other second order long-range interaction, the London dispersion energy, is caused by an attraction between the instantaneous charge distributions of the two molecules. It may be described semi-classically as an interaction between the instantaneous moments of a fluctuating charge distribution in one molecule and the moments induced thereby in the other molecule, which are always in phase and thus produce a net attractive force despite the fact that the average moment vanishes. The leading term is the instantaneous dipole-induced dipole interaction, which is r^{-6} dependent. The accurate expression for the dispersive interaction may be approximated to allow calculation from experimentally measurable quantities.

At short range, repulsion (valence) interactions between the two molecules become important—these are caused by overlap of the electronic distributions resulting in a large distortion, with a consequent decrease in charge density between the two nuclei involved. These forces first appear in a first order perturbation calculation with symmetrized wave functions, whereas long-range forces appear in perturbation calculations with unsymmetrized wave functions. Thus at intermediate separations, where the interactions are comparable, it is a poor approximation simply to add them. Since, in any case, the short-range interaction has not, in general, been calculated, the repulsion term is approximated using an intermolecular potential function, normally the Lennard-Jones (6–12) potential.

Thus the intermolecular potential energy V may be expressed approximately as the sum of four terms

$$V = V_{elec} + V_{ind} + V_{dis} + V_{rep} \tag{4.6}$$

Defining the vibrational shift for the ($v \leftarrow 0$) band as

$$\Delta_v v = v_{matrix} - v_{gas} \tag{4.7}$$

so that shifts to lower frequency are negative, and shifts to higher frequency are positive, then from first order perturbation theory

$$\Delta_v v = \frac{1}{hc}\{[V]_v - [V]_0\} \tag{4.8}$$

where

$$[V]_v = \int \psi_v^* V \psi_v \, d\tau$$

Thus

$$\Delta_v v = \Delta_v v_{elec} + \Delta_v v_{ind} + \Delta_v v_{dis} + \Delta_v v_{rep} \tag{4.9}$$

The contribution from each of the terms in equation (4.9) will be considered in turn, the solute molecule being assumed to occupy a substitutional site in the undistorted matrix lattice, and solute–solute interactions being ignored. Only diatomic molecules will be considered, thus allowing V, μ, α, etc. to be expanded as Taylor series, in the inter-nuclear configuration of the solute molecule, for evaluation.

4.2.3 RTC model[9]

Whereas previous models of the rotation of diatomic molecules in matrices had been based on a hindered rotor model, Friedmann and Kimel's rotation-translation coupling (RTC) model used a cell model treatment, in which the solute molecule is assumed to be undergoing a constrained translational motion, based on the following assumptions:

(i) A trapped molecule occupies a substitutional site in the undistorted matrix lattice, and is confined to a 'cage' formed by its rigidly fixed neighbours.

(ii) Interactions between solute molecules, in highly dilute matrices, are negligible.

(iii) The potential energy V_v, governing the rotational and translational motions of the molecule (in the vth vibrational level) in its cell, depends only on the positions of the molecular centres of interaction (the intermolecular potential energy is most conveniently expressed as a function of the co-ordinates of the centre of interaction, which is a position fixed with respect to the charge distribution of the molecule—the centre of symmetry for a symmetrical molecule and, for an unsymmetrical molecule, chosen so as to minimize the average angular dependence of the intermolecular interactions), thus retaining only the isotropic term in the expansion of the intermolecular potential energy in spherical harmonics.

(iv) V_v depends only on the absolute values of the co-ordinates of the centre of interaction with respect to the cell centre (i.e. the site symmetry group must contain D_{2h} as a sub-group).

(v) Virtually only the ground state of the translational motion of the solute molecule is populated at the low temperatures considered.

(vi) No transitions between translational energy levels (or between phonon states of the matrix lattice) are induced by the radiation process.

From the repeated application of the Born–Oppenheimer approximation, Friedmann and Kimel show that the intermolecular potential energy depends only on the vibrational state of the molecules. The coupling between the rotational motion of the solute molecule and its constrained translational motion is treated as a perturbation.

The gas phase frequencies are determined by adding the unperturbed rotational energy to the unperturbed vibrational energy, there being no constraint on the molecule. In a matrix, where the solute molecule is undergoing constrained translational motion, the band centre of the $(v \leftarrow 0)$

absorption is shifted because of the dependence of this translational energy on the vibrational state of the molecule.

$$\Delta_v v = \frac{1}{hc}(E_{v,0}^{tr} - E_{0,0}^{tr})$$ (4.10)

The value of $E_{v,0}^{tr}$ depends on the intermolecular potential energy $V_v(r)$

$$V_v(r_a) = \sum_b 4\varepsilon(x)_v \left\{ \left[\frac{\sigma(x)_v}{r_{ab}}\right]^{12} - \left[\frac{\sigma(x)_v}{r_{ab}}\right]^{6} \right\}$$ (4.11)

assuming the Lennard-Jones (6–12) potential,[10] where $\varepsilon(x)_v$ and $\sigma(x)_v$ are v-dependent parameters representing an energy and a length, respectively, characteristic of the pair of interacting molecules (Figure 4.1) for a fixed

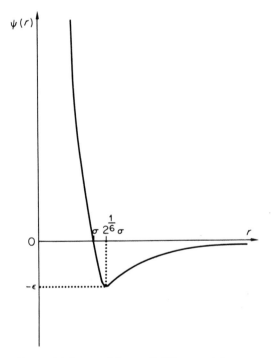

Figure 4.1 Lennard-Jones (6–12) potential curve

inter-nuclear configuration $x = (r - r_e)$ of the solute molecule a. The matrix molecules b are assumed to be non-polar, and thus there is no electrostatic interaction to be considered.

Using an idealization of the Lennard-Jones–Devonshire cell model— either the harmonic oscillator cell model,[11] or the spherical box model[12]—

Table 4.1 Lennard-Jones parameters

Molecule	σ (Å)	ε/k (K)
Ne	2·76	35·0
Ar	3·406	120·8
Kr	3·62	170
Xe	4·03	223
N_2	3·70	95·5
O_2	3·52	117·8
Cl_2	[4·26][a]	[307][a]
CO	3·76	100·2
NO	3·17	131
CO_2	4·27	199
N_2O	4·57	191
CH_4	3·813	148·2
CF_4	4·70	152·3
$CClF_3$	4·92	222
CCl_2F_2	5·16	286
SF_6	5·7	195
C_4F_8	7·03	222·6
C_2H_4	4·52	199
C_2H_2	[4·22][a]	[185][a]
HF	2·84[b]	—
HCl	3·60[c]	250[c]
HBr	3·80[c]	281[c]
HI	[4·12][a]	[324][a]

a. Data derived from second virial coefficient
   data, except for those in brackets which are
   from viscosity measurements (ref. 17).
b. J. H. Jaffe, A. Rosenberg, M. A. Hirshfeld,
   and N. M. Gailar, *J. Chem. Phys.*, **43**, 1525,
   (1965).
c. From critical constants (ref. 9).

$E^{tr}_{v,0}$ is given by

$$E^{tr}_{v,0} = V_v(0) + \omega_v(r)_0 \qquad (4.12)$$

where the potential energy of the solute molecule fixed at a lattice site is
(from equation 4.11)

$$V_v(0) = 4\varepsilon(x)_v \left\{ N_{12}\left[\frac{\sigma(x)_v}{d_0}\right]^{12} - N_6\left[\frac{\sigma(x)_v}{d_0}\right]^6 \right\} \qquad (4.13)$$

N_6 and N_{12} are the effective numbers of nearest neighbours in the matrix for
r^{-6} and r^{-12} dependences (14·454 and 12·132, respectively, in a f.c.c. lattice).[13]
d_0 is the spacing between nearest neighbours in the matrix lattice and
$\omega_v(r)_0$ is the zero-point energy of the translational motion of the solute

molecule in its cell, given by

$$\omega_v(r)_0 = \frac{3h}{2\pi d_0}\left(\frac{Z\varepsilon(x)_v}{m}\right)^{1/2}\left\{44\left[\frac{\sigma(x)_v}{d_0}\right]^{12} - 10\left[\frac{\sigma(x)_v}{d_0}\right]^{6}\right\}^{1/2} \tag{4.14}$$

in the harmonic oscillator cell model, and by

$$\omega_v(r)_0 = \frac{h^2}{8mr_f^2} \tag{4.15}$$

in the spherical box model, where

Z is the number of nearest neighbours (12 for a f.c.c. lattice)
m is the reduced mass of the solute molecule
r_f is the radius of the box in which the centre of the molecule (considered as a hard sphere) is free to move; it is determined from the relation $V_v(r_f) = 0$

From equations (4.10) and (4.12), the vibrational frequency shift is given by

$$\Delta_v v = \frac{1}{hc}[V_v(0) - V_0(0) + \omega_v(r)_0 - \omega_0(r)_0] \tag{4.16}$$

Assuming that the attractive term in the Lennard-Jones potential is due to the sum of inductive and dispersive forces, the static part of the shift may be separated as follows

$$\frac{1}{hc}[V_v(0) - V_0(0)] = \Delta_v v_{\text{ind}} + \Delta_v v_{\text{dis}} + \Delta_v v_{\text{rep}} \tag{4.17}$$

The various terms are given by the expressions:

$$\Delta_v v_{\text{ind}} = -\frac{4}{hc}N_6\varepsilon(x)_0\left[\frac{\sigma(x)_0}{d_0}\right]^6 \mathfrak{F}_v(\mu) \tag{4.18}$$

$$\Delta_v v_{\text{dis}} = -\frac{4}{hc}N_6\varepsilon(x)_0\left[\frac{\sigma(x)_0}{d_0}\right]^6 \mathfrak{F}_v(\alpha) \tag{4.19}$$

$$\Delta_v v_{\text{rep}} = \frac{4}{hc}N_{12}\varepsilon(x)_0\left[\frac{\sigma(x)_0}{d_0}\right]^{12}[\mathfrak{F}_v(\mu) + \mathfrak{F}_v(\alpha)]\left(\frac{2+y}{1+y}\right) \tag{4.20}$$

where

$$\mathfrak{F}_v(\mu) = \frac{8\mu_a(x)_0[\mu_a(x)_v - \mu_a(x)_0]}{3W_{ab}\alpha_a(x)_0} \tag{4.21}$$

$$\simeq \frac{v\mu_a(x)_0\alpha_b(d\mu/dx)_0\langle 0|x|0\rangle}{\varepsilon(x)_0\sigma^6(x)_0} \tag{4.22}$$

μ_a is the dipole moment of the solute molecule
α_b is the polarizability of the matrix molecules
W_{ab} is the excitation energy

$\langle 0|x|0 \rangle$ is the mean displacement from the equilibrium internuclear distance in the ground state of the solute molecule (for low vibrational levels[14] $\langle v|x|v \rangle - \langle 0|x|0 \rangle \simeq 2v\langle 0|x|0 \rangle$).

$$\mathfrak{F}_v(\alpha) = \frac{\alpha_a(x)_v - \alpha_a(x)_0}{\alpha_a(x)_0} \tag{4.23}$$

$$\simeq \frac{\alpha_a(x)_0 - \alpha'_a(x)_0}{\alpha_a(x)_0} \frac{2v(m')^{1/2}}{(m')^{1/2} - (m)^{1/2}} \tag{4.24}$$

the prime designating an isotopic molecule. The parameter y in equation (4.20) is given approximately by

$$y \simeq \frac{1}{2} \frac{\sigma_a(x)_0 + \sigma_b}{\sigma_a(x)_0} [1 \cdot 93(\log 8/\log N) - 1] \tag{4.25}$$

where N is the number of electrons in the valence shell of the solute molecule. This crude empirical estimate is a good enough approximation since $(2 + y)/(1 + y)$ varies only from 2 to 1 as y increases from 0 to ∞.

4.2.4 Electrostatic interaction

The electrostatic term in the potential energy represents the interaction between the permanent charge distributions of the solute and matrix molecules. For the noble gas matrices this term is obviously zero, but this is not the case for other matrices. Of the commonly used matrix materials, only carbon monoxide and nitrous oxide have dipole moments, while nitrogen, carbon monoxide, nitric oxide and carbon dioxide have quadrupoles. The symmetrical tetrahedral molecules, such as methane, have an octopole as their lowest moment, and the octahedral sulphur hexafluoride molecule has only a hexadecapole moment.

The 2^l moment may be defined as the following tensor[15,16]

$$M_{\alpha,\beta,\ldots\lambda} = \frac{(-1)^l}{l!} \sum_i e_i r_i^{2l+1} \frac{\partial^l}{\partial r_{i\alpha} \partial r_{i\beta} \ldots \partial r_{i\lambda}} \left(\frac{1}{r_i}\right) \tag{4.26}$$

and thus

$l = 1$ dipole

$$\mu_\alpha = \sum_i e_i r_{i\alpha} \tag{4.27a}$$

2 quadrupole

$$\Theta_{\alpha\beta} = \frac{1}{2} \sum_i e_i [3r_{i\alpha} r_{i\beta} - r_i^2 \delta_{\alpha\beta}] \tag{4.27b}$$

3 octopole

$$\Omega_{\alpha\beta\gamma} = \frac{1}{2} \sum_i e_i [5r_{i\alpha} r_{i\beta} r_{i\gamma} - r_i^2 (r_{i\alpha} \delta_{\beta\gamma}$$

$$+ r_{i\beta} \delta_{\alpha\gamma} + r_{i\gamma} \delta_{\alpha\beta})] \tag{4.27c}$$

4 hexadecapole $\quad \Phi_{\alpha\beta\gamma\delta} = \frac{1}{8}\sum_i e_i[35r_{i\alpha}r_{i\beta}r_{i\gamma}r_{i\delta} - 5r_i^2(r_{i\alpha}r_{i\beta}\delta_{\gamma\delta}$

$$+ r_{i\alpha}r_{i\gamma}\delta_{\beta\delta} + r_{i\alpha}r_{i\delta}\delta_{\beta\gamma} + r_{i\beta}r_{i\gamma}\delta_{\alpha\delta}$$

$$+ r_{i\beta}r_{i\delta}\delta_{\alpha\gamma} + r_{i\gamma}r_{i\delta}\delta_{\alpha\beta}) + r_i^4(\delta_{\alpha\beta}\delta_{\gamma\delta}$$

$$+ \delta_{\alpha\gamma}\delta_{\beta\delta} + \delta_{\alpha\delta}\delta_{\beta\gamma})] \qquad (4.27d)$$

etc.

where δ_{xy} is the Kronecker delta, and α, β, γ, δ, ... permute x, y and z.

For charge distributions possessing axial ($C_{\infty v}$) symmetry, each multipole moment may be defined by a single scalar quantity μ, Θ, Ω, etc. Then for example

$$\Theta_{\alpha\beta} = \tfrac{1}{2}\Theta(3l_\alpha l_\beta - \delta_{\alpha\beta}) \qquad (4.28)$$

where l_α and l_β are the components of unit vectors along the molecular axis.

Only the lowest moment possessed by a molecule is independent of origin—values quoted are usually referred to the centre of mass, the co-ordinate system used having the z axis as the molecular axis, in the direction of the atom of lower mass when the molecule does not have a centre of symmetry—but the moments are easily transformed to a different point on the molecular axis as follows:

$$\mu'_z = \mu_z \qquad (4.29a)$$

$$\Theta'_{zz} = \Theta_{zz} - 2\mu_z z \qquad (4.29b)$$

$$\Omega'_{zzz} = \Omega_{zzz} - 3\Theta_{zz}z + 3\mu_z z^2 \qquad (4.29c)$$

etc.

where z is the displacement of the origin.

Expanding the charge distributions of both the solute (a) and the matrix (b) molecules as sums of multipole moments, the energy of interaction between two molecules is given by[17]

$$\psi_{elec} = \frac{\mu_a\mu_b}{r_{ab}^3}[-2\cos\theta_a\cos\theta_b + \sin\theta_a\sin\theta_b\cos(\phi_a - \phi_b)]$$

$$+ \frac{3\mu_a\Theta_b}{2r_{ab}^4}[\cos\theta_a(3\cos^2\theta_b - 1) - 2\sin\theta_a\sin\theta_b\cos\theta_b\cos(\phi_a - \phi_b)]$$

$$- \frac{3\mu_b\Theta_a}{2r_{ab}^4}[\cos\theta_b(3\cos^2\theta_a - 1) - 2\sin\theta_a\sin\theta_b\cos\theta_a\cos(\phi_a - \phi_b)]$$

$$+ \frac{3\Theta_a\Theta_b}{4r_{ab}^5}[1 - 5\cos^2\theta_a - 5\cos^2\theta_b + 17\cos^2\theta_a\cos^2\theta_b$$

$$- 16\sin\theta_a\cos\theta_a\sin\theta_b\cos\theta_b\cos(\phi_a - \phi_b)$$

$$+ 2\sin^2\theta_a\sin^2\theta_b\cos^2(\phi_a - \phi_b)] + \cdots \qquad (4.30)$$

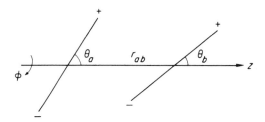

Figure 4.2 Relative orientation of two molecules

where r_{ab} is the intermolecular distance, and θ, ϕ define the relative orientation of the two molecules (Figure 4.2).

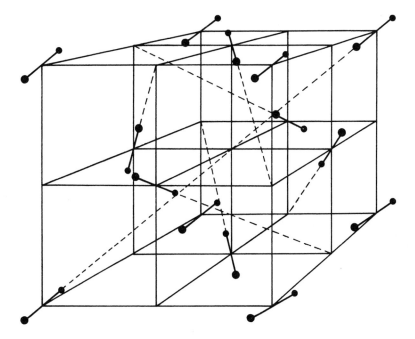

Figure 4.3 Face-centred cubic (f.c.c.) lattice of diatomic molecules

Assuming that the solute molecule occupies the [111] direction in the f.c.c. matrix lattice (Figure 4.3), the total energy of interaction

$$V_{elec} = \mu_a \left\{ \frac{\mu_b}{d_0^3} \left[\sum_b d_0^3 \frac{-2 \cos \theta_a \cos \theta_b + \sin \theta_a \sin \theta_b \cos (\phi_a - \phi_b)}{r_{ab}^3} \right] \right.$$
$$\left. + \frac{\Omega_b}{d_0^5} [\ldots] + \ldots \right\}$$

$$
\begin{aligned}
+\; \Theta_a \left\{ \frac{\Theta_b}{d_0^5} \left[\sum_b \frac{3}{4} d_0^5 \frac{\left(\begin{array}{c} 1 - 5\cos^2\theta_a - 5\cos^2\theta_b + 17\cos^2\theta_a\cos^2\theta_b \\ - 4\sin 2\theta_a \sin 2\theta_b \cos(\phi_a - \phi_b) \\ + 2\sin^2\theta_a \sin^2\theta_b \cos^2(\phi_a - \phi_b) \end{array} \right)}{r_{ab}^5} \right] \right. \\[2mm]
\left. + \frac{\Phi_b}{d_0^7}[\ldots] + \cdots \right\} + \Omega_a\{\ldots\} + \cdots \qquad (4.31)
\end{aligned}
$$

terms of the type $\mu\Theta$ vanishing because of the symmetry of the lattice.

These sums have been evaluated by Melhuish and Scott,[18] for the purpose of calculating the total electrostatic interaction energy in the pure solid lattices. If the quadrupole moments of the solute and matrix molecules are Q_a and Q_b, respectively, measured with respect to the centre of mass as origin, the values of the various multipole moments evaluated at the lattice point are as follows:

$$
\mu = \mu
$$

$$
\Theta = Q - 2\mu z
$$

$$
\Omega \simeq -3Qz + 3\mu z^2
$$

where z is the displacement of the centre of mass of the molecule from the lattice point. In both cases molecular electric moments of order octopole and above are neglected, since their values are virtually unknown. Thus

$$
V_{elec} \simeq -1\cdot533\frac{\mu_a\mu_b}{d_0^3} + 7\cdot05\frac{\mu_a(Q_b z_b - \mu_b z_b^2)}{d_0^5}
$$

$$
-10\cdot58\frac{(Q_a - 2\mu_a z_a)(Q_b - 2\mu_b z_b)}{d_0^5} + 7\cdot05\frac{(Q_a z_a - \mu_a z_a^2)\mu_b}{d_0^5} \qquad (4.32)
$$

and hence the shift is given by

$$
\Delta_v\nu_{elec} \simeq -\frac{1\cdot533\mu_b}{hcd_0^3}\Delta\mu_a
$$

$$
-\frac{[(7\cdot05z_a^2 + 42\cdot31z_a z_b + 7\cdot05z_b^2)\mu_b - (21\cdot15z_a + 7\cdot05z_b)Q_b]}{hcd_0^5}\Delta\mu_a
$$

$$
-\frac{(10\cdot58Q_b - 7\cdot05\mu_b z_a - 21\cdot15\mu_b z_b)}{hcd_0^5}\Delta Q_a \qquad (4.33)
$$

In deriving this expression it has been assumed that the arrangement of dipoles in the matrix lattice is ordered (Figure 4.4). However it appears likely from the residual entropies[19] that the dipoles of carbon monoxide and

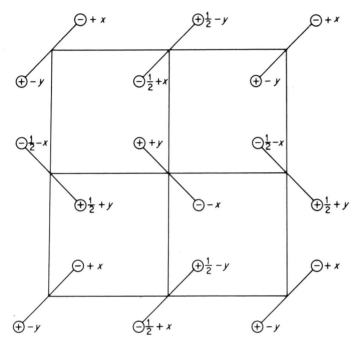

Figure 4.4 Ordered arrangement of dipoles in a f.c.c. lattice

nitrous oxide are randomly orientated in their respective lattices, and a neutron diffraction study[20] of nitrous oxide also indicated that the structure is disordered. If the orientation of the dipoles is, in fact, completely random all the terms involving μ_b vanish in equations (4.31), (4.32) and (4.33) since they average to zero. The shift would then be the same as for a non-dipolar lattice:

$$\Delta_v v_{\text{elec}} \simeq \frac{(21 \cdot 15 z_a + 7 \cdot 05 z_b) Q_b}{h c d_0^5} \Delta \mu_a - \frac{10 \cdot 58 Q_b}{h c d_0^5} \Delta Q_a \tag{4.34}$$

Nevertheless it is possible that, under the conditions used to deposit matrices, partial or complete ordering of the matrix dipoles around the solute dipole may occur.

The electrostatic shift for lithium fluoride[21] and for hydrogen chloride[22] in a nitrogen matrix has been estimated by assuming maximum dipole-quadrupole interaction between the solute dipole and all 25·338 effective nearest neighbours;[13] such estimates must, however, be about an order of magnitude too large, if the solute molecule does in fact occupy a substitutional site in the matrix lattice, since the symmetry of the lattice causes the dipole-quadrupole interaction to vanish—leaving only a second order effect due to displacement of the molecules from the lattice points. Even if the restriction of an ordered lattice around the solute molecule is removed, the electrostatic interaction cannot be as high as these estimates would imply, since the maximum interaction is obtained only when the quadrupole is on the dipole axis.

4.2.5 Inductive interaction

The inductive effect arises from the interaction between the permanent charge distribution of one molecule and the moments induced in the other molecule. The intermolecular potential energy for two linear molecules (orientated as in Figure 4.2) is given by[23]

$$\psi_{ind} = -\frac{\mu_a^2 \alpha_b (3 \cos^2 \theta_a + 1)}{2r_{ab}^6} - \frac{6\mu_a \Theta_a \alpha_b \cos^3 \theta_a}{r_{ab}^7} - \cdots$$

$$-\frac{\mu_a^2 (\alpha_\| - \alpha_\perp)_b \left[\begin{array}{c} 12 \cos^2 \theta_a \cos^2 \theta_b + 3 \sin^2 \theta_a \sin^2 \theta_b \cos^2 (\phi_a - \phi_b) \\ - 3 \cos^2 \theta_a - 1 \\ - 12 \sin \theta_a \sin \theta_b \cos \theta_a \cos \theta_b \cos (\phi_a - \phi_b) \end{array}\right]}{6r_{ab}^6}$$

$$+\frac{3\mu_a^2 A_{\| b} \left[\begin{array}{c} 6 \cos^2 \theta_a \cos^3 \theta_b - 2 \cos^2 \theta_a \cos \theta_b \\ - 7 \sin \theta_a \cos \theta_a \sin \theta_b \cos^2 \theta_b \cos (\phi_a - \phi_b) \\ + \sin \theta_a \cos \theta_a \sin \theta_b \cos (\phi_a - \phi_b) \\ + 2 \sin^2 \theta_a \sin^2 \theta_b \cos \theta_b \cos^2 (\phi_a - \phi_b) \end{array}\right]}{2r_{ab}^7}$$

$$-\frac{2\mu_a^2 A_{\perp b} \left[\begin{array}{c} 6 \cos^2 \theta_a \cos^3 \theta_b - 5 \cos^2 \theta_a \cos \theta_b \\ - 7 \sin \theta_a \cos \theta_a \sin \theta_b \cos^2 \theta_b \cos (\phi_a - \phi_b) \\ + 2 \sin \theta_a \cos \theta_a \sin \theta_b \cos (\phi_a - \phi_b) \\ + 2 \sin^2 \theta_a \sin^2 \theta_b \cos \theta_b \cos^2 (\phi_a - \phi_b) - \cos \theta_b \end{array}\right]}{r_{ab}^7}$$

$$+ \cdots \tag{4.35}$$

where μ_a and Θ_a are the dipole and quadrupole moments, respectively, of the solute molecule, α_b is the polarizability of the matrix molecule, A_b describes the polarization of the matrix molecule, determining the dipole induced by a field gradient or the quadrupole induced by a static field.

This simplifies to

$$\psi_{ind} = -\frac{\mu_a^2 \alpha_b (3 \cos^2 \theta_a + 1)}{2r_{ab}^6} - \frac{6\mu_a \Theta_a \alpha_b \cos^3 \theta_a}{r_{ab}^7}$$

$$-\frac{9\Theta_a^2 \alpha_b (1 - 2 \cos^2 \theta_a + 5 \cos^4 \theta_a)}{2r_{ab}^8} - \cdots \tag{4.36}$$

for the interaction of a dipolar linear molecule with a spherical atom.

The potential energy of interaction, given by equation (4.36), depends on the relative orientation of the two molecules. The average over all orientations[17]

$$\bar{\psi}_{ind} = - \frac{\mu_a^2 \alpha_b}{r_{ab}^6} - \frac{6\Theta_a^2 \alpha_b}{r_{ab}^8} - \ldots \tag{4.37}$$

Point polarizable atom model

If the dipole-induced dipole interaction is considered to be the only important term in equation (4.37), effectively regarding the interaction as being between a point non-polarizable dipole and a point polarizable atom, the potential energy due to the inductive interaction for a solute trapped in a f.c.c. lattice may be evaluated, by summing over the 14·454 effective nearest neighbours for an r^{-6} dependence,[13] as[21,22]

$$V_{ind} = - \frac{14 \cdot 454 \mu_a^2 \alpha_b}{d_0^6} \tag{4.38}$$

Thus

$$\Delta_v V_{ind} = - \frac{14 \cdot 454 \alpha_b}{hcd_0^6} \Delta_v(\mu_a^2) \tag{4.39}$$

for spherical matrix molecules, but for matrices such as nitrogen the anisotropy in the polarizability of the matrix molecule must be taken into account

$$\Delta_v V_{ind} = - \frac{14 \cdot 454}{hcd_0^6} [\alpha_b - 0 \cdot 5(\alpha_\parallel - \alpha_\perp)_b] \Delta_v(\mu_a^2) \tag{4.40}$$

Expanding the dipole moment as a Taylor series in $x = (r - r_e)$

$$\mu = \mu_e + \mu' x + \tfrac{1}{2}\mu'' x^2 + \ldots \tag{4.41}$$

where

$$\mu' = \left(\frac{\partial \mu}{\partial x}\right)_0, \mu'' = \left(\frac{\partial^2 \mu}{\partial x^2}\right)_0, \quad \text{etc.}$$

Thus

$$\Delta_v(\mu_a^2) = [\mu_a^2]_v - [\mu_a^2]_0 = 2\mu_{ea}\mu_a'\{[x]_v - [x]_0\}$$
$$+ (\mu_{ea}\mu_a'' + \mu_a'^2)\{[x^2]_v - [x^2]_0\}$$
$$+ (\tfrac{1}{3}\mu_{ea}\mu_a''' + \mu_a'\mu_a'')\{[x^3]_v - [x^3]_0\} + \ldots \tag{4.42}$$

Neglecting higher terms in equation (4.42)

$$\Delta_v(\mu_a^2) \simeq 2\mu_a \left(\frac{\partial \mu}{\partial x}\right)_0 \{[x]_v - [x]_0\}$$

$$\simeq 4v\mu_a \left(\frac{\partial \mu}{\partial x}\right)_0 \langle 0|x|0 \rangle \tag{4.43}$$

RTC model

From equations (4.18) and (4.22), the shift is given by

$$\Delta_v v_{ind} = -\frac{4vN_6\mu_a(x)_0\alpha_b(\partial\mu/\partial x)_0\langle 0|x|0\rangle}{hcd_0^6} \tag{4.44}$$

which is exactly the same expression as that obtained by combining equation (4.39) and (4.43).

Cavity model

The point polarizable atom model takes no account of further polarization arising from the induced moments themselves, and also ignores the relatively important short-range interaction with the nearest neighbours actually in contact with the solute molecule. If a point dipole is assumed to lie at the centre of a cavity, of radius $d_0/2$, in a uniform dielectric[24] (cf. solvent shift theories)

$$V_{ind} = -\frac{8\mu_a^2}{d_0^3}\frac{(\varepsilon'-1)}{(2\varepsilon'+1)}\frac{1}{1-\frac{2(\varepsilon'-1)}{(2\varepsilon'+1)}\frac{8}{d_0^3}\alpha_a}$$

$$\simeq -\frac{8\mu_a^2}{d_0^3}\frac{(\varepsilon'-1)}{(2\varepsilon'+1)} \tag{4.45}$$

where ε' is the dielectric constant of the matrix medium, and μ_a is the polarizability of the solute molecule. Thus

$$\Delta_v v_{ind} = -\frac{8(\varepsilon'-1)}{(2\varepsilon'+1)hcd_0^3}\Delta_v(\mu_a^2) \tag{4.46}$$

This model, with its basic assumption that the molecules surrounding a particular solute molecule may be considered to behave as a uniform dielectric, is just as unrealistic as the point polarizable atom model. On the molecular scale, the cavity is no longer a well-defined concept and the choice of radius is arbitrary. The cavity model gives values of the shift two to three times greater than the PPA model, and the two models may well, in fact, give upper and lower bounds, respectively, of $|\Delta_v v_{ind}|$, since the cavity model over-estimates the short-range contribution.

A further objection to both models is that they are based on a model which assumes that the dipole moment, inducing the reaction field, is stationary. If the solute molecule is actually rotating, as for hydrogen chloride in the noble gas matrices, for example—it cannot produce a reaction field which will instantaneously follow its motion, and thus the models used are inadequate.

Effect of including higher terms

In both the models considered above, the poles of higher order than the dipole have been ignored—the effect being to treat the solute as a point dipole at the centre of a substitutional site in the matrix, whereas, in fact, it approximately fills the cavity and is not necessarily centred about the lattice point. In principle, at least, the magnitude of the dipole moment changes between the vibrational states can be derived from gas phase intensity measurements, but the information required to calculate the changes in higher moments during a vibration is virtually non-existent. Thus, in general, the terms involving higher moments have been assumed to be small compared with the dipole-induced dipole term. In the case of matrices such as nitrogen, the terms involving the anisotropy of the matrix molecule have also been neglected.

Considering first the noble gas matrices, on the PPA model, the first two terms in the interaction energy are given in equation (4.37), and thus, neglecting the octopole and higher moments

$$\Delta_v v_{ind} = -\frac{14 \cdot 454 \alpha_b}{hcd_0^6} \Delta_v(\mu_a^2) - \frac{76 \cdot 811 \alpha_b}{hcd_0^8} \Delta_v(Q_a^2) \qquad (4.47)$$

For matrices such as nitrogen, further terms must be included to take account of the anisotropy in the polarizability of the matrix molecule. The terms in $A_{\parallel}$ and $A_{\perp}$ in equation (4.35) are difficult to evaluate; these and higher terms will be neglected. Thus

$$\Delta_v v_{ind} = -\frac{14 \cdot 454}{hcd_0^6}[\alpha_b - 0 \cdot 5(\alpha_{\parallel} - \alpha_{\perp})_b] \Delta_v(\mu_a^2) - \frac{76 \cdot 811 \alpha_b}{hcd_0^8} \Delta_v(Q_a^2) \quad (4.48)$$

where the angular-dependent term has been approximated by summing over the first few shells, assuming that the solute molecule occupies the [111] direction.

Using the cavity model, the potential, due to the reaction field, in the cavity in the dielectric medium is given by[25]

$$\psi = -\sum_{l=0}^{\infty} \left[\frac{(l+1)(\varepsilon'-1)}{(l+1)\varepsilon'+l} b_l \frac{r^l}{(d_0/2)^{2l+1}} \right] P_l(\cos \theta) \qquad (4.49)$$

where b_l represents the original charge distribution (the values of the moments M_l evaluated at the centre of the cavity), r is the distance from the centre of the cavity, and $P_l(\cos \theta)$ are Legendre functions. Since $P_l(\cos \theta) \propto r^l$, the terms decrease as $(2r/d_0)^2$. Assuming that the charge distribution has cylindrical symmetry, the potential field in a varying electric field ψ is[15]

$$V = M_0 \psi_0 + M_1 \left(\frac{d\psi}{dz} \right)_0 + \frac{1}{2!} M_2 \left(\frac{d^2\psi}{dz^2} \right)_0 + \dots \qquad (4.50)$$

where

$$M_l = \sum_i e_i r_i^l P_l(\cos \theta_i)$$

Thus the extra potential energy of the charge distribution in its own (this introduces a factor of $\frac{1}{2}$) reaction field is found by combining equations (4.49) and (4.52), equation (4.49) being differentiated the appropriate number of times with respect to z, and then z set equal to zero, since $(d^l\psi/dz^l)$ are required to be evaluated at the centre of the cavity. Thus the total potential energy is given by[24]

$$V_{\text{ind}} = -\sum_{l=0}^{\infty} \frac{1}{2} \frac{(l+1)(\varepsilon'-1)}{(l+1)\varepsilon'+l}\left(\frac{2}{d_0}\right)^{2l+1}\left[\sum_i e_i r_i^l P_l(\cos \theta_i)\right]^2 \quad (4.51)$$

and hence

$$\Delta_v v_{\text{ind}} = -\frac{8(\varepsilon'-1)}{(2\varepsilon'+1)hcd_0^3}\Delta_v(\mu_a^2) - \frac{48(\varepsilon'-1)}{(3\varepsilon'+2)hcd_0^5}\Delta_v(Q_a^2) - \dots \quad (4.52)$$

$\Delta_v(Q_a^2)$ may be derived, in a similar manner to the corresponding dipole moment change, by expanding Q_a as a Taylor series in x. Then

$$\Delta_v(Q_a^2) = 4vQ_a\left(\frac{\partial Q}{\partial x}\right)_0 \langle 0|x|0\rangle \quad (4.53)$$

Since however, in general, no quadrupole (or higher moments) data are available, McKean[24] uses an effective charge model—placing effective charges on the nuclei, which move with them in all their motions—to calculate values of the multipole factor, i.e. the multiplier of the shift calculated for dipole-induced dipole interactions only. For the group IV hydrides, he derives multipole factors ~ 3.5 for the stretching vibrations and ~ 1.8 for the bending motions, thus showing that, at least in these cases, the contribution from these higher moments is significant.

4.2.6 Dispersive interaction

The dispersive interaction between two molecules is due to a net attractive force produced by the instantaneous electronic configurations of the two molecules. In the case of a linear molecule and a spherical atom, it is given approximately by[26]

$$\psi_{\text{dis}} = -\frac{3E_a E_b \alpha_b}{2(E_a + E_b)}[\alpha_a + \tfrac{1}{6}(\alpha_\parallel - \alpha_\perp)_a(3\cos^2\theta_a - 1)]\frac{1}{r_{ab}^6}$$

$$- \frac{45E_a E_b \alpha_a \alpha_b}{8e^2}\left[\frac{\alpha_a E_a}{2E_a + E_b} + \frac{\alpha_b E_b}{E_a + 2E_b}\right]\frac{1}{r_{ab}^8} - \dots \quad (4.54)$$

where α_a, α_b are the mean polarizabilities of the two molecules, and E_a, E_b

are the first ionization energies of the two molecules. Thus, for solute molecules trapped in noble gas matrices

$$V_{dis} \simeq -14.454 \frac{3E_a E_b \alpha_a \alpha_b}{2(E_a + E_b)d_0^6} \tag{4.55}$$

since the mean value of $(3\cos^2\theta - 1) = 0$. Ignoring the change in E_a between the two vibrational states (which is only a few per cent.), the shift is given by

$$\Delta_v v_{dis} = -\frac{21.681 E_a E_b \alpha_b}{(E_a + E_b)hcd_0^6} \Delta_v \alpha_a \tag{4.56}$$

$\Delta_v \alpha_a$ may be evaluated by expanding the polarizability as a Taylor series in x, giving

$$\Delta_v \alpha_a = 2v \left(\frac{\partial \alpha}{\partial x}\right)_0 \langle 0|x|0 \rangle \tag{4.57}$$

or alternatively, from data for isotopic molecules, using equation (4.24).

The interaction between two linear molecules displays a strong angular dependence at large values of r_{ab}, and, at closer range, deviates markedly from the relationship implied by equation (4.54). The interaction now includes a r^{-7} dependent term[23,27]

$$\psi_{dis} = -\frac{3E_a E_b \alpha_a \alpha_b}{2(E_a + E_b)r_{ab}^6} \left[1 + \frac{(\alpha_{\parallel} - \alpha_{\perp})_a}{6\alpha_a}(3\cos^2\theta_a - 1) \right.$$

$$+ \frac{(\alpha_{\parallel} - \alpha_{\perp})_b}{6\alpha_b}(3\cos^2\theta_b - 1) \bigg]$$

$$+ \frac{3E_a E_b \alpha_a}{(E_a + E_b)r_{ab}^7} [A_{\parallel b}\cos^3\theta_b + \tfrac{2}{3}A_{\perp b}(3\cos\theta_b - 2\cos^3\theta_b)]$$

$$- \frac{3E_a E_b \alpha_b}{(E_a + E_b)r_{ab}^7} [A_{\parallel a}\cos^3\theta_a + \tfrac{2}{3}A_{\perp a}(3\cos\theta_a - 2\cos^3\theta_a)] - \ldots \tag{4.58}$$

However, the symmetry of the f.c.c. lattice of nitrogen, and similar substances, is such that equation (4.58) reduces, to a first approximation, to the same value of the shift as above (equation (4.56)).

RTC model

From equations (4.19) and (4.23), the shift is given by

$$\Delta_v v_{dis} = -\frac{4N_6 \varepsilon(x)_0 \sigma^6(x)_0 \, \Delta_v \alpha_a}{hc\alpha_a d_0^6} \tag{4.59}$$

and thus, for a f.c.c. lattice

$$\Delta_v v_{dis} = -\frac{57 \cdot 816 \varepsilon(x)_0 \sigma^6(x)_0}{hc\alpha_a d_0^6} \Delta_v \alpha_a \tag{4.60}$$

4.2.7 Repulsive interaction

Few attempts have been made to calculate the shift due to the repulsive interaction, reflecting the lack of an adequate theory for the short-range valence forces. The calculations that have been performed are based on the assumption that the repulsive potential is described by the r^{-12} term in the Lennard-Jones (6–12) potential[10]

$$\psi_{ab} = 4\varepsilon_{ab} \left[\left(\frac{\sigma_{ab}}{r_{ab}}\right)^{12} - \left(\frac{\sigma_{ab}}{r_{ab}}\right)^6 \right] \tag{4.61}$$

where ε_{ab} is the intermolecular potential well depth, σ_{ab} is the distance at which attractive and repulsive forces cancel, and r_{ab} is the intermolecular distance. The intermolecular force constants are obtained by assuming the combining law

$$\varepsilon_{ab}\sigma_{ab}^{12} = (\varepsilon_a \sigma_a^{12})^{1/2}(\varepsilon_b \sigma_b^{12})^{1/2} \tag{4.62}$$

Charles and Lee[28] have calculated the shifts due to the repulsive interaction for carbon monoxide in various matrices on a geometrical model, using the above formulae, and Whyte[29] has performed similar calculations for hydrogen chloride in an argon matrix. Such calculations are, however, strongly dependent on the assumptions made to obtain the parameters.

RTC model

Friedmann and Kimel[9] derive the following formula for the repulsive shift, based similarly on the Lennard–Jones (6–12) potential (§4.2.3)

$$\Delta_v v_{rep} = \frac{4}{hc} N_{12}\varepsilon(x)_0 \left[\frac{\sigma(x)_0}{d_0}\right]^{12} [\mathfrak{F}_v(\mu) + \mathfrak{F}_v(\alpha)]\left(\frac{2+y}{1+y}\right) \tag{4.20}$$

where $\mathfrak{F}_v(\mu)$, $\mathfrak{F}_v(\alpha)$ and y are given by equations (4.21)–(4.22), (4.23)–(4.24) and (4.25) respectively, and the effective number of nearest neighbours for a r^{-12} law (N_{12}) is 12·132 for a f.c.c. lattice.[13] Thus

$$\Delta_v v_{rep} = \frac{48 \cdot 528}{hcd_0^{12}} \varepsilon(x)_0 \sigma^{12}(x)_0 [\mathfrak{F}_v(\mu) + \mathfrak{F}_v(\alpha)]\left(\frac{2+y}{1+y}\right) \tag{4.63}$$

4.2.8 Dynamic contribution

The dynamic contribution[9] to the frequency shift

$$\Delta_v v_{dyn} = \frac{1}{hc}[\omega_v(r)_0 - \omega_0(r)_0] \tag{4.64}$$

is derived, on the harmonic oscillator cell model, from equation (4.14)

$$\Delta_v v_{\text{dyn}} = \tfrac{9}{2}\rho(x)_v v_0 \left\{ y + \frac{88[\sigma(x)_0/d_0]^{12} - 10[\sigma(x)_0/d_0]^6}{44[\sigma(x)_0/d_0]^{12} - 10[\sigma(x)_0/d_0]^6} \right\} \quad (4.65)$$

where

$$\rho(x)_v = \frac{\mathfrak{F}_v(\mu) + \mathfrak{F}_v(\alpha)}{6(1 + y)} \quad (4.66)$$

or, on the spherical box model, from equation (4.15)

$$\Delta_v v_{\text{dyn}} = \frac{h}{4mc} \frac{\rho(x)_v \sigma(x)_0}{d_0^3} \left(\frac{r_f}{d_0}\right)^{-3} \frac{\partial(r_f/d_0)}{\partial(\sigma(x)_0/d_0)} \quad (4.67)$$

The dynamic part of the shift—arising, of course, from the constrained translational motion of the solute molecule—is small compared with the static part of the shift, and thus the inaccuracies of the model used are unimportant. Typical values, on the two models, for hydrogen chloride and hydrogen bromide in noble gas matrices are given in Table 4.2.

Table 4.2 Dynamic contributions (in cm^{-1}) to the frequency shifts of the hydrogen halides in noble gas matrices[9]

Solute	Matrix	Harmonic oscillator cell model	Spherical box model
HCl	Ar	1·5	0·4
	Kr	1·3	0·2
	Xe	1·0	0·1
HBr	Ar	0·8	0·3
	Kr	0·7	0·1
	Xe	0·5	0·0

4.2.9 Comparison with experimental results for diatomic solute molecules

The contribution from each of the interactions discussed in the preceding sections may be calculated to give a value of the overall shift for any diatomic solute molecule in any matrix, provided that all the relevant physical properties of the solute and matrix molecules are known. The properties of matrix molecules are detailed in Chapter 2, and it can be seen that the lattice parameter at 4 K or 20 K—a very important number, since it is involved in all the equations as a large inverse power—is often not available for matrices other than the noble gases, nitrogen, etc., and has to be estimated by extrapolation from data at higher temperatures. Thus X-ray crystallographic examination of materials such as sulphur hexafluoride, carbon dioxide, etc., at liquid hydrogen or liquid helium temperatures would be of great value. Other data required for matrix materials are normally available, with the exception that the values of the polarizability are generally determined in the gas phase at room temperature using light of visible wavelengths, rather than in the solid phase at cryogenic temperatures using infrared radiation, and consequently are not completely satisfactory. The usefulness of McKean's cavity model for the calculation of the inductive shift is limited by the complete lack of any data on the dielectric constants of the matrix materials at very low temperatures, the only available values (for but a few of the solid substances) being at liquid nitrogen, or higher, temperatures.

The data required on the properties of diatomic solute molecules are only partially available. In particular, satisfactory values of the Lennard–Jones parameters (i.e. derived from thermodynamic properties rather than from transport properties) are not available for all the hydrogen halides, and values of the change in polarizability with vibrational state are scanty and of dubious accuracy. The only information on the changes in electrical

Table 4.3 Physical properties of some diatomic solutes

	HF	HCl	HBr	HI	CO	NO
Bond length r_e (Å)	0.9181	1.2747	1.4146	1.6091	1.1282	1.1506
$\langle 0\|x\|0\rangle$ (Å)[a]	0.0159	0.0162	0.0167	0.0175	0.0041	0.0045
μ_0 (e.s.u. cm) $\times 10^{18}$	1.819	1.093	0.834	0.451	−0.112	−0.153
$(\partial\mu/\partial x)_0$ (e.s.u.) $\times 10^{10}$	1.50[b]	0.925[c]	$\begin{pmatrix}0.455\\0.341\end{pmatrix}$[d]	−0.055[e]	3.10[f]	2.14[g]
$\Delta\mu$ (e.s.u. cm) $\times 10^{18}$	0.0476	0.0304	$\begin{pmatrix}0.0163\\0.0116\end{pmatrix}$	0.0010	0.0244	0.0146
$\Delta\mu^2$ (e.s.u.^{2} cm^2) $\times 10^{36}$	0.1836	0.0754	$\begin{pmatrix}0.0299\\0.0208\end{pmatrix}$	0.0009	0.0164	0.0060
α_0 (Å^3)	2.46	2.60	3.61	5.45	1.98	1.74
$(\partial\alpha/\partial x)_0$ (Å^2)	0.72[h]	1.0[j]	1.2[j]	—	—	—
$\Delta\alpha$ (Å^3)	0.0228	0.0323	0.0401	[0.06][k]	0.0200[l]	—
IP E (eV)	16.05	12.74	11.67	10.38	14.01	9.26
Gas phase ν_0 (cm^{-1})	3961.4	2885.5	2558.7	2229.6	2143.3	1876.0

a. Expressions for $\langle 0\|x^n\|0\rangle$ in terms of Dunham potential constants are given in R. M. Herman, R. H. Tipping and S. Short, *J. Chem. Phys.*, **53**, 595, (1970).
b. R. J. Lovell and W. F. Herget, *J. Opt. Soc. Amer.*, **52**, 1374, (1962).
c. E. W. Kaiser, *J. Chem. Phys.*, **53**, 1686 (1970).
d. R. H. Tipping and R. M. Herman, *J. Mol. Spectroscopy*, **36**, 404, (1970).
e. W. Benesch, *J. Chem. Phys.*, **39**, 1048, (1963).
f. R. A. Toth, R. H. Hunt and E. K. Plyler, *J. Mol. Spectroscopy*, **32**, 85, (1969).
g. B. Schurin and R. E. Ellis, *J. Chem. Phys.*, **45**, 2528, (1966).
h. B. Oksengorn, *Spectrochim. Acta*, **20**, 99, (1964).
j. E. J. Stansbury, M. F. Crawford and H. L. Welsh, *Can. J. Phys.*, **31**, 954, (1953).
k. Estimated value.
l. Ref. 9.

moments, of order higher than dipole, with vibrational state is a theoretical calculation by Nesbet[30] for hydrogen chloride, which gives $Q_0 = 3.96 \times 10^{-26}$ e.s.u. cm^2 (experimental value[16] $+ 3.8 \times 10^{-26}$ e.s.u. cm^2) and $(\partial Q/\partial x)_0 = 4.54 \times 10^{-18}$ e.s.u. cm. The best available values, estimated where necessary, for the various physical properties of diatomic solute molecules are summarized in Table 4.3.

In the absence of data for ΔQ_a, the electrostatic shifts calculated from the first term of equation (4.34), which gives small positive or negative values, are negligible. Thus they will be assumed to be zero in the calculations below, although it should not be forgotten that they *may* provide an important contribution to the total shift in polar matrices.

The inductive shifts calculated on the PPA model, equation (4.39), and on the cavity model, equation (4.46), for hydrogen chloride in argon, krypton and xenon matrices (the only matrices for which any values of the dielectric constant are available) are compared in Table 4.4. It can be seen that the

Table 4.4 Inductive shift (in cm^{-1}) of HCl in noble gas matrices on different models

Matrix	Dipole term only		Dipole + quadrupole terms	
	PPA	cavity	PPA	cavity
Ar	-3.1	-8.4	-24.6	-48.7
Kr	-3.3	-8.8	-23.2	-46.1
Xe	-3.3	-8.2	-20.2	-38.1

cavity model gives values 2.5–3 times larger than the PPA model and thus, since the data required to calculate the shifts on the cavity model are not available for most matrices, the inductive shift will be taken as twice the value given by the PPA model, i.e. approximately the mean of the two models. In fact, inductive shifts are generally rather small and thus the exact value used is of little importance. Also given in Table 4.4 are values of the inductive shifts calculated including the quadrupole term, from equation (4.47) and equation (4.52), with Nesbet's value of ΔQ_a. It can be seen that the second (quadrupole) term swamps the first (dipole) term but, since ΔQ_a is not an experimental value but derived from a theoretical calculation, this effect may not be meaningful. In the absence of further data, the quadrupole, and higher, terms will be assumed to be negligible.

The various contributions to the total vibrational shift—the inductive term calculated as described above, and the dispersive and the repulsive terms calculated from equation (4.60) and (4.63) respectively—for the hydrogen halides in a variety of matrices are compared with the experimental

Table 4.5 Comparison of calculated and observed frequency shifts (in cm^{-1}) for hydrogen halides in matrices

Solute	Matrix	Δv_{ind}	Δv_{dis}	Δv_{rep}	Total Δv_{calc}	Δv_{exp}
HF	Ne	−10·6	−25·3	+23·3	−12·6	−8·8
	Ar	−15·3	−29·3	+17·7	−26·9	−40·9
	Kr	−16·1	−29·5	+15·4	−30·2	−39·5
	Xe	−16·0	−28·5	+12·1	−32·4	−58·9
	CH_4	−13·1	−25·2	+11·7	−26·6	—
	CF_4	−8·9	−24·3	+9·8	−23·4	—
	SF_6	−3·5	−19·4	+5·0	−17·9	—
	N_2	−9·1	−23·6	+12·5	−20·2	−81·1
	CO_2	−11·9	−63·6	+58·1	−17·4	—
	CO	−11·0	−25·7	+14·7	−22·0	—
HCl	Ne	−4·3	−51·4	+74·8	+19·1	+14·5 (non-rot.) +0·5 (rot.)
	Ar	−6·3	−56·3	+49·6	−13·0	−15·1
	Kr	−6·6	−55·8	+41·4	−21·0	−30·6
	Xe	−6·6	−52·5	+30·9	−28·2	−43·8
	CH_4	−5·4	−47·1	+30·9	−21·6	−38·2[a]
	CF_4	−3·6	−42·8	+24·4	−22·0	−18·0 (non-rot.) −38·2 (rot.)
	SF_6	−1·4	−32·5	+12·0	−21·9	−20·8
	N_2	−3·7	−44·3	+34·0	−14·0	−31·7
	CO_2	−4·9	−114·8	+153·7	+34·0	−42·7
	CO	−4·5	−48·2	+39·2	−13·5	−79·5
HBr	Ne	−1·7	−58·5	+100·1	+39·9	—
	Ar	−2·5	−63·1	+63·8	−1·8	+0·9
	Kr	−2·6	−62·1	+52·6	−12·1	−16·7
	Xe	−2·6	−58·1	+38·6	−22·1	—
	CH_4	−2·1	−52·2	+39·0	−15·3	−25·7
	CF_4	−1·4	−46·7	+30·1	−18·0	—
	SF_6	−0·6	−34·9	+14·5	−21·0	−19·9
	N_2	−1·5	−49·3	+43·3	−7·5	−12·8
	CO_2	−1·9	−126·1	+192·9	+64·9	—
	CO	−1·8	−53·5	+49·7	−5·6	−53·9
HI	Ne	−0·1	−82·8	+186·7	+103·8	—
	Ar	−0·1	−87·2	+113·0	+25·7	+16
	Kr	−0·1	−85·2	+91·6	+6·3	−6·6
	Xe	−0·1	−78·9	+65·8	−13·2	—
	CH_4	−0·1	−71·2	+67·2	−4·1	−8·8
	CF_4	−0·0	−62·1	+49·5	−12·6	—
	SF_6	−0·0	−45·4	+23·0	−22·4	+2
	N_2	−0·0	−67·4	+75·2	+7·8	+7·6
	CO_2	−0·1	−169·5	+325·3	+155·7	+30·0
	CO	−0·1	−73·0	+86·0	+12·9	−22[a]

a. Derived from DX value (HX spectrum unobtainable due to matrix absorption).

values of the shift in Table 4.5. It can be seen that there is a reasonable correlation between the calculated shifts and the experimental results for the matrices offering spherical cavity sites, but that the agreement is very poor in the case of the polar, non-spherical cavity site, matrices nitrogen, carbon monoxide, and carbon dioxide. The fact that approximately the correct trends are predicted for the hydrogen halides in spherical cavity site matrices lends some support to the assumption that the solute molecules occupy substitutional sites in the matrix, and the agreement is surprisingly good in view of the many assumptions and approximations involved.

In the case of the matrices nitrogen, carbon monoxide, and carbon dioxide, two additional assumptions have been made in these calculations. The first is that the repulsive shift of a molecule in a 'cylindrical' cavity may be approximately represented by equation (4.63), using the mean cavity diameter as d_0; this is clearly a gross over-simplification. Secondly, it has been assumed that there is no electrostatic contribution to the frequency shift; from the trend of the observed shifts in these matrices it is obvious that there is a considerable specific interaction between the small, highly polar solute molecule and the polar matrix molecules, which is particularly marked in the case of hydrogen fluoride as a solute or carbon monoxide as a matrix. This second effect is not, at present, calculable, but the problem of cavity shape in calculating the repulsive shift may be overcome by using a geometrical model, summing the contributions of the interactions between each pair of atoms. This approach has been used by Charles and Lee[28] to predict quite accurate shifts for carbon monoxide as a solute in various matrices (Table 4.6).

Table 4.6 Comparison of calculated and observed frequency shifts (in cm^{-1}) for CO in matrices[28]

Solute	Matrix	Δv_{ind}	Δv_{dis}	Δv_{rep}	Total Δv_{calc}	Δv_{exp}
CO	Ar	−0·9	−7·2	+10·5	+2·4	+5·6
	Kr	−1·0	−8·1	+10·5	+1·4	+1·9
	Xe	−1·0	−7·3	+6·3	−2·0	−1·9
	N$_2$	−0·7	−5·8	+0·7	−5·8	−3·7
$^{12}C^{17}O$						−4·0
$^{12}C^{18}O$	CO	−0·8	−6·5	+1·1	−6·2	−3·7
$^{13}C^{16}O$						−3·9

An alternative approach to the entire problem is the empirical method of comparing the relative frequency shifts with a standard, which has been used with great success in the (liquid) solution field[2] to demonstrate the importance of short-range specific interactions between solute and solvent. Hydrogenic stretching frequencies yield linear Bellamy–Hallam–Williams (BHW) plots with no observable discontinuity between non-polar and highly polar solvents, each solute giving a characteristic slope depending on the polarity of the R—H bond. When plotted against $v(N-H)$ of pyrrole as a reference solute, the shifts of the hydrogen halide stretching frequencies in liquid solvents yield reasonable lines,[2] whose slopes reflect the electronegativity of the halogen. When limited to a particular class of solvents, such as aromatic hydrocarbons, the linear relationships are greatly improved, but the lines do not pass through the origin. BHW plots of the hydrogen halides in various liquid solvents and solid matrices are shown[31] in Figure 4.5, using hydrogen chloride as the reference solute. The values used of $\Delta v/v = (v_m - v_g)/v_g$ are calculated using the band centre $[R(0) + P(1)]/2$ for rotating HX solutes, and the strongest monomer band for non-rotating solutes. It can be seen that, with some exceptions, the matrix points fall fairly close to the solvent lines,

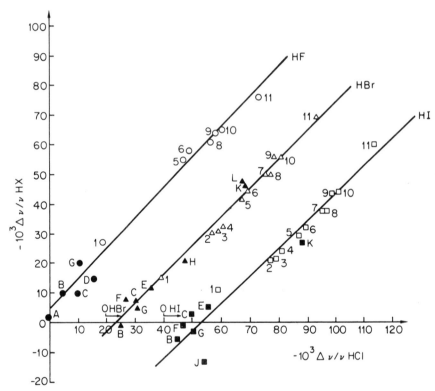

Figure 4.5 BHW plots of the hydrogen halides in liquid solvents (open points $1 = CCl_4$, $2 = C_6H_5Cl$, $3 = C_2H_4Cl_2$, $4 = C_6H_5Br$, $5 = C_6H_6$, $6 = C_6H_5Me$, $7 = pC_6H_4Me_2$, $8 = mC_6H_4Me_2$, $9 = 1,2,4C_6H_3Me_3$; $10 = 1,3,5C_6H_3Me_3$. and $11 = C_6Me_6$ and in solid matrices (solid points) $A = Ne$, $B = Ar$, $C = Kr$, $D = Xe$, $E = CH_4$, $F = SF_6$, $G = N_2$, $H = CO$, $J = CO_2$, $K = C_2H_4$. and $L = HBr$ against HCl as standard (ref. 31)

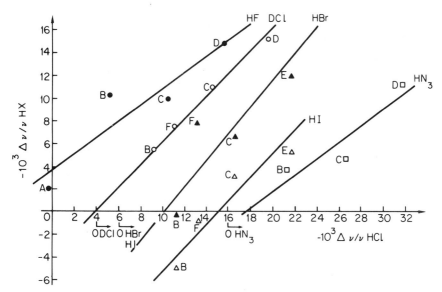

Figure 4.6 BHW plots of H—X type molecules in solid matrices with spherical cavities A = Ne, B = Ar, C = Kr, D = Xe, E = CH$_4$, and F = SF$_6$ against HCl as standard (data from refs 31, 32, and 66)

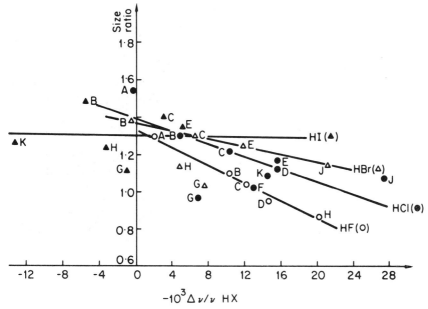

Figure 4.7 Relative shift plotted against cavity diameter for hydrogen halides in matrices A = Ne, B = Ar, C = Kr, D = Xe, E = CH$_4$, F = CF$_4$, G = SF$_6$, H = N$_2$, J = CO, and K = CO$_2$ (ref. 31)

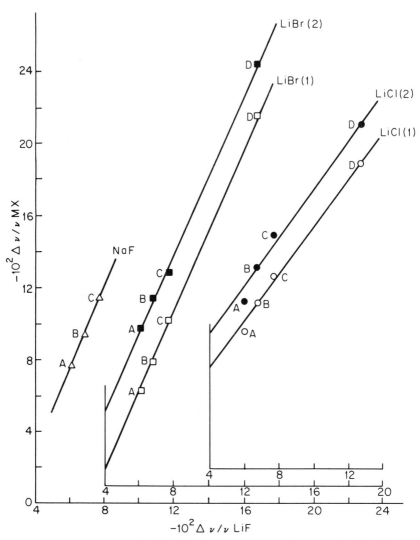

Figure 4.8 BHW plots of alkali halides in matrices A = Ar, B = Kr, C = Xe, and D = N₂ against LiF as standard (data from A. Snelson and K. S. Pitzer, *J. Phys. Chem.*, **67**, 882 (1963) and S. Schlick and O. Schnepp, *J. Chem. Phys.*, **41**, 463 (1964)).

which is particularly significant in the case of the very strong $HX-C_2H_4$ matrix interaction. The plots for the less polar matrices only are shown on a larger scale in Figure 4.6, together with those of related solute molecules. Although the data are incomplete, it would appear that the noble gases give straight line plots which differ slightly from the liquid solution lines and whose slopes are in the reverse order. This suggests that repulsive effects are important, and the 'cylindrical' cavity matrices show considerable deviations from these lines. A plot of relative shift against size of matrix cavity (Figure 4.7) might be expected to be more successful, but an effective size is required for matrices, such as nitrogen, possessing asymmetric cavities—in Figure 4.7, the mean diameter is used.

The alkali halide molecules, which give very large shifts in matrices such as the noble gases, provide a more critical test of the applicability of BHW plots to matrices, but unfortunately they have only been studied in a limited range of matrices. Their relative shifts are plotted, against lithium fluoride as an internal standard, in Figure 4.8 and it can be seen that—although the situation is complicated by multiple trapping sites—quite good straight lines are obtained. Thus it would appear that BHW plots and relative shift vs cavity size plots are potentially of value in matrix isolation spectroscopy, but that the data is, at present, too limited for their usefulness to be properly evaluated.

4.2.10 Polyatomic solute molecules

The intermolecular potential energy V for a diatomic solute molecule may be expanded as a Taylor series in $x = (r - r_e)$

$$V = V_e + \left(\frac{\partial V}{\partial x}\right)_0 x + \frac{1}{2}\left(\frac{\partial^2 V}{\partial x^2}\right)_0 x^2 + \dots \tag{4.68}$$

For a polyatomic solute molecule the expansion is more complicated

$$V = V_e + \sum_j \left(\frac{\partial V}{\partial Q_j}\right)_0 Q_j + \frac{1}{2}\sum_j \sum_k \left(\frac{\partial^2 V}{\partial Q_j \partial Q_k}\right)_0 Q_j Q_k + \dots \tag{4.69}$$

and thus quantitative evaluation of frequency shifts becomes impracticable. However, Pimentel and Charles have given a qualitative discussion[32] of the factors which influence vibrational shifts of polyatomic solute molecules in matrices. Buckingham[5] derives the expression

$$\Delta v = v_{\text{solution}} - v_{\text{gas}} = \frac{B_e}{hc\omega_e}\left[V'' - \frac{3A}{\omega_e}V'\right] \tag{4.70}$$

Thus lowering the frequency, reduced mass, or force constant of the vibration all have the effect of increasing the value of $|\Delta v|$. However the sign of the shift is determined by the expression in brackets in equation (4.70). Since the anharmonicity A is negative, it can be seen from Figure 4.9 that for $R_{CM} < R_e$ a positive shift is always obtained, while for $R_{CM} > R_1$ a negative shift is always obtained. In the terminology of Pimentel and Charles these situations are 'tight cage' and 'loose cage' respectively. They argue that the site occupied by a polyatomic solute molecule has to be a compromise between the optimum positions (R_e) for each of the points of contact between the molecule

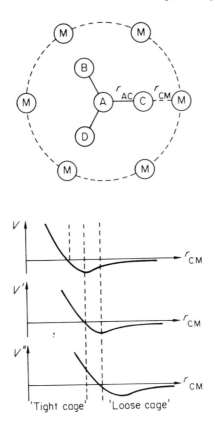

Figure 4.9 Potential curves for a polyatomic solute molecule in a matrix
cavity $V' = (\partial V/\partial r_{AC})$ and $V'' = (\partial^2 V/\partial r_{AC}^2)$ (ref. 32)

Table 4.7 Experimental frequency shifts (in cm^{-1}) of HN_3 in matrices[32]

Mode	v_1 (NH str.) 3335·6	v_2 (NNN a.str.) 2140·4	v_3 (NNN s.str.) 1269·0	v_4 (NNH bend) 1152·5
Matrix				
Ar	−12	−2	−5	−6
Kr	−15	0	−3	−4
Xe	−37	−8	−6	−7
N_2	−12	+10	+4	+15
CO	−41	—	+15	+25
CO_2	−31	+26	+18	+35

and the cage, thus leading to a situation where low force constant co-ordinates are more likely to be in a 'tight cage' environment than high force constant co-ordinates. The observed trend for polyatomic solute molecules in matrices is for high frequency stretching vibrations to display negative shifts, and for low frequency stretching, bending, or rocking vibrations to give positive shifts—for example HN_3 (Table 4.7). Note that the effect is more extreme in the matrices with unsymmetrical cavities.

4.2.11 Multiple trapping sites

A solute molecule trapped in two (or more) distinguishable sites in a matrix will obviously display more than one band, in general, for each of its vibrations since the intermolecular potential energy for the sites will differ. However multiple bands can arise from other causes—for example, rotation, or aggregation. The reversible temperature dependence of rotational lines enables these to be readily characterized, but very careful studies are required to differentiate multiple trapping site effects from aggregation effects. In many cases, multiple trapping site effects have been incorrectly invoked to explain additional bands—for example, the structure observed[33] for bands of formaldehyde in argon and nitrogen matrices is almost certainly due to aggregation.[34] There are, however, a few cases where complex bands may be assigned with reasonable certainty to multiple trapping sites, such as the hydrogen halides in nitrogen, carbon monoxide, or carbon dioxide matrices[31] (Figure 3.1). Unfortunately, since these are matrices with unsymmetrical cavities, they do not lend themselves to calculation. The alkali halides also display multiple trapping site effects, and Figure 4.8 shows BHW plots for both sites in the cases of lithium chloride and bromide.

4.3 Rotational shift

4.3.1 Rotation in matrices

Evidence has been found for the rotation of a number of small molecules in noble gas matrices from the appearance of multiplets in the infrared spectra, which show *reversible* temperature dependence. The molecules which have been reported to exhibit rotational fine structure to their vibrational bands are hydrogen fluoride,[35] hydrogen chloride,[36-43] hydrogen bromide,[37,39,40,43] hydrogen iodide,[37] water,[44-49] ammonia[49-52] and methane,[49,53-56] including their deuterium analogues. (The spectra of ammonia are complicated by inversion effects, and those of water and of methane by nuclear spin species conversion). In the cases of hydrogen fluoride, hydrogen chloride and water confirmatory evidence has been found from the far IR pure rotational spectra.[57-59] Certain small radicals also appear to rotate in noble gas matrices, for example[60] NH_2. Matrices other than the noble gases have not been used to any great extent, but the hydrogen halides have been found to show rotational structure in matrices with spherical cavity sites, such as sulphur hexafluoride.[31,61] For all these solute molecules, discrete rotational lines are observed although the relative spacings vary from those in the gas phase, apart from the superimposed overall vibrational shift (§4.2).

Rotation does not seem to occur in matrices such as nitrogen[32]—although this has been disputed for hydrogen chloride and hydrogen bromide,[62] water[63] and ammonia,[50] later evidence[31,58,64,65] confirms the non-rotation hypothesis. It would appear that, in general, rotational motion of small molecules may take place in any matrix which provides a spherical cavity, but not in matrices with unsymmetrical cavities. Exceptionally rotation appears not to occur in methane matrices.[66] Other solute molecules, such as carbon monoxide,[67,68] and nitrogen dioxide and sulphur dioxide,[69] exhibit apparent rotational effects, but the evidence is contradictory.

Numerous attempts have been made to calculate the perturbation of the rotational levels of a diatomic solute molecule, in noble gas matrices, on the basis of the anisotropy of the crystal field and/or on the basis of a rotation-translation coupling (RTC) model. These two approaches will be discussed in the subsequent paragraphs, and the predictions compared with the experimental data for the hydrogen halides as solutes in noble gas matrices. The perturbation of the rotational levels of tetrahedral solute molecules will be discussed in paragraph 4.3.5.

4.3.2 Crystal field model

The problem of hindered rotation of molecules in the solid state was first analysed by Pauling,[70] who assumed a model of a diatomic molecule rotating in a potential well possessing cylindrical symmetry. This gave the barrier to rotation of a plane rotor in the lattice as $V_0(1 - \cos 2\theta)$, where V_0 is the barrier to end-for-end rotation of the molecule and θ is the angle of rotation about an axis through the centre of mass perpendicular to the molecular axes. Schoen, Mann, Knobler, and White[36] attempted to analyse the vibration-rotation spectrum of hydrogen chloride in an argon matrix on a similar model, using a cylindrically symmetric potential of the form $C(1 - \cos^2 \theta)$.

Devonshire[71] investigated the Schrödinger equation for the rotation of a linear molecule in a potential field of octahedral symmetry, which is the appropriate symmetry for a substitutional site in a face-centred cubic lattice, and thus calculated energy levels as a function of the barrier to rotation (Figure 4.10). Flygare[72] investigated the nature of this barrier to the rotation of molecules in noble gas matrices, on the following assumptions:
(i) Solute-solute interactions may be ignored (Armstrong[73] showed that solute-solute dipolar interactions for concentrations <1 per cent. perturb the rotational levels only very weakly).

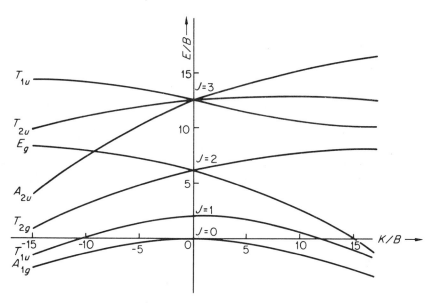

Figure 4.10 Rotational energy levels of a diatomic molecule in a cavity of octahedral symmetry as a function of the barrier to rotation (ref. 71)

(ii) The solute molecule occupies a substitutional site in the undistorted matrix lattice.

(iii) The centre of mass of the solute molecule is fixed at the lattice point.

(iv) At the low temperatures considered, perturbation due to lattice vibrations may be ignored.

(v) Only electrostatic interactions between the lattice atoms and the solute molecule need to be considered, exchange interactions being ignored (Armstrong[74] has examined the effect of repulsive interactions on the rotational energy levels of the solute molecule, and has found that for hydrogen chloride in noble gas matrices the effect is small compared with that of the long-range interactions).

As a consequence of the octahedral site symmetry of the noble gas face-centered cubic lattices, only terms involving the moments of order 2^L, where $L = 0, 4, 6, 8, 9, 10, \ldots$, of the solute molecule's charge distribution can have a non-zero effect on the rotational energy of the trapped molecule. The monopole ($L = 0$) term represents the orientation of the centre of mass of the solute molecule, and thus has no effect on the rotational energy as a consequence of assumption (iii). Hence the first term in the potential barrier to rotation is due to the solute hexadecapolar charge distribution interacting with the fourth gradient of the electric potential, due to the induced dipoles in the lattice atoms, at the solute molecule's centre of mass. A further contribution arises from the quadrupole-induced quadrupole interaction.[39] The total electrostatic interaction energy is given by

$$V = -K[P_4^0(\cos \theta) + \tfrac{1}{168}P_4^4(\cos \theta) \cos 4\phi] + K' \qquad (4.71)$$

where the coupling constant K depends on the hexadecapole moment and the square of the quadrupole moment of the solute molecule. Ignoring the constant K', equation (4.71) is the Devonshire[71] interaction of a diatomic molecule with an octahedral lattice. From the variation of energy levels with K/B (where B is the gas phase rotational constant), it can be seen that the effect of increasing the barrier is to reduce the spacings $R(1) - R(0)$ and $R(0) - P(1)$, and also to split $R(1)$ due to the lifting of the degeneracy of the $J = 2$ rotational level (Figure 4.10).

Determination of the barriers to rotation from the observed spectra of the hydrogen halides in noble gas matrices leads to rather high values for the electrical moments of the solute molecules (e.g. $3\cdot9 \times 10^{-42}$ e.s.u. cm^4 for the hexadecapole moment[72] of hydrogen chloride). The crystal field model also predicts that

$$(K/B)_{DX} \simeq 2(K/B)_{HX} \qquad (4.72)$$

whereas the values of K/B calculated from the experimental data for hydrogen chloride are about three times greater than those of deuterium chloride, and similarly the barrier for hydrogen bromide is significantly larger than that for deuterium bromide—Table 4.8. Thus this model is unsatisfactory, and it is not

Table 4.8 Barriers to rotation of hydrogen and deuterium halides in noble gas matrices at 20 K[39]

Solute	Matrix	K/B
HCl	Ar	15
	Kr	14
DCl	Ar	5
	Kr	4
HBr	Kr	26
DBr	Kr	18

significantly improved by assuming that rotation occurs about a centre of interaction rather than about the centre of mass. The rotation-translation coupling model, discussed in the next section, which neglects any effect due to the anisotropy of the crystal field but removes the constraint of rotation about a fixed point, gives a much closer agreement with the experimental results for the hydrogen and deuterium halides in noble gas matrices.

4.3.3 Rotation-translation coupling (RTC) model

The assumptions on which the RTC model is based have already been detailed in paragraph 4.2.3. The rotational shifts are interpreted by treating the coupling between the rotation of the solute molecule and its constrained translational motion (arising from the assumption that rotation takes place about the centre of interaction, which is not in general coincident with the centre of mass) as a perturbation. Neglecting, as a first approximation, the orientational forces arising from the anisotropy of the crystal field, so that the oscillational motion of the centre of interaction may be described by means of the harmonic oscillator cell model, the Hamiltonian governing the rotational and translational motions of the solute molecule takes the form[9,75,76]

$$H = H^0 + aH_{\mathrm{I}} + a^2 H_{\mathrm{II}} \tag{4.73}$$

$$= \frac{p^2}{2m} + \frac{P^2}{2I} + V(r) - \frac{a}{I}\sum_F (i\hbar p_F \Phi_{Fz} + p_F \Phi_{Fx}P_y - p_F \Phi_{Fy}P_x)$$

$$+ \frac{a^2}{2I}\left(p^2 - \sum_{F,F'} p_F p_{F'} \Phi_{Fz}\Phi_{F'z} \right) \tag{4.74}$$

where p is the linear momentum operator (independent of origin) and P is the angular momentum operator about the centre of interaction, a is the distance between the centre of mass and the centre of interaction, $F = X, Y,$ or Z represents a co-ordinate system fixed in space, and $f = x, y,$ or z represents a co-ordinate system fixed with respect to the molecule. The Φ_{Ff} are the direction cosines between the space-fixed co-ordinates F and the molecule-fixed co-ordinates f.

To order a^2, treating the RTC operator $(aH_{\mathrm{I}} + a^2 H_{\mathrm{II}})$ as a perturbation, and assuming that the potential energy $V(r)$ is harmonic when the translational motion is not highly excited, the change in the rotational energy levels is

$$\frac{\Delta E(J)}{hcB} = \frac{1}{2}b\xi\left[1 - \frac{2\xi(J^2 + J + 1) - 4J(J + 1)}{(\xi - 2J)(\xi + 2J + 2)} \right] \tag{4.75}$$

where $b = ma^2/I$, m being the mass, I the moment of inertia, and B the rotational constant of the solute molecule, $\xi = v/B$, where v is the frequency of oscillation of the solute molecule in its cell. The frequency v is given by (cf. equation (4.14))

$$v = \frac{1}{\pi d_0 c}\left(\frac{12\varepsilon}{m}\right)^{1/2}\left[44\left(\frac{\sigma}{d_0}\right)^{12} - 10\left(\frac{\sigma}{d_0}\right)^6 \right]^{1/2} \tag{4.76}$$

The rotational shift can be expressed by

$$\Delta_J = \pm 2bB\mathfrak{F}(\xi, J) \tag{4.77}$$

where

$$\mathfrak{F}(\xi, J) = \frac{(J + 1)\xi^4}{(\xi - 2J - 2)(\xi - 2J)(\xi + 2J + 2)(\xi + 2J + 4)} \tag{4.78}$$

the $+$ and $-$ signs in equation (4.77) referring to the lines $P(J + 1)$ and $R(J)$ respectively. The total shift of a rotation-vibration line from its gas phase frequency may be obtained by adding Δ_J to the vibrational shift Δ_v obtained in paragraph 4.2. In particular

$$\Delta[R(0) - P(1)] = -\frac{4bB\xi^4}{(\xi + 4)(\xi + 2)\xi(\xi - 2)} \tag{4.79}$$

$$\Delta[R(1) - R(0)] = -\frac{2bB\xi^4}{(\xi + 4)(\xi + 2)\xi(\xi - 2)} \left(\frac{\xi^2 + 2\xi + 24}{\xi^2 + 2\xi - 24}\right) \tag{4.80}$$

To first order in a, the states $J \pm 1$ and $n \pm 1$ (where n is the translational quantum number) are mixed.[76] Thus, in addition to the normal rotational transitions, there will be an RTC spectrum obeying the selection rules

$$\Delta n = \pm 1$$

$$\Delta J = 0, \pm 2$$

At very low temperatures, where only the rotational state $J = 0$ and the translational state $n = 0$ are occupied, only three transitions should be observed

(a) $R_Q(00) \equiv R(0)$ from $Jn = 00$ to 10
(b) $Q_R(00)$ from $Jn = 00$ to 01
(c) $S_R(00)$ from $Jn = 00$ to 21

Since the energy separation between the levels 01 and 10 is smaller than that between the levels 21 and 10, the intensity of $S_R(00)$ is very much less than the intensity of the first two transitions. To first order in the parameter b, the frequency difference $Q_R(00) - Q_Q(00)$ (where $Q_Q(00)$ is the band centre frequency) which equals $n(1 \leftarrow 0)$, the frequency of the RTC line in the far infrared, is given by

$$n(1 \leftarrow 0) = v\left[1 + \frac{b\xi^2}{3(\xi + 2)(\xi - 2)}\right] \tag{4.81}$$

and, to the same order in b, the intensity of the RTC line $Q_R(00)$ relative to the rotational line $R(0)$ is given by

$$\frac{\text{Int. } Q_R(00)}{\text{Int. } R(0)} = \frac{4b\xi^3}{3(\xi^2 - 4)^2} \frac{\omega_{Q_R(00)}}{\omega_{R(0)}} \tag{4.82}$$

At somewhat higher temperatures, further transitions arise from the state 10.

These are

(a) $P_Q(10) \equiv P(1)$ from $Jn = 10$ to 00
(b) $R_Q(10) \equiv R(1)$ from $Jn = 10$ to 20
(c) Weaker transitions arising from levels mixed with 10 and 20 by the RTC operator

Figure 4.11 illustrates the vibration–rotation spectral region of hydrogen chloride and deuterium chloride in argon matrices at several temperatures.

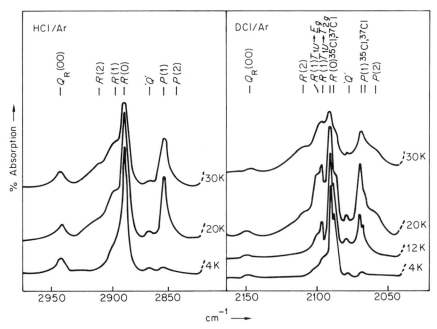

Figure 4.11 Vibration–rotation spectra of HCl and DCl in argon matrices at various temperatures (refs 31 and 66)

Equations (4.75) and (4.77)–(4.82) break down when $\xi \simeq 2J$ or $2(J + 1)$, i.e. under conditions of near-resonance when the separation between two rotational levels approaches that between two translational levels. Keyser and Robinson[38] overcame this problem by using a variational procedure, and solving the resulting secular equations numerically, with simplifying assumptions. Friedmann and Kimel[77] apply a generalized perturbation theory, taking advantage of the spherical symmetry of the cell, to obtain the energy levels in the case of near-resonance.

The parameter $\xi = v/B$ has been calculated[9] using the Lennard–Jones–Devonshire cell model, and it may also be obtained from the observed frequencies of the lines $Q_R(00)$, $R(0)$, and $P(1)$, i.e. from the separation

$Q_R(00) - Q_Q(00)$. The distance a (not strictly a molecular parameter since the position of the centre of interaction depends on the experimental conditions, e.g. the environment) can, in principle, be obtained from an independent quantum mechanical calculation, but is in fact only available for hydrogen chloride in argon.[78] This method calculates the interaction between an hydrogen chloride molecule and one argon atom, and thus probably gives too high a value due to neglect of the other atoms surrounding the hydrogen chloride molecule.[79] Again the parameter a may be obtained empirically from the frequencies of the observed lines. It is thus possible, from the observed positions of $R(0)$, $P(1)$ and $Q_R(00)$ for HX in a noble gas matrix,* to obtain values of the parameters ξ and a. Consequently predicted values for the spacing $R(1) - R(0)$ for HX may be obtained, and for the positions of the lines in the DX matrix spectrum, using the relations

$$\xi_{DX} = \xi_{HX}(B_{HX}/B_{DX})(m_{HX}/m_{DX})^{1/2} \tag{4.83}$$

$$a_{DX} = a_{HX} - r_0[(m_D/m_{DX}) - (m_H/m_{HX})] \tag{4.84}$$

where r_0 is the inter-nuclear distance (a is changed only as a result of the shift in the centre of mass from HX to DX).

It can be seen from Table 4.9 that the agreement between the observed and calculated parameters is satisfactory, and also that the variation in a is sufficiently small that it may be considered to be a molecular parameter.

Table 4.9 RTC parameters for HCl in noble gas matrices[77]

Matrix	ξ (exp.)	ξ (calc.)	b (exp.)	a (exp.) (Å)	a (calc.)[78] (Å)
Ne	8·25	9·10	0·21	0·098	—
Ar	6·47	6·49	0·19	0·093	0·109
Kr	5·50	5·42	0·20	0·095	—
Xe	4·08	3·81	0·18	0·090	—

$\xi_{DCl} = 1·910\, \xi_{HCl}$
$a_{DCl} = a_{HCl} - 0·034$.

If the perturbation due to orientational forces is small, it is possible to combine rotation-translation coupling with the crystal field theory.[9] To a first approximation, the two effects are simply additive:

$$\Delta_J = 2bB\mathfrak{F}(\xi, J) + \Delta_J^{cr.f.}(K/B) \tag{4.85}$$

A third factor affecting the rotational energy levels is the effect of lattice

---

* A trivial correction has to be applied for the effect of finite line width, which leads to a small shift of $P(1)$ and $Q_P(01)$ to higher frequency.[77] This effect is particularly important in liquid solution, where the $Q_P(01)$ line of hydrogen in liquid neon or argon[80] is shifted by about 70 cm^{-1}.

vibrations on the crystal field interaction. Pandey[79] expands the instantaneous interaction between the molecule and a noble gas atom in powers of the displacement of the lattice atom from its equilibrium position, and then averages over all neighbouring atoms. By treating the resulting potential energy of the solute molecule as a perturbation, he calculates the displacement of the rotational energy levels. The resulting shift is comparatively small, the change in $R(0) - P(1)$ for hydrogen chloride, deuterium chloride or hydrogen bromide in noble gas matrices being approximately 0.7 cm^{-1}.

4.3.4 Comparison with experimental results for diatomic solute molecules

Hydrogen chloride (HCl) *and deuterium chloride* (DCl)

In Table 4.10 the observed and calculated values for the various rotational spacings of hydrogen chloride and deuterium chloride are compared, and it can be seen that there is reasonable agreement. The expected relative intensities of the several rotational lines (calculated from the Boltzmann

Table 4.10 Comparison of calculated and observed rotational spacings (in cm^{-1}) of H^{35}Cl and D^{35}Cl in noble gas matrices

Solute	Rotational spacing		Gas	Ne	Ar	Kr	Xe
HCl	$R(0) - P(1)$	exp.	41·16	28	34·4	34·8	35·6
		calc.[a]	—	34·8	34·8	35·0	34·9
	$R(1) - R(0)$	exp.	19·64	11	~10	12	13·5
		calc.[b]	—	16	16	15	13
	$R(2) - R(0)$	exp.	38·66	—	25	—	—
	$P(2) - P(1)$	exp.	21·47	—	~10	—	—
	$J(1 \leftarrow 0)$	exp.	20·88	—	18·6	—	—
		calc.[a]	—	17·4	17·4	17·5	17·5
	$Q_R(00) - Q_Q(00)$	exp.	—	—	74	62	45
	$n(1 \leftarrow 0)$	exp.	—	93	73	62	49
		calc.[a]	—	102	73·3	61·6	43·9
DCl	$R(0) - P(1)$	exp.	21·34	19·6	20·2	21·2	21
		calc.[b]	—	20·5	20·6	20·6	20·6
	$R(1) - R(0)$	exp.	10·33	—	7·7	8	7
		calc.[b]	—	9·3	9·3	9·3	8·9
	$R(2) - R(0)$	exp.	20·43	—	19	—	—
	$P(2) - P(1)$	exp.	11·01	—	9	—	—
	$J(1 \leftarrow 0)$	exp.	10·78	—	10·9	—	—
		calc.[b]	—	10·3	10·3	10·3	10·3
	$Q_R(00) - Q_Q(00)$	exp.	—	—	70	58·6	—
	$n(1 \leftarrow 0)$	exp.	—	80	72	58	—
		calc.[b]	—	81	68	58	40

a. From calculated values of ξ (ref. 9 and 76).
b. From fitted values of ξ for $R(0)$, $P(1)$, and $Q_R(00)$ of HCl (ref. 77). Experimental values are derived from ref. 31, 36–43, 58, and 66.

distribution) and of the RTC line $Q_R(00)$ (calculated from equation (4.82)) are listed in Table 4.11 for hydrogen chloride and deuterium chloride in argon matrices at various temperatures, and comparison with Figure 4.11 shows that the absorptions are correctly assigned.

Table 4.11 Calculated relative intensities of vibration-rotation lines of HCl and DCl in argon matrices

Solute	Transition	4 K	10 K	20 K	30 K	40 K
HCl	$Q_R(00)^a$	0·056	0·056	0·056	0·056	0·056
	$R(2)$	0	0·005	0·134	0·363	0·617
	$R(1)$	0·006	0·171	0·599	0·883	1·084
	$R(0)$	1	1	1	1	1
	$P(1)$	0·006	0·168	0·590	0·870	1·067
	$P(2)$	0	0·005	0·131	0·355	0·604
DCl	$Q_R(00)^a$	0·005	0·005	0·005	0·005	0·005
	$R(2)$	0	0·053	0·418	0·787	1·102
	$R(1)$	0·061	0·464	0·980	1·233	1·393
	$R(0)$	1	1	1	1	1
	$P(1)$	0·061	0·458	0·966	1·216	1·374
	$P(2)$	0	0·052	0·408	0·769	1·077

a. Ref. 76.

Qualitatively, the RTC model does predict correctly that the rotational spacings for the deuterated species should decrease less than those for the hydrogen halides, and is thus clearly more realistic than the crystal field model in explaining the main features of the spectra. However, the $R(1)$ line of deuterium chloride in noble gas matrices has been observed to be split by about 3 cm^{-1}, which is not predicted by the RTC model. Such a splitting is to be expected from the lifting of the degeneracy of the $J = 2$ level in an octahedrally symmetric field,[71] and the separation of the T_{2g} and E_g sublevels can be accounted for by a barrier parameter $K \simeq -2B_{DCl} \simeq -10\,\text{cm}^{-1}$ (the sign of the coupling constant is expected to be negative[37]). Since the orientational forces are due, to a first approximation, to the interactions of the solute quadrupole with the induced quadrupoles in the lattice atoms,[39] and of the solute hexadecapole with the induced dipoles in the lattice atoms,[72] a value of 0.5×10^{-42} e.s.u. cm^4 is derivable for the hexadecapole moment of the hydrogen chloride molecule.[77,79] A barrier parameter of this order results in a decrease in the $R(0) - P(1)$ spacing for hydrogen chloride or deuterium chloride in noble gas matrices of about 0·7 cm^{-1},[79] which should strictly be allowed for in the calculation of the rotational spacings using the RTC model.

The sharpness of $R(1)$ of deuterium chloride, compared with the diffuse absorption observed for hydrogen chloride, may be explained by the fact that $J(2 \leftarrow 1)$ of hydrogen chloride falls within the region of the first maximum in the phonon states density of the noble gas matrix, whereas the corresponding transition of deuterium chloride is at lower frequency. A more elaborate theory is necessary to describe the interaction of the solute impurity molecules with the lattice vibrations of the matrix crystal.

Hydrogen bromide (HBr) *and deuterium bromide* (DBr)

Hydrogen and deuterium bromide give spectra similar to those of hydrogen chloride and deuterium chloride, but the experimental data are sparse and complicated by splittings of the various spectral features, so that comparison with calculated rotational spacings is of little value. From the splitting of $R(1)$, a value of 0.2×10^{-42} e.s.u. cm^4 has been derived for the hexadecapole moment of hydrogen bromide.[79]

Hydrogen fluoride (HF) *and deuterium fluoride* (DF)

Spectra of hydrogen fluoride and deuterium fluoride in noble gas matrices in both the vibration–rotation and the far infrared pure rotation regions have been reported in which the spacing of the $J = 0$ and $J = 1$ rotational levels *increases* (Table 4.12). This phenomenon cannot be explained on the

Table 4.12 Observed rotational spacings (in cm^{-1}) of HF and DF in noble gas matrices[57]

Solute	Spacing	Gas	Ne	Ar	Kr	Xe
HF	$J(1 \leftarrow 0)$	41.9	39.8	43.4	45.4	50.5
	$[R(0) - P(1)]/2$	40.3	—	43.5	44.5	50.5
DF	$J(1 \leftarrow 0)$	22.0	21.6	19.8	24.6	—
	$[R(0) - P(1)]/2$	21.5	—	18.5	15.0	—

crystal field model, but is qualitatively accounted for by the RTC theory.[35,81] From equation (4.79) it can be seen that the $R(0) - P(1)$ spacing (or the $J(1 \leftarrow 0)$ frequency in the far IR) increases when $\xi < 2$, and such a low value for ξ is to be expected for hydrogen fluoride because of the large value of the rotational constant and the small value of the Lennard–Jones parameter σ. The RTC transition $n(1 \leftarrow 0)$ will have a frequency less than $2B_{\mathrm{HF}}$ (41 cm^{-1}), from equation (4.81). Deuterium fluoride, on the other hand, would be more likely to display a decrease in the spacing of its rotational levels since $\xi_{\mathrm{DF}} = 1.847\xi_{\mathrm{HF}}$. If $\xi_{\mathrm{DF}} > 2$, then the rotational transition $J(1 \leftarrow 0)$ would be below $2B_{\mathrm{DF}}$ (22 cm^{-1}) while the RTC transition $n(1 \leftarrow 0)$ would be above 22 cm^{-1}. As can be seen from Table 4.12, an increase in rotational spacing for hydrogen fluoride and a decrease for deuterium fluoride is in fact observed in neon and argon matrices. In a krypton matrix there is some disagreement in the rotational spacings derived from the vibration–rotation regions and from the pure rotational transitions. In the spectrum of hydrogen fluoride in argon,[35] a line was observed 34 cm^{-1} above the band centre which could be the RTC transition $Q_R(00)$; this would give $\xi \simeq 1.7$. Since the frequencies of these transitions for hydrogen fluoride lie within the phonon bands of the noble gas matrices, a quantitative explanation of the data is not possible without a more elaborate theory. A recent theoretical treatment,[57] assuming *uncoupled* rotation-translation behaviour, gives good agreement with the observed spectra.

4.3.5 Tetrahedral solute molecules

King and Hornig[82] carried out calculations for a tetrahedral molecule in a cell of octahedral symmetry based on the anisotropy of the crystal field interacting with the molecule, rotating with its centre of mass fixed at a lattice point. As a consequence of the site symmetry, the angular dependent interaction involves only moments of order 2^4, 2^6, etc. The rotational fine structure of the v_3 band of CH_4 and CD_4 could be interpreted[53] in a semi-quantitative manner on this basis, retaining only the octopole term, with a rotational barrier of the order of 50 cm^{-1}. However, the fine structure of the v_4 band could only be interpreted with a dominant hexadecapole term[56] and a rotational barrier of the order of 400 cm^{-1}. The effect of rotation-translation coupling (RTC) on the energy levels of a tetrahedral molecule in an octahedral cell has been considered by Friedmann, Shalom, and Kimel.[83]

The Hamiltonian governing the rotational and translational motion of a tetrahedral solute molecule in a rigid matrix may be written as

$$H = \frac{p^2}{2m} + \frac{P^2}{2I} + V(r) + V(\omega) + V_{RTC}(r, \omega) \qquad (4.86)$$

where p is the linear momentum operator, P is the angular momentum operator about the centre of mass, m is the mass and I the moment of inertia of the solute molecule. $r \equiv r, \theta, \phi$ denotes the radius vector of the centre of mass of the solute molecule relative to the centre of the cell, and $\omega \equiv \alpha, \beta, \gamma$ denotes the Eulerian angles describing the orientation of the molecule.

In the harmonic oscillator cell model the term $V(r)$ in equation (4.86) reduces to

$$V(r) = V(0) + 2\pi^2 c^2 v^2 m r^2 \qquad (4.87)$$

where v is the frequency of the translational motion of the molecule in the cell.

The symmetry of an octahedral cell restricts the values which the total angular momentum can take to 0, 4, 6, 8, 9, etc. Consequently equation (4.86) may be expressed as

$$H = H_0 + V_{RTC}^{(0)} + V^{(4)}(\omega) + V_{RTC}^{(4)} \qquad (4.88)$$

where

$$H_0 = \frac{p^2}{2m} + \frac{P^2}{2I} + 2\pi^2 c^2 v^2 mr^2 + V(0) \qquad (4.89)$$

In a spherical cell model only $H_0 + V_{\text{RTC}}^{(0)}$ are retained in equation (4.88), non-sphericity giving rise to the terms $V^{(4)}(\omega) + V_{\text{RTC}}^{(4)}$ whose relative importance depends on the molecular hexadecapolar and octopolar (respectively) charge distributions. Provided that the last three terms in equation (4.88) are small, compared with H_0, perturbation theory may be applied. The first order corrections vanish and, to second order, the perturbations due to $V_{\text{RTC}}^{(0)}$, $V^{(4)}(\omega)$ and $V_{\text{RTC}}^{(4)}$ are additive.

The rotational shifts calculated[83] in this manner for the v_3 band of methane in noble gas matrices, neglecting $V^{(4)}(\omega)$ and fitting the overall calculated shift to the experimental value, are given in Table 4.13. $V_{\text{RTC}}^{(0)}$ splits the $J = 2$ level into two sub-levels, according to the representations $\bar{F}$ and $\bar{E}$ of the molecular group $\bar{T}$. As a result of the action of $V_{\text{RTC}}^{(4)}$ further splitting into the sub-levels $\bar{F}\bar{F}$, $\bar{F}\bar{E}$, $\bar{E}\bar{F}$ and $\bar{E}\bar{E}$ occurs, which is ignored in Table 4.13. With more detailed data for deuteromethanes, the relative importance of the $V^{(4)}(\omega)$ term could be estimated.

Table 4.13 Rotational shifts (in cm^{-1}), relative to $Q(1)$, in the v_3 band of methane in noble gas matrices[83]

		Ar	Kr	Xe
$\Delta P(1)$	exp.	-3.1	-1.9	-0.4
	calc. $V_{\text{RTC}}^{(0)}$	-1.8	-1.2	-0.4
	calc. $V_{\text{RTC}}^{(4)}$	-1.3	-0.7	0
$\Delta R(1)_{\bar{F}}$	exp.	-8.3	-8.5	-7.2
	calc. $V_{\text{RTC}}^{(0)}$	-37.9	-21.0	-7.2
	calc. $V_{\text{RTC}}^{(4)}$	$+29.8$	$+12.7$	0
$\Delta R(1)_{\bar{E}}$	exp.	—	—	—
	calc. $V_{\text{RTC}}^{(0)}$	$+125$	$+69.0$	$+24.2$
	calc. $V_{\text{RTC}}^{(4)}$	$+41$	$+17.7$	0
v		116	98	105

4.4 References

1. H. E. Hallam, p. 245 *in* M. J. Wells (ed.), *Spectroscopy, Proc. 3rd Inst. Petroleum Hydrocarbon Research Group Conference*, Inst. of Petroleum, London, 1962.
2. L. J. Bellamy, H. E. Hallam and R. L. Williams, *Trans. Faraday Soc.*, **54**, 1120, (1958); ibid., **55**, 14, (1959); ibid., **55**, 220, (1959).
3. J. G. Kirkwood, *J. Chem. Phys.*, **2**, 351, (1934).
4. W. West and R. T. Edwards, *J. Chem. Phys.*, **5**, 14, (1937); E. Bauer and M. Magat, *J. Phys. Rad.*, **9**, 319, (1938), and *Physica*, **5**, 718, (1938).

5. A. D. Buckingham, *Proc. Roy. Soc.*, **A248**, 169, (1958); ibid., **A255**, 32 (1960); *Trans. Faraday Soc.*, **56**, 753, (1960).
6. J. G. David and H. E. Hallam, *Spectrochim. Acta*, **23A**, 593, (1967).
7. G. L. Caldow and H. W. Thompson, *Proc. Roy. Soc.*, **A254**, 1, (1960).
8. A. van der Avoird and F. Hofelich, *J. Chem. Phys.*, **45**, 4664, (1966).
9. H. Friedmann and S. Kimel, *J. Chem. Phys.*, **43**, 3925, (1965).
10. J. E. Lennard-Jones, *Proc. Roy. Soc.*, **A106**, 463, (1924).
11. I. Prigogine, *The Molecular Theory of Solutions*, North-Holland Publishing Co., Amsterdam, 1957.
12. S. D. Hamann, *Trans. Faraday Soc.*, **48**, 303, (1952).
13. J. E. Lennard-Jones and A. E. Ingham, *Proc. Roy. Soc.*, **A107**, 636, (1925).
14. R. P. Bell, *Trans. Faraday Soc.*, **38**, 422, (1942).
15. A. D. Buckingham, *Quart. Rev.*, **13**, 183, (1959).
16. D. E. Stogryn and A. P. Stogryn, *Mol. Phys.*, **11**, 371, (1966).
17. J. O. Hirschfelder, C. F. Curtiss and R. B. Bird, *Molecular Theory of Gases and Liquids*, Wiley, New York, 1954.
18. M. W. Melhuish and R. L. Scott, *J. Phys. Chem.*, **68**, 2301, (1964).
19. J. O. Clayton and W. F. Giauque, *J. Amer. Chem. Soc.*, **54**, 2610, (1932); R. W. Blue and W. F. Giauque, ibid., **57**, 991, (1935).
20. W. C. Hamilton and M. Petrie, *J. Phys. Chem.*, **65**, 1453, (1961).
21. M. J. Linevsky, *J. Chem. Phys.*, **34**, 587, (1961).
22. H. F. Shurvell, Ph.D. Thesis, University of British Columbia, 1964.
23. A. D. Buckingham, *Disc. Faraday Soc.*, **40**, 232, (1965); *Adv. Chem. Phys.*, **12**, 107, (1967).
24. D. C. McKean, *Spectrochim. Acta*, **23A**, 2405, (1967).
25. C. J. F. Bottcher, *The Theory of Electric Polarization*, Elsevier, 1952.
26. F. London, *Z. Physik. Chem.*, **B11**, 222, (1930); *J. Phys. Chem.*, **46**, 305, (1942).
27. A. D. Buckingham, *J. Chem. Phys.*, **48**, 3827, (1968).
28. S. W. Charles and K. O. Lee, *Trans. Faraday Soc.*, **61**, 2081, (1965).
29. T. E. Whyte, jr., Ph.D. Thesis, Howard University, 1965.
30. R. K. Nesbet, *J. Chem. Phys.*, **41**, 100, (1964).
31. A. J. Barnes, H. E. Hallam and G. F. Scrimshaw, *Trans. Faraday Soc.*, **65**, 3159, (1969).
32. G. C. Pimentel and S. W. Charles, *Pure Appl. Chem.*, **7**, 111, (1963).
33. K. B. Harvey and J. F. Ogilvie, *Can. J. Chem.*, **40**, 85, (1962).
34. J. J. Smith and B. Meyer, *J. Chem. Phys.*, **50**, 456, (1969).
35. M. T. Bowers, G. I. Kerley and W. H. Flygare, *J. Chem. Phys.*, **45**, 3399, (1966).
36. L. J. Schoen, D. E. Mann, C. Knobler and D. White, *J. Chem. Phys.*, **37**, 1146, (1962).
37. M. T. Bowers and W. H. Flygare, *J. Chem. Phys.*, **44**, 1389, (1966).
38. L. F. Keyser and G. W. Robinson, *J. Chem. Phys.*, **44**, 3225, (1966).
39. D. E. Mann, N. Acquista and D. White, *J. Chem. Phys.*, **44**, 3453, (1966).
40. H. F. Shurvell and K. B. Harvey, *Can. Spectroscopy*, **14**, 32, (1969).
41. J. C. Kwok, Ph.D. Thesis, California Institute of Technology, 1964.
42. J. M. P. J. Verstegen, H. Goldring, S. Kimel and B. Katz, *J. Chem. Phys.*, **44**, 3216, (1966).
43. L. C. Brunel and M. Peyron, *Compt. Rend.*, **262C**, 1297, (1966).
44. E. Catalano and D. E. Milligan, *J. Chem. Phys.*, **30**, 45, (1959).
45. J. A. Glasel, *J. Chem. Phys.*, **33**, 252, (1960).
46. T. Miyazawa, *Bull. Chem. Soc. Japan*, **34**, 202, (1961).
47. R. L. Redington and D. E. Milligan, *J. Chem. Phys.*, **37**, 2162, (1962).
48. R. L. Redington and D. E. Milligan, *J. Chem. Phys.*, **39**, 1276, (1963).

49. H. P. Hopkins, jr., R. F. Curl, jr. and K. S. Pitzer, *J. Chem. Phys.*, **48**, 2959, (1968).
50. D. E. Milligan, R. M. Hexter and K. Dressler, *J. Chem. Phys.*, **34**, 1009, (1961).
51. J. F. Hanlan, Ph.D. Thesis, University of California, 1961.
52. R. E. Meredith, Ph.D. Thesis, University of Michigan, 1963.
53. A. Cabana, G. B. Savitsky and D. F. Hornig, *J. Chem. Phys.*, **39**, 2942, (1963).
54. A. Cabana, A. Anderson and R. Savoie, *J. Chem. Phys.*, **42**, 1122, (1965).
55. F. H. Frayer and G. E. Ewing, *J. Chem. Phys.*, **46**, 1994, (1967).
56. F. H. Frayer and G. E. Ewing, *J. Chem. Phys.*, **48**, 781, (1968).
57. D. W. Robinson and W. G. von Holle, *J. Chem. Phys.*, **44**, 410, (1966); M. G. Mason, W. G. von Holle and D. W. Robinson, *J. Chem. Phys.*, **54**, 3491, (1971).
58. B. Katz, A. Ron and O. Schnepp, *J. Chem. Phys.*, **46**, 1926, (1967); A. J. Barnes, J. B. Davies, H. E. Hallam, G. F. Scrimshaw, G. C. Hayward and R. C. Milward, *Chem. Comm.*, 1089, (1969); B. Katz and A. Ron, *Chem. Phys. Letters*, **7**, 357, (1970).
59. D. W. Robinson, *J. Chem. Phys.*, **39**, 3430, (1963).
60. D. E. Milligan and M. E. Jacox, *J. Chem. Phys.*, **43**, 4487, (1965).
61. R. Ranganath, T. E. Whyte, jr., T. Theophanides and G. C. Turrell, *Spectrochim. Acta*, **23A**, 807, (1967).
62. K. B. Harvey and H. F. Shurvell, *Chem. Comm.*, 490, (1967); *Can. J. Chem.*, **45**, 2689, (1967).
63. K. B. Harvey and H. F. Shurvell, *J. Mol. Spectroscopy*, **25**, 120, (1968).
64. A. J. Tursi and E. R. Nixon, *J. Chem. Phys.*, **52**, 1521, (1970).
65. G. C. Pimentel, M. O. Bulanin and M. van Thiel, *J. Chem. Phys.*, **36**, 500, (1962).
66. J. B. Davies and H. E. Hallam, *J.C.S. Faraday II*, **67**, 3176, (1971).
67. G. E. Leroi, G. E. Ewing and G. C. Pimentel, *J. Chem. Phys.*, **40**, 2298, (1964).
68. S. W. Charles and K. O. Lee, *Trans. Faraday Soc.*, **61**, 614, (1965).
69. M. Allavena, R. Rysnik, D. White, V. Calder and D. E. Mann, *J. Chem. Phys.*, **50**, 3399, (1969).
70. L. Pauling, *Phys. Rev.*, **36**, 430, (1930).
71. A. F. Devonshire, *Proc. Roy. Soc.*, **A153**, 601, (1936).
72. W. H. Flygare, *J. Chem. Phys.*, **39**, 2263, (1963).
73. R. L. Armstrong, *J. Chem. Phys.*, **36**, 2429, (1962).
74. R. L. Armstrong, *J. Chem. Phys.*, **44**, 530, (1966).
75. H. Friedmann and S. Kimel, *J. Chem. Phys.*, **41**, 2552, (1964).
76. H. Friedmann and S. Kimel, *J. Chem. Phys.*, **44**, 4359, (1966).
77. H. Friedmann and S. Kimel, *J. Chem. Phys.*, **47**, 3589, (1967).
78. R. M. Herman, *J. Chem. Phys.*, **44**, 1346, (1966).
79. G. K. Pandey, *J. Chem. Phys.*, **49**, 1555, (1968).
80. G. E. Ewing and S. Trajmar, *J. Chem. Phys.*, **41**, 814, (1964); ibid., **42**, 4038, (1965).
81. G. K. Pandey and S. Chandra, *J. Chem. Phys.*, **45**, 4369, (1966).
82. H. F. King and D. F. Hornig, *J. Chem. Phys.*, **44**, 4520, (1966).
83. H. Friedmann, A. Shalom and S. Kimel, *J. Chem. Phys.*, **50**, 2496, (1969).

5 Infrared spectra of free radicals and chemical intermediates trapped in low-temperature matrices

LESTER ANDREWS

Contents

5.1 Introduction

Free radicals have played a significant role in the development of Chemistry during the twentieth century. Since Gomberg's preparation in 1900 of the first free radical, triphenylmethyl,[1] and along with it a challenge to the understanding of valence, free radicals have been used increasingly to explain reaction mechanisms. As early as 1918, Nernst suggested[2] that the photochemical reaction of hydrogen and chlorine involved the individual atoms as intermediates. In the famous metal mirror experiments, Paneth postulated[3] in 1929 that alkyl free radicals, produced by thermal decomposition of metal alkyls, were responsible for removing a metal film. Following this early work, the appearance of free radical intermediates in the literature became itself a chain reaction. Free radicals were subsequently proposed as reaction intermediates by numerous prominent scientists. Steacie has reviewed much of this work[4] in his book *Atomic and Free Radical Reactions*.

Free radicals are most simply defined as species containing one or two unpaired electrons. Examples include methyl, CH_3, the chlorine atom, Cl, and the lithium atom, Li, as well as the stable molecules nitric oxide, NO, and nitrogen dioxide, NO_2, each of which contains a single unpaired electron. Diradicals, such as triplet CH_2, contain two unpaired electrons. In this discussion we will use the term radical to refer to species containing a single unpaired electron. Other reactive intermediates such as carbenes, believed to have singlet ground states, and nitrenes with triplet ground states are also of interest here.

Recently Pryor[5] has reviewed the mechanistic, synthetic, and industrial significance of free radicals in contemporary organic chemistry. Obviously, the validity of a proposed mechanism is improved by the direct observation of the free radical from a related chemical process. Furthermore, free radicals provide unique tests for theories of chemical bonding. The presence of an unpaired electron in a half-filled orbital supplies an excellent opportunity for electronic interaction with neighbouring atoms and groups. Knowledge of the vibrational spectrum of simple free radicals provides insight into the structure and bonding of these interesting chemical species.

Since most free radicals by nature are highly reactive and thus chemically unstable, direct observation cannot be made using conventional infrared techniques. In order to observe the gas phase infrared spectrum of free radicals with lifetimes typically near one millisecond, a spectral scan would have to be produced in a fraction of a millisecond. With the recent development of doped semiconductor detectors, spectrophotometers with this capability have been developed. Herr and Pimentel[6] have described such an instrument capable of scanning $1000\ cm^{-1}$ in 0·1 ms. A second approach to the problem is to immobilize the free radical of interest in an inert medium or matrix, thus minimizing radical reaction and preserving the radical for conventional infrared analysis. The matrix isolation technique was first applied to free radicals by Lewis and Lipkin.[7] These workers obtained visible spectra of large organic free radicals produced by photolysis and suspended in a glassy matrix at low temperature. The matrix isolation technique as we know it today was proposed independently by Norman and Porter[8] and Whittle, Dows and Pimentel[9] in 1954. Matrix isolation using the rigid inert matrix materials nitrogen and argon was subsequently developed in the laboratory of Pimentel at Berkeley. This technique has been successfully used by numerous subsequent workers for the preparation and stabilization of free radicals and chemical intermediates in sufficient concentration for direct spectroscopic observation.

5.2 Matrix isolation technique

Any species trapped in an inert matrix will sustain perturbation by the matrix. These 'matrix shifts' are usually of the order of several wave numbers for nonpolar species and are negligible for most purposes. Theories of matrix shifts have been discussed earlier in this monograph (Chapter 4) and elsewhere.[10] Frequently doubling of infrared absorptions are observed depending upon the relative sizes of the guest molecule and host cavity. Such splittings are attributed to different orientations of a guest species in a host cage and can be identified by the use of two different matrix materials such as krypton and argon. Further splittings of a spectral line can be caused by guest molecules in nearest or next-nearest neighbour sites in the host matrix. These splittings are obviously concentration dependent, and can be removed by decreasing the concentration of the guest species.

Rotational structure is quenched by the matrix for all but the simplest hydride molecules (see Chapter 3). This has the obvious advantage of sharpening spectral lines so that nearby fundamentals or isotopic splittings can be resolved with the accompanying disadvantages of loss of structural data from rotational line spacings and symmetry designations from vibration-rotation band envelopes. Even with the loss of band contour, structural information may be found in the observed spectrum. Gas phase selection rules are expected to be obeyed in the matrix. The infrared observation of a symmetric fundamental vibration indicates a molecular species of lower symmetry. However, the absence of such a frequency could be due to selection rules or to the very low intensity of the symmetric vibrational mode. The low-temperature matrix provides for the thermal occupation of only ground vibrational states. For molecules having low frequency vibrational fundamentals, the observed spectrum may thus be simplified by the removal of complications due to 'hot bands'.

With the above-mentioned limitations, the matrix isolation technique has been uniquely applied to spectroscopic studies of numerous chemical systems where its advantages have been thoroughly exploited. Since matrix absorptions are sharp, matrix isolation has been used to study hydrogen stretching modes of hydrogen-bonded species. Large band widths characteristic of solution spectra prevent identification of frequencies due to hydrogen-bonded aggregates of various sizes. Characteristic O—H stretching frequencies of methanol dimers, trimers and tetramers have been observed in nitrogen matrices.[11] In recent work, hydrogen-bonding between hydrazoic acid dimers has been interpreted to indicate that the dimer has an open rather than a cyclic structure;[12] these aspects have been fully discussed in Chapter 3.

Matrix isolation provides a means of studying photolysis. The photolytic process can be terminated without allowing secondary reactions in some chemical systems, thus providing direct information on the primary act of photolysis. The photolysis of nitromethane in solid argon at 20 K yields methyl nitrite. Further photolysis of methyl nitrite was shown by Brown and Pimentel[13] to produce formaldehyde and a new chemical intermediate, HNO. Photolysis of matrix isolated molecules can also lead to new molecular species by secondary reaction of the photolytic products. The work of Moore, Pimentel and Goldfarb[14] indicates that diazomethane photolyses to give methylene which reacts with the diazomethane precursor to produce methyleneimine and hydrogen cyanide; the vibrational frequencies of CH_2 itself, however, have yet to be identified in any matrix study.

The matrix technique can be applied to the study of rate processes since it provides a means of bringing reactive species together at very low temperature as well as isolating them. Pimentel has described the usefulness of the matrix technique in reaction kinetics.[15] A particularly interesting study is the infrared photochemical induced isomerization of nitrous acid. This first reported example of an infrared photochemical reaction[16] produced a measure of the potential barrier to isomerization.

Perhaps the most significant application of the matrix isolation technique has been for the production, stabilization, and spectroscopic observation of free radicals. Free radicals are readily stabilized by an inert matrix which minimizes recombination or further reaction. Even though inert gas matrix samples scatter light due to their polycrystalline nature, these samples are reasonably transparent throughout the entire infrared region. Numerous physical and chemical means have been developed for the production of free radicals for spectroscopic study using the matrix isolation technique. Some of these methods will now be described in more detail.

5.3 Techniques for producing radicals

To date photolysis has been the most prolific method for producing free radicals for matrix study. The medium-pressure mercury arc has been widely used as an ultra-violet source for the photolysis of such molecules as HI,[17] HN_3[18] and H_2CN_2.[14] The later development of microwave-powered discharge lamps[19] using hydrogen, xenon or inert gas mixtures has provided a source of vacuum ultra-violet radiation. Numerous molecules have been photolysed with this source including NH_3,[20] CH_4[21] and CH_3F.[22]

The photolysis of a suitable precursor isolated in a matrix at low temperature is greatly dependent upon the cage effect. The *in situ* photolysis of methyl iodide likely produced some methyl radical and iodine atom pairs in the same matrix site which subsequently recombined to form methyl iodide. However, if the photodecomposition produces a nonreactive fragment which remains trapped in an adjacent matrix site, the free radical of interest can be stabilized. In the photolysis of HN_3 and the XN_3 molecules by Milligan and Jacox,[23,24] the triplet nitrenes HN and XN are stabilized since back reaction with N_2 does not occur. The photolysis of CF_2N_2 likewise produces good yields of difluorocarbene CF_2 which does not react with the N_2 product.[25] This contrasts with the photolysis of diazomethane where the photoproducts CH_2 and N_2 do recombine as has been shown by Moore and Pimentel[26] using $CH_2^{15}N_2$ in a $^{14}N_2$ matrix.

The matrix cage can be escaped if one of the photodecomposition products is light and small enough to diffuse away from the photolysis site. The molecules HCN, FCN, ClCN and BrCN all photolyse in a matrix environment; however, the experiments of Milligan and Jacox[27] show that the CN radical concentration produced decreases rapidly in the series H > F > Cl > Br. Even though only a small amount of CN is produced upon photolysis of ClCN and BrCN, photodissociation of these precursors is shown by their rearrangement to ClNC and BrNC. These results suggest that the larger, heavier atoms not unexpectedly diffuse through the matrix much less readily than smaller, lighter atoms.

In early work, the photolysis of HI and HBr provided a source of hydrogen atoms which subsequently diffuse through argon or nitrogen matrices to react with carbon monoxide[17] or oxygen.[28] Similarly the *in situ* photolysis of molecular fluorine produces fluorine atoms which also diffuse through the matrix and react with such molecules as carbon monoxide[29] and oxygen.[30] In recent work Milligan and Jacox[31] have developed a photolytic carbon atom source, N_3CN, which first photolyses to NCN and then to give carbon atoms. The photolytically produced carbon atoms find such reaction partners as chlorine,[32] fluorine,[33] hydrogen chloride[34] and carbon monoxide[35] also present in the matrix sample. Later, vacuum–ultra-violet photolysis provided the capability for photodissociation of stronger X—H bonds leaving as products new free radicals and hydrogen atoms which diffuse away through the matrix. Milligan and Jacox have photolysed NH_3, CH_4 and C_2H_2 in this manner producing observable quantities of the NH_2, CH_3 and C_2H free radicals.[20,21,36] Thus, photodecomposition of suitable precursors trapped in matrices has shown a great deal of versatility and productivity.

Pimentel has outlined[37] some of the early attempts to produce free radicals by electron impact and gamma radiolysis. These techniques have the common disadvantage of high input energy and the resulting low selectivity in free radical production. Gamma radiolysis has been used primarily by electron spin resonance spectroscopists. Recently, Current[38] has attempted to revive the electron impact technique for free radical production for infrared matrix study. Again, excessive fragmentation and the possibility of negative ion formation has complicated this work.

Early attempts to stabilize the methyl radical by gas-phase pyrolysis of dimethyl mercury in argon, azomethane in argon, and di-tert-butyl peroxide in CF_2Cl_2 provided evidence only of stable products.[39] Recent revival of the pyrolysis technique using furnace zones inside of the cryostat vacuum vessel have been successful. Current[40] has produced CCl_3 by pyrolysis of CCl_3Br and Snelson[41] has succeeded in stabilizing methyl from thermal decompositions of methyl iodide and dimethyl mercury.

Furnace flow reactions provide a source of reactive species for matrix study. Margrave's group[42] has succeeded in producing silicon difluoride by passing SiF_4 over a heated bed of elemental silicon. The SiF_2 so produced is trapped in an argon matrix for spectral study. Recently, Milligan and Jacox[43] have studied a series of transition metal dichlorides prepared by the reaction of a stream of chlorine gas over the heated transition metal surface. This technique should have further applications in the infrared study of high temperature chemical species (see Chapter 6).

The immediate condensation of gases flowing through a microwave discharge has provided a number of interesting new chemical species. In early work, Rice and Freamo[44] passed hydrazoic acid through a glow discharge and reported the formation of a blue, paramagnetic solid upon condensation. The glow discharge also has the disadvantage of being non-selective which limits its application to simple chemical systems. Harvey and Brown[45] produced hydrogen atoms by microwave discharge of H_2 and observed their reaction with nitric oxide to form HNO. Recently, Steudel[46] has produced an infrared absorption which turned out to be the trichloromethyl radical by condensing a discharged stream of CCl_4 or $HCCl_3$. In later work, Nelson and Pimentel have stabilized the trichloride radical[47] using a microwave discharge of a krypton-chlorine stream and xenon dichloride[48] in like manner from xenon and chlorine. More recently microwave discharge of a HCl, Cl_2, Ar gas mixture by Noble and Pimentel[49] has produced the hydrogen dichloride radical for spectral study.

Shirk and Bass[50] have used microwave discharged gases to 'sputter' metal atoms from filaments into matrices. This work suggests that users of microwave discharges be aware of the possible presence of metal atoms, particularly copper, in their matrix samples.

The use of molecular beam evaporation from an oven or Knudsen cell as a source of new chemical species for matrix isolation study was suggested by Pimentel[37] in 1960. Subsequently, Linevsky observed the infrared spectrum of lithium fluoride[51] molecules in an argon matrix using the Knudsen cell technique. Since that time, numerous workers have studied molecules in matrices deposited from high temperature ovens. This subject is covered in the next chapter of this monograph.

As a logical extension of these evaporative methods, Andrews and Pimentel[52] developed a technique for providing lithium atoms as reaction partners in matrices. This technique employs the co-deposition of an atomic beam of lithium from a Knudsen cell with a suitable reactive species at high dilution in argon. During the 15 K condensation of these gaseous species, diffusion takes place on the surface of the matrix which produces collisions between lithium atoms and reactive molecules and provides an opportunity for reaction with subsequent product stabilization by the matrix. The matrix deposition of alkali metal atoms has been used to synthesize a number of unique inorganic molecules for infrared analysis including LiON,[52] LiO$_2$[53] and NaO$_2$.[54] These reactions feature co-deposition of the alkali metal atom and nitric oxide or oxygen molecules prediluted in argon. Secondary reactions are also observed leading to the peroxide molecules LiO$_2$Li and NaO$_2$Na.

The early sodium flame work of Polanyi[4] suggests that this new matrix method might be capable of stabilizing free radicals by halogen abstraction from a suitable alkyl halide. The first work of Andrews and Pimentel[39] with methyl iodide and bromide produced what turned out to be the methyl radical perturbed by a lithium halide or a weakly-bonded methyl alkali-halide[55] complex at least partially stabilized by the cage effect. After 'unperturbed' methyl was reported by Milligan and Jacox,[21] re-examination of the spectra from the alkali metal reaction with methyl iodide indeed shows a sizeable yield of the isolated methyl radical as well. Further work of Tan and Pimentel[55] and Carver and Andrews[56] has shown that comparison of spectra produced by the reactions of lithium and sodium are necessary to discriminate between isolated radicals and perturbed radicals. The heavier alkali halide forms more perturbed radical relative to isolated radical since it is less likely to diffuse away from the reaction site during the condensation of gases forming the matrix sample.[56] Yields of radicals produced by the lithium reaction exceed those from the sodium reaction, presumably because the lighter lithium atom diffuses farther during condensation and thus has more reactive collisions than a sodium atom would have under similar conditions.

Andrews and Carver have used the alkali metal technique to produce the trichloromethyl[57,58] and tribromomethyl[59] radicals by halogen abstraction from the appropriate carbon tetra-halide. Secondary reaction of the trihalomethyl radicals with alkali metal atoms in these experiments have yielded the appropriate dihalocarbene[59,60] for spectral study. In more recent work, dihalomethyl[57,61] and monohalomethyl[62] radicals have been stabilized for matrix observation using this technique.

The alkali-metal matrix reaction is ideally suited for synthesis of interesting alkali-metal inorganic molecules which contain one or two alkali atoms. As a method for producing free radicals for spectral study, the halogen-abstraction reaction has the disadvantage of stabilizing radicals perturbed by the alkali halide product in addition to isolated radicals; however, with due care the two can be identified. Furthermore, this method is capable of stabilizing unusually large concentrations of certain free radicals, such as trichloromethyl.[58] The alkali-metal technique has the unique application of studying relative rates of abstraction of two different halogen atoms from the same precursor. Unlike the sodium flame and crossed molecular beam reactions, the matrix technique readily provides positive identification for each different radical and alkali halide product.

5.4 Identification of new molecular species

The identification of free radicals and chemical intermediates from vibrational spectra necessarily depends heavily upon isotopic substitution and vibrational analysis.[63] In nearly all of the work cited in the following section of this chapter, two or more isotopic molecules were studied. The presence of a certain atom can be verified by observing an isotopic shift. By using isotopic mixtures, frequently the molecular formula can be determined. For example, in the reaction of lithium atoms with oxygen molecules,[53] an equimolar mixture of lithium-6 and -7 isotopes produced two prominent doublet and two triplet absorptions. The doublet absorptions are clearly due to a vibrational mode involving a single lithium atom whereas the triplet bands are caused by the vibration of two equivalent lithium atoms. If the symmetry of the molecule is such that each lithium atom participates equally in these vibrations, then the molecules can be identified to contain a single lithium atom and two equivalent lithium atoms, respectively. Furthermore, observation of the same isotopic pattern for all absorptions of a particular molecule adds strong support for its identification. An interesting contrast is seen in the reaction of hydrogen atoms and lithium atoms with mixtures of $^{16}O_2$, $^{16}O^{18}O$ and $^{18}O_2$ in argon matrices. Milligan and Jacox[28] identified three absorptions of a bent molecule HOO in their experiments; a doublet at 3414 cm^{-1} and 3402 cm^{-1}, a quartet at 1389 cm^{-1}, 1386 cm^{-1}, 1384 cm^{-1} and 1380 cm^{-1}, and a quartet at 1101 cm^{-1}, 1072 cm^{-1}, 1069 cm^{-1} and 1040 cm^{-1}. The high frequency mode involved the vibration of hydrogen and primarily *one* oxygen atom whereas the two lower frequency modes involved *two* unequivalent oxygen atoms. Thus, the molecule is unsymmetrical and the two oxygen atoms are not equivalent. The magnitude of the isotopic shifts can be shown by vibrational analysis to indicate that the absorption near 1400 cm^{-1} is primarily a valence angle bending vibration and the 1100 cm^{-1} band principally an oxygen–oxygen stretching vibration. Figure 5.1 illustrates triplets at 1097 cm^{-1}, 1067 cm^{-1} and 1036 cm^{-1}, 744 cm^{-1}, 741 cm^{-1} and 738 cm^{-1}, and 507 cm^{-1}, 499 cm^{-1} and 492 cm^{-1} assigned by Andrews to a symmetrical triangular $^6LiO_2$ molecule. (The triplets near 830 cm^{-1} and 470 cm^{-1} are due to LiO_2Li.) In the latter case, all three vibrational modes are triplets indicating the presence of two equivalent oxygen atoms in the LiO_2 molecule. The magnitude of these isotopic shifts indicates their correct vibrational assignments. If a complete assignment of all vibrations (or all vibrations of a given symmetry) is possible, the value

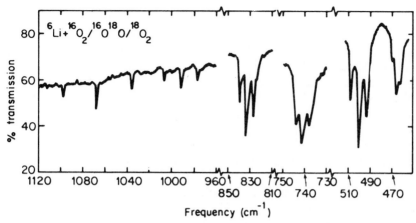

Figure 5.1 Infrared spectra showing oxygen isotopic splittings for ^{6}Li deposited with a 2/5/3 mixture of $^{16}O_2/^{16}O^{18}O/^{18}O_2$ in argon at 15 K, $Ar/O_2 = 130$, $Li/O_2 = \frac{1}{2}$. (From ref. 53, used with permission)

of normal co-ordinate analysis cannot be underestimated. In the LiO$_2$ work, eighteen vibrational frequencies were fit to a four constant potential function with an average difference between calculated and observed frequencies of 0.5 cm^{-1}. The excellent frequency fit confirms the vibrational assignments to LiO$_2$ and adds support to the view that matrix-isolation spectroscopy is ideally suited for vibrational analysis.

The usefulness of isotopes in making vibrational assignments is illustrated by three further examples. Unusual isotope shifts and intensity changes result when two vibrations of the same symmetry fall nearby in frequency. Carver and Andrews[61] observed a weak band at 1226 cm^{-1} and a very intense absorption at 902 cm^{-1} for HCCl$_2$ whereas a very intense band at 974 cm^{-1} and a weak one at 814 cm^{-1} were found for DCCl$_2$. The antisymmetric D—C—Cl valence angle bend is almost accidentally degenerate with the antisymmetric C—Cl stretching mode. Intensity alternation and a large splitting result from the interaction of these two vibrations of the same symmetry. Absorptions of 1316 cm^{-1} and 1175 cm^{-1} were found for HCF$_2$ which contrast 934 cm^{-1} and 1216 cm^{-1} for DCF$_2$. The former frequency in each pair was assigned to the antisymmetric H—(or D)—C—F bend and the latter to the antisymmetric C—F stretching vibration. This unusual shift to higher frequency following deuterium substitution can be attributed to the interaction of vibrations of the same symmetry class. For HCF$_2$ interaction with the bending mode forces the C—F stretch to lower frequency whereas in DCF$_2$ the bend has shifted below the C—F stretch and mutual interaction forces the C—F vibration to higher frequency. Isotopic splittings are also useful for determining vibrational assignments to modes of different symmetry. Two absorptions were observed for dichlorocarbene in the carbon–

chlorine stretching region each showing chlorine isotopic splittings appropriate to a species containing two equivalent chlorine atoms.[60] For a molecule of C_{2v} symmetry and an obtuse valence angle, the symmetry coordinate G-matrix element for the symmetric stretch is more dependent upon the reduced mass of chlorine than is the G-matrix element for the antisymmetric stretch. Thus, the band with a chlorine isotope splitting of 3 cm^{-1} was assigned to v_1, and the absorption with a 2 cm^{-1} splitting was attributed to v_3 of CCl_2.

Whenever possible, additional information for identifying new free radical species can be provided by using two or more chemical precursors for the same free radical. In early photolysis experiments,[17] both HBr and HI were used as a source of hydrogen atoms. Later work of Milligan and Jacox[65] utilized four different chemical routes for production of the free radical NCO: photolysis of the precursors HNCO, HN_3 in carbon monoxide, HCN and N_2O, and N_3CN with NO in an argon matrix. In recent halogen-abstraction work, the precursors H_2CClF, H_2CCl_2, H_2CClBr and H_2CClI have yielded the same absorptions which are attributed to the H_2CCl free radical.[62] Identification of the products of the reactions of alkali atoms with carbon tetrahalides was facilitated by the use of polyhalogen starting materials. Figure 5.2 contrasts argon matrix spectra for the reaction of CCl_4, CCl_3Br, CCl_2Br_2, $CClBr_3$ and CBr_4 with lithium atoms.[58] The band labelled A_1 is produced from CCl_4 and CCl_3Br and not any of the other precursors whereas the B_1 and B_2 bands arise only from the reaction of CCl_3Br and CCl_2Br_2. Likewise the C and D bands are produced from two and only two different precursors. That the resulting species must be chemically common to the two precursors from which it is produced leads to the identification of the A band as CCl_3 and the D band as CBr_3. The presence of B_1 and B_2 bands for CCl_2Br and C_1 and C_2 bands for $CClBr_2$ further indicates that the A and D bands are doubly degenerate since two bands result when the degeneracy is broken in the mixed chlorine–bromine radicals. The production of four distinct products from the five reactions represented in Figure 5.2 indicates that three equivalent halogens are present in the species of higher symmetry, consistent with the conclusions formed from comparing experiments using precursors of the same radical.

Sample warming experiments are helpful in identifying free radicals, since upon molecular diffusion two free radicals can react with little or no activation energy to form a stable molecule. However, great care must be taken in the interpretation of these experiments since other small molecules can diffuse and aggregate, thus causing a decrease in the intensity of an absorption which may be due to aggregation rather than reaction to form a new chemical species (see Chapter 3). The diffusion behaviour of carbon dioxide and methylene chloride illustrate this point. The doublet at 662 cm^{-1} and 664 cm^{-1} due to trace impurities of CO_2 in an argon matrix disappears completely after sample warming to 35 K and recooling to 14 K. The hydrogen

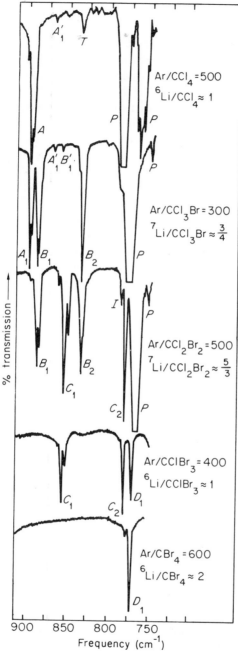

Figure 5.2 Infrared spectra illustrating the effect of bromine substitution for chlorine in the starting materials CCl_4, CCl_3Br, CCl_2Br_2, $CClBr_3$ and CBr_4 in argon deposited with lithium atoms at 15 K. (From ref. 58, used with permission)

deformation mode v_7 of H_2CCl_2 appears as a sharp, intense band at 895 cm^{-1} in an argon matrix. Following a similar cycling of the sample temperature, the sharp, intense band is reduced to a weak, broad band while other fundamentals of H_2CCl_2 retain their original intensity. After the diffusion disappearance of a new band tentatively associated with a free radical, it is particularly helpful to observe the concomitant growth of the radical dimer by detecting one or more infrared absorptions of this stable molecule. In argon matrix samples containing CCl_2 and CCl_3, Andrews[58] observed the complete disappearance of CCl_2 and appearance of the most intense band of C_2Cl_4 after warming to 40 K, and a decrease of CCl_3 with the appearance of an absorption due to C_2Cl_6 following a warm-up to 47 K. The incomplete disappearance of CCl_3 at this 'high' temperature for an argon matrix is attributed to its bulk and high molecular weight which retard the diffusion process, rather than to a lack of reactivity. A similar failure to disappear upon sample warming has been noted for $SiCl_3$.[66] In more recent work, Carver and Andrews[61] have observed HCl_2C-CCl_2H after diffusion and partial disappearance of the $HCCl_2$ free radical.

5.5 Infrared spectra of free radicals and chemical intermediates

We now turn our attention to a brief review of the matrix infrared spectra of free radicals and chemical intermediates in noble gas matrices. These studies are tabulated in seven groups along with the method of production of each species. Only certain selected examples from each group will be discussed in any detail.

5.5.1 Methyl and halomethyl radicals

The methyl radical is certainly one of the most important chemical intermediates. Recently Andrews and Pimentel[39] have mentioned some of the early frustrating attempts to observe the infrared spectrum of methyl. These later workers present convincing evidence for the infrared detection of methyl in the spectrum recorded after matrix reaction of methyl bromide and iodide with lithium atoms. Later work of Milligan and Jacox[21] using the vacuum ultra-violet photolysis of methane assigned another feature to the out-of-plane bending mode of CH_3. The following work of Tan and Pimentel[55] reacting sodium and potassium atoms with methyl bromide and iodide showed that the absorption reported earlier by Andrews and Pimentel[39] was due to methyl perturbed by the alkali halide or a methyl alkali-halide complex. This later work also identified the band observed by Milligan and Jacox and confirmed their assignment to v_2 of CH_3. The 454 cm^{-1} band assigned to v_2 of CD_3 in an argon matrix agrees well with Herzberg's $2v_2''$ assignment of a hot band 894 cm^{-1} to the red of the 2140 Å 0—0 band of CD_3 in the gas phase. It is interesting to note that the isotopic

Table 5.1 Methyl and halomethyl radicals

Radical	Method of production	Reference
CH_3	photolysis of CH_4 in Ar, N_2	Milligan and Jacox[21]
	reaction of CH_3X with Li, Na, K	Tan and Pimentel[55]
	pyrolysis of CH_3I, $(CH_3)_2Hg$	Snelson[41]
CF_3	photolysis of CF_3Br, HCF_3 in Ar, N_2	Milligan, Jacox and
	photolysis of N_2F_2 and N_3CN or CF_2N_2	Comeford[83]
	photolysis of N_3CN and F_2	Milligan and Jacox[33]
	reaction of CF_3Cl, CF_3I with Li	Tan and Pimentel[84]
	pyrolysis of CF_3I	Snelson[85]
CCl_3	reaction of CCl_4 with Li, Na, K	Andrews[57,58]
	reaction of CCl_3Br, CCl_3I with Li	
	pyrolysis and discharge of CCl_4, $HCCl_3$, C_2Cl_6	Steudel[46]
	electron impact on CCl_4	Current[38]
	photolysis of $HCCl_3$	Rogers, et al.[58a]
	pyrolysis of CCl_3Br	Current[40]
	pyrolysis of $C_6H_5HgCCl_3$	Maltsev, et al. [111]
CBr_3	reaction of CBr_4 with Li, Na	Andrews and Carver[59]
	reaction of CBr_3Cl with Li, Na	
	photolysis of $HCBr_3$	Rogers, et al.[58a]
CI_3	reaction of CI_4 with Li, Na	Smith and Andrews[112]
	pyrolysis of CI_4	Smith and Andrews[112]
HCF_2	reaction of HCF_2Br, HCF_2Cl with Li, Na	Carver and Andrews[64]
	photolysis of H_2CF_2	Milligan and Jacox[87]
$HCCl_2$	reaction of $HCCl_3$, $HCCl_2Br$ with Li, Na	Carver and Andrews[61]
$HCBr_2$	reaction of $HCBr_3$, $HCBr_2Cl$ with Li, Na	Carver and Andrews[56]
HCI_2	reaction of HCI_3 with Li, Na	Smith and Andrews[112]
H_2CF	photolysis of CH_3F in Ar, N_2	Milligan and Jacox[22]
	reaction of H_2CFCl, H_2CFBr with Li, Na	Raymond and Andrews[88]
H_2CCl	reaction of H_2CClF, H_2CCl_2, H_2CClBr, H_2CClI with Li, Na	Andrews and Smith[113]
	photolysis of H_3CCl_3 in Ar, N_2	Jacox and Milligan[86]
H_2CBr	reaction of H_2CBr_2, H_2CBrF with Li	Smith and Andrews[62]
H_2CI	reaction of H_2CI_2 with Li	Smith and Andrews[114]

data of Milligan and Jacox for methyl suggests considerable anharmonicity in the out-of-plane bend opposite in sign to that usually contributed by cubic terms in a potential function. However, by symmetry only even powers of the out-of-plane vibration can contribute to the vibrational potential function. The quartic term, in contrast to cubic, may be of either sign.

More recently, Snelson[41] has pyrolyzed streams of CH_3I and CD_3I near 1200 °C and trapped the resulting products in neon matrices. The out-of-plane bending mode of the methyl radical identified by Snelson agrees with the assignment of Milligan and Jacox. Two additional weak bands were observed in the pyrolysis work and attributed to the antisymmetric bending and stretching modes of methyl.

The trichloromethyl radical was first observed by Andrews[57] following the matrix reaction of CCl_4 with alkali metals. Interest in this free radical is due, at least partially, to its electronic stabilization determined from bond dissociation energies. In addition to the exceedingly intense antisymmetric C—Cl stretching vibration, Andrews[58] also reported a weaker absorption assigned to the symmetric C—Cl mode. However, subsequent workers[58a] have not observed the weaker band in photolysis studies. Additional alkali metal reactions[58b] have shown that the disputed absorption is due to the carbene perturbed by a metal chloride molecule. Carbon–chlorine force constants for chloromethyl radicals are higher than those for stable chloro-carbon molecules; this observation and electronic stabilization in chloro-methyl radicals lead to interesting conclusions about the bonding present, which has been recently discussed by Carver and Andrews[56,61] and Smith and Andrews.[62]

5.5.2 Carbenes and nitrenes

The infrared detection of methylene has eluded workers for more than a decade. Moore et al.[14] have reviewed some of the earlier unsuccessful

Table 5.2 Carbenes and nitrenes

Intermediate	Method of production	Reference
CH_2	photolysis of CH_2CO, CH_2N_2	Milligan and Pimentel[89], Moore, Pimentel and Goldfarb[14]
	photolysis of C_3O_2 and H_2 in Ar	Moll and Thompson[67]
	photolysis of CH_4 in Ar, N_2	Milligan and Jacox[21]
	reaction of H_2CBr_2, H_2Cl_2 with Li in Ar	Andrews and Smith[62]
CF_2	photolysis of CF_2N_2	Milligan et al.[25]
	photolysis of F_2 and N_3CN	Milligan and Jacox[33]
	pyrolysis of C_2F_4	Snelson[85]
CCl_2	photolysis of Cl_2 and N_3CN in Ar, N_2	Milligan and Jacox[32]
	reaction of CCl_4, CCl_3Br with Li, Na	Andrews[60]
	pyrolysis of $C_6H_5HgCCl_3$	Maltsev et al.[111]
CBr_2	reaction of CBr_4, CBr_3Cl with Li, Na	Andrews and Carver[59]
HCF	photolysis of CH_3F in Ar, N_2	Jacox and Milligan[22]
	photolysis of HF and N_3CN	Jacox and Milligan[22]
HCCl	photolysis of HCl and N_3CN	Jacox and Milligan[34]
NH	photolysis of HN_3	Milligan and Jacox[23] Rosengren and Pimentel[18]
	photolysis of NH_3	Milligan and Jacox[20]
NF	photolysis of FN_3 in Ar, N_2	Milligan and Jacox[24]
	photolysis of NF_2 in Ar, Kr	Comeford and Mann[78]
NCl	photolysis of ClN_3 in Ar, N_2	Milligan and Jacox[24]
NBr	photolysis of BrN_3 in Ar, N_2	Milligan and Jacox[24]

attempts to observe matrix isolated methylene. The recent attempts to add carbon atoms[67] to H_2, photolyse off hydrogen atoms[21] from CH_4, and to abstract bromine atoms[62] from H_2CBr_2 in order to stabilize methylene have yielded chemical processes involving CH_2 but no absorption which can be attributed to this elusive intermediate. The comparative ease of diffusion and high reactivity of CH_2 may account for these failures, as well as possibly low infrared extinction coefficients. However, the more sensitive electron paramagnetic resonance technique succeeded[67a] in observing triplet methylene following photolysis of diazirine in solid xenon at 4 K. Subsequently, Herzberg and Johns[67b] have pointed out that the vacuum ultra-violet spectrum can be interpreted in terms of a bent triplet methylene species, consistent with the ESR spectrum.

Dichlorocarbene was first observed by Milligan and Jacox,[32] while the following work of Andrews[60] produced identical absorptions within two wave numbers. Such excellent agreement using completely different methods of production illustrates the success of the matrix isolation technique. The potential constants for dichlorocarbene were shown by Andrews to be near those for carbon tetrachloride which suggests similar $C—Cl$ bond strengths in these molecules. This fact is surprising in view of earlier proposed pi bonding in this chemical intermediate, which suggests that pi bonding may not be appreciable in CCl_2.

5.5.3 Diatomic molecule-atom addition radicals and intermediates

The first free radical stabilized in sufficient concentration for matrix infrared detection was formyl, HCO. Ewing, Thompson, and Pimentel[17] photolysed HI or HBr in a carbon monoxide matrix and identified the two low-frequency fundamentals of HCO. Milligan and Jacox[68] improved the yield by photolysing H_2S in a carbon monoxide matrix and assigned a weak band at $2488\ cm^{-1}$ to the $C—H$ stretch with a deuterium counterpart at $1937\ cm^{-1}$. Later work of Ogilvie[69] and these authors[68] substantiate the earlier results. Even though the large amount of anharmonicity associated with the weak $C—H$ bond has complicated the normal co-ordinate analysis of HCO, the use of a relatively long $C—H$ bond length in these calculations has led to improved results.[70] The weak $C—H$ bond deduced from infrared spectra of HCO is consistent with the contribution of a nonbonded excited state suggested to explain the high positive proton coupling constant from the electron spin resonance spectrum.[71]

Spratley and Pimentel[72] have suggested that bonding in XO_2 molecules is $(s - \pi^*)$ sigma or $(p - \pi^*)$ sigma depending upon whether hydrogen or a halogen is bonded to the diatomic molecule. The chemical properties of the atom are reflected in the diatomic bond strength since the electron density forming the sigma $X—O$ bond also contributes to the occupancy of a π^* molecular orbital on the oxygen molecule. As the electronegativity of the

Table 5.3 Diatomic molecule-atom addition radicals and intermediates

Radical	Method of production	Reference
HCO	photolysis of HI, HBr in CO	Pimentel, et al.[17]
	photolysis of H_2S in CO	Milligan and Jacox[68]
	photolysis of HI and CO in Ar	Ogilvie[69]
	photolysis of HCl, H_2O, CH_4 in CO and with CO in Ar	Milligan and Jacox[68]
FCO	photolysis of OF_2, NF_2, N_2F_2 in CO and with CO in Ar	Milligan and Jacox[29]
ClCO	photolysis of HCl, Cl_2, Cl_2CO, $(Cl_2CO)_2$ in CO and Cl_2, CO in Ar	Jacox and Milligan[90]
CCO	photolysis of N_3CN, CO in Ar	Jacox, et al.[35]
	photolysis of C_3O_2 in Ar	Moll and Thompson[67]
NCO	photolysis of HNCO in Ar, CO, N_2	Milligan and Jacox[65]
	photolysis of HN_3 in CO	Milligan an Jacox[65]
	photolysis of HCN, N_2O in Ar	Milligan and Jacox[65]
	photolysis of NO, N_3CN in Ar, N_2	Milligan and Jacox[65]
HOO	photolysis of HBr, HI, O_2 in Ar	Milligan and Jacox[28]
	photolysis of HI, O_2 in Ar	Ogilvie[69]
	glow discharge of H_2O in Ar	Smith and Andrews[115]
FOO	photolysis of OF_2, O_2 in Ar, N_2	Arkell[91]
	photolysis of F_2, O_2 in Ar, N_2	Arkell,[91] and Spratley Turner, Pimentel[30]
	photolysis of F_2 in O_2	Noble and Pimentel[92]
ClOO	photolysis of Cl_2 in O_2	Arkell and Schwager[73]
	photolysis of ClO_2 in Ar	Arkell and Schwager[73]
LiO_2	reaction of Li and O_2 in Ar	Andrews[53]
NaO_2	reaction of Na and O_2 in Ar	Andrews[54]
KO_2	reaction of K and O_2 in Ar	Andrews[54a]
RbO_2	reaction of Rb and O_2 in Ar	Andrews[54a]
CsO_2	reaction of Cs and O_2 in Ar	Andrews[54a]
HNO	photolysis of CH_3NO_2 in Ar	Brown and Pimentel[13]
	discharged H_2 and NO in Ar	Harvey and Brown[45]
	photolysis of HI, NO in Ar	Ogilvie[69]
LiON	reaction of Li and NO in Ar	Andrews and Pimentel[52]
LiN_2	reaction of Li in N_2	Spiker, et al.[116]
LiOF	reaction of OF_2 with Li in Ar	Andrews and Raymond[121]

added atom decreases, the sigma bonding electrons contribute more electron density to the antibonding molecular orbital of the diatomic molecule and thus reduce its frequency and force constant. The O—O force constants for the molecules FOO,[30] $ClOO$,[73] HOO[28] and LiO_2[53] are 10·5 mdyn/Å, 9·7 mdyn/Å, 6·1 mdyn/Å and 5·6 mdyn/Å, respectively in agreement with the theory of Spratley and Pimentel. The bonding in LiO_2 is suggested to be approximately completely ionic[53] since the force constant for isolated O_2^- itself is also 5·6 mdyn/Å.[53] Apparently, the valence electron of lithium has been completely transferred to the anti-bonding molecular orbital of the oxygen molecule. The bonding in $M^+O_2^-$ species has recently been described[73a] in terms of a polarizable ion pair model.

5.5.4 Inorganic radicals and intermediates

Bassler, Timms and Margrave[42] have produced the intermediate SiF_2 by passing SiF_4 over silicon at 1150 °C. After deposition of this reaction mixture in an argon matrix, numerous new absorptions were noted in the 800–860 cm^{-1} region. Due to the complexity of the spectrum, a vibrational assignment to SiF_2 was not possible. Later work of Khanna[74] et al., suggested two prominent features in the argon matrix work at 843 cm^{-1} and 855 cm^{-1} for the two Si—F fundamentals due to their proximity to band centres near 855 cm^{-1} and 872 cm^{-1} assigned to gas phase SiF_2. Recent vacuum ultraviolet photolysis of H_2SiF_2 and D_2SiF_2 in argon by Milligan and Jacox[75] produced prominent absorptions at 843 cm^{-1} and 855 cm^{-1} which they assigned to SiF_2. This agreement between the infrared spectra reported by two independent research groups using different techniques of production for SiF_2 is excellent and it adds confirmation to the interpretation of their results.

Table 5.4 Inorganic radicals and intermediates

Intermediate	Method of production	Reference
SiH_2, SiH_3	photolysis of SiH_4 in Ar	Milligan and Jacox[117]
SiF_2	high temperature reaction of Si, SiF_4 deposited in Ar	Bassler, Timms, Margrave[42]
	photolysis of H_2SiF_2 in Ar	Milligan and Jacox[75]
SiF_3	photolysis of $HSiF_3$ in Ar	Milligan,[93] et al.
$SiCl_2$	photolysis of H_2SiCl_2 in Ar	Milligan and Jacox[94]
$SiCl_3$	photolysis of $HSiCl_3$ in Ar	Jacox and Milligan[66]
GeH_2, GeH_3	photolysis of GeH_4 in Ar	Smith and Guillory[125]
GeF_2	evaporation of solid GeF_2 in Ne	Margrave, et al.[95]
$GeCl_2$	evaporation from solid $GeCl_2$ in Ar	Andrews and Frederick[96]
	photolysis of H_2GeCl_2 in Ar	Guillory and Smith[97]
$GeCl_3$	photolysis of $HGeCl_3$ in Ar	Guillory and Smith[97]
$SnCl_2$	evaporation from solid $SnCl_2$ in Ar	Andrews and Frederick[96]
$PbCl_2$	evaporation from solid $PbCl_2$ in Ar	Andrews and Frederick[96]
PF_2	photolysis of PF_2H in Ar photolysis of P_2F_4 in Ar	J. K. Burdett et al.[118]
PCl_2	reaction of PCl_3 with Li, Na in Ar	Andrews and Frederick[98]
PBr_2	reaction of PBr_3 wjth Li, Na in Ar	Andrews and Frederick[98]
H_2GeCl	photolysis of H_2GeCl_2 in Ar	Isabel and Guillory[119]
LiO	reaction of Li with N_2O in N_2	Spiker and Andrews[120]
CsO	reaction of Cs with N_2O in N_2	Spiker and Andrews[120]

An entirely different type of inorganic free radical, the ozonide ion in the species $M^+O_3^-$, has recently been studied in argon matrices. Spiker and Andrews[75a] employed matrix reactions of ozone and alkali metal atoms to yield the orange ozonide species which produced a very intense v_3 near $800\,cm^{-1}$ and a weaker v_2 near $600\,cm^{-1}$; these features exhibited small alkali metal effects and appropriate oxygen-18 shifts for a C_{2v} species. Jacox and Milligan[75b] synthesized the ozonide species by photolysing matrix samples containing alkali metal atoms, N_2O and O_2; these workers observed v_3 and an electronic band from 5100 Å to 3700 Å. In very recent work from this laboratory,[75c] the ozonide sample has been studied by argon plasma laser excitation. The observed emission spectrum consisted of a progression in v_1 of the ground electronic state of O_3^- which has been attributed to resonance fluorescence. This new band, observed $1010\,cm^{-1}$ lower than the exciting lines, showed alkali metal effects and the oxygen-18 isotopic shift for a pure oxygen vibrational mode.

5.5.5 Other nitrogen containing species

Colburn[76] has discussed the significance of the difluoroamino radical in the chemistry of the fluorides of nitrogen. The difluoroamino radical exists in equilibrium with tetrafluorohydrazine just as nitrogen dioxide does with dinitrogen tetroxide. As might be expected, the first attempt to stabilize

Table 5.5 Other nitrogen containing species

Species	Method of production	Reference
NH_2	photolysis of NH_3 in Ar, N_2	Milligan and Jacox[20]
NF_2	thermal decomposition of N_2F_4	Harmony and Myers[77]
		Comeford and Mann[78]
	photolysis of F_2, NH_3 in Ar, N_2	Milligan and Jacox[79]
HNNH	photolysis of HN_3 in Ar, N_2	Rosengren and Pimentel[18]
HCN_2	photolysis of CH_2N_2 in Ar, Kr, N_2	Ogilvie[99]
F_2CN	photolysis of CF_2N_2, FCN in Ar, N_2	Jacox and Milligan[100]
HNF	photolysis of HN_3, F_2 in Ar	Jacox and Milligan[79]
HNSi	photolysis of SiH_3N_3 in Ar	Ogilvie and Cradock[101]
NCN	photolysis of N_3CN in Ar, N_2, CO	Milligan, et al.[31]
	photolysis of C_3O_2 in N_2	Moll and Thompson[67]
CNN	photolysis of N_3CN in Ar, N_2	Milligan and Jacox[102]
	photolysis of C_3O_2 in N_2	Moll and Thompson[67]
CH_2NH	photolysis of CH_3N_3 in Ar	Milligan[103]
	photolysis of CH_2N_2 in N_2	Moore, et al.[14]
HNC	photolysis of CH_3N_3 in Ar	Milligan and Jacox[104]
CN	photolysis of HCN in Ar, N_2	Milligan and Jacox[27]
FNC	photolysis of FCN in Ar, N_2	Milligan and Jacox[27]
ClNC	photolysis of ClCN in Ar, N_2	Milligan and Jacox[27]
BrNC	photolysis of BrCN in Ar, N_2	Milligan and Jacox[27]

NF_2 by Harmony and Myers[77] employed the thermal decomposition of tetrafluorohydrazine followed by trapping in a nitrogen matrix. Coupled with their own gas phase results, these workers assigned bands at 1069·6 cm^{-1} to v_1 and 930·7 cm^{-1} to v_3 of NF_2. The observation of $v_1 > v_3$ is also found for OF_2 and is conceivably a mechanical consequence of the atoms having nearly the same mass. Later work by Comeford and Mann[78] used higher decomposition temperatures and obtained more complete dissociation to NF_2. These workers photolysed NF_2 and observed the appearance of NF and NF_3. The photolysis of HN_3 and F_2 by Jacox and Milligan[79] produced bands at 1069 cm^{-1} and 929 cm^{-1} which they attributed to NF_2. By using ^{15}N enriched HN_3, these workers were also able to observe $^{15}NF_2$.

5.5.6 Other hydrogen and halogen containing species

Recently, Noble and Pimentel[80] have stabilized hypofluorous acid by photolysing mixtures of fluorine and water suspended in a nitrogen matrix. These workers suggest that atomic fluorine formed by photolysis abstracts hydrogen from H_2O giving HF, also observed as a by-product, and OH radical which reacts with an additional fluorine atom. The frequencies of HOF are discussed in further detail in paragraph 3.6.3 where they are compared with those of matrix-isolated HOCl and HOBr.

Table 5.6 Other hydrogen and halogen containing species

Species	Method of production	Reference
OH	photolysis of H_2O in Ar	Acquista et al.[105]
OF	photolysis of OF_2 in Ar	Arkell, et al.[106]
	photolysis of OF_2, N_2O in Ar	Arkell[107]
	reaction of Li, Na, K and OF_2 in Ar	Andrews and Raymond[121]
HOF	photolysis of F_2, H_2O in Ar, Ne	Noble and Pimentel[80]
ClO	reaction of Li, Na, K and Cl_2O in Ar	Andrews and Raymond[122]
ClClO	photolysis of Cl_2O in Ar, N_2	Rochkind and Pimentel[108]
$(ClO)_2$	photolysis of Cl_2O in Ar, N_2	Rochkind and Pimentel[108]
		Alcock and Pimentel[109]
ClF_2	photolysis of ClF_3 in Ar	Mamantov, et al.[123]
	photolysis of ClF, F_2 in Ar	
HCl_2	glow discharge HCl, Cl_2, Ar	Noble and Pimentel[49]
HBr_2	glow discharge HBr, Br_2, Ar	Bondybey, et al.[124]
Cl_3	glow discharge Cl_2, Kr	Nelson and Pimentel[47]
Br_2Cl_2	glow discharge of Cl_2, Br_2 in Ar, Kr, Xe	Nelson and Pimentel[110]
HC_2	photolysis of C_2H_2 in Ar, N_2, Ne	Milligan and Jacox[36]

In other recent work, the infrared detection of the Cl_3 radical has been provided by Nelson and Pimentel[47] from the glow discharge of one to two per cent. chlorine in krypton mixtures. Since the presence of krypton in the discharge is necessary for production of the new band near 370 cm^{-1}, these authors attempted unsuccessfully to fit the observed isotopic pattern to several krypton chloride molecules. However, the spectrum is readily explained as the antisymmetric stretching fundamental of linear and slightly asymmetric Cl_3. The force constant for Cl_3 is slightly higher than that for Cl_3^- since the extra electron in Cl_3^- is nonbonding and its only effect on the bond strength might be to weaken it slightly due to increased electron repulsions.

5.5.7 Noble gas compounds

Turner and Pimentel[81] have outlined early attempts to prepare noble gas compounds by matrix methods. These workers photolysed argon-fluorine mixtures and found no evidence for argon fluorides. Upon photolysis of Ar, Kr, F_2 mixtures, two new absorptions appeared which were assigned to

Table 5.7 Noble-gas compounds

Compound	Method of production	Reference
KrF_2	photolysis of Kr, F_2 in Ar	Turner and Pimentel[81]
XeF_2	photolysis of Xe, F_2 in Ar	Turner and Pimentel[82]
XeF_4	photolysis of Xe, F_2 in Ar	Turner and Pimentel[82]
$XeCl_2$	glow discharge of Xe, Cl_2	Nelson and Pimentel[48]

KrF_2. Photolysis[82] of Ar, Xe, F_2 mixtures gave absorptions characteristic of XeF_2 and XeF_4. More recently Nelson and Pimentel[48] have passed mixtures of xenon and chlorine through a microwave discharge and condensed the products on an optical window at 20 K and thus identified $XeCl_2$. These are discussed in more detail along with other halogenated molecules in paragraph 3.6.2.

5.6 Conclusions

The large body of work using the matrix isolation technique as a means to produce free radicals and chemical intermediates for infrared spectral study certainly speaks for itself. The studies listed here have made a significant contribution to the understanding of bonding, structure, and chemical behaviour of free radicals. Further work in these areas is expected to produce equally interesting results.

It is fitting to conclude this chapter with a comment of Professor George C. Pimentel made at the 1968 Gordon Research Conference on Infrared

Spectroscopy, since a large amount of the thought and development behind the work described here was pioneered in his laboratory at Berkeley.

'Early in our use of the matrix technique, I began to have high hopes for the application of matrix isolation to a variety of chemical problems. It is gratifying that the success of the technique has far surpassed my most optimistic expectations.'

5.7 References

1. M. Gomberg, *J. Am. Chem. Soc.*, **22**, 757, (1900).
2. W. Nernst, *Z. Elektrochem.*, **24**, 335, (1918).
3. F. Paneth and W. Hofeditz, *Ber.*, **62**, 1335, (1929).
4. E. W. R. Steacie, *Atomic and Free Radical Reactions*, 2nd ed., Reinhold, New York, 1954.
5. W. A. Pryor, 'Organic Free Radicals,' *Chem. Eng. News*, 15 January, 70, (1968).
6. K. C. Herr and G. C. Pimentel, *Appl. Opt.*, **4**, 25, (1965).
7. G. N. Lewis and D. Lipkin, *J. Am. Chem. Soc.*, **64**, 2801, (1942).
8. I. Norman and G. Porter, *Nature*, **174**, 508, (1954).
9. E. Whittle, D. A. Dows and G. C. Pimentel, *J. Chem. Phys.*, **22**, 1943, (1954).
10. G. C. Pimentel and S. W. Charles, *Pure Appl. Chem.*, **7**, 111 (1963).
11. E. D. Becker, G. C. Pimentel and M. Van Thiel, *J. Chem. Phys.*, **26**, 145, (1957).
12. G. C. Pimentel, S. W. Charles and Kj. Rosengren, *J. Chem. Phys.*, **44**, 3029, (1966).
13. H. W. Brown and G. C. Pimentel, *J. Chem. Phys.*, **29**, 883, (1958).
14. C. B. Moore, G. C. Pimentel and T. D. Goldfarb, *J. Chem. Phys.*, **43**, 63, (1965).
15. G. C. Pimentel, *J. Am. Chem. Soc.*, **80**, 62, (1958).
16. R. T. Hall and G. C. Pimentel, *J. Chem. Phys.*, **38**, 1889, (1963).
17. G. E. Ewing, W. E. Thompson and G. C. Pimentel, *J. Chem. Phys.*, **32**, 927, (1960).
18. Kj. Rosengren and G. C. Pimentel, *J. Chem. Phys.*, **43**, 507, (1965).
19. P. Warnek, *Appl. Opt.*, **1**, 721, (1962).
20. D. E. Milligan and M. E. Jacox, *J. Chem. Phys.*, **43**, 4487, (1965).
21. D. E. Milligan and M. E. Jacox, *J. Chem. Phys.*, **47**, 5146, (1967).
22. M. E. Jacox and D. E. Milligan, *J. Chem. Phys.*, **50**, 3252, (1969).
23. D. E. Milligan and M. E. Jacox, *J. Chem. Phys.*, **41**, 2838, (1964).
24. D. E. Milligan and M. E. Jacox, *J. Chem. Phys.*, **40**, 2461, (1964).
25. D. E. Milligan, M. E. Jacox and R. A. Mitsel, *J. Chem. Phys.*, **41**, 1199, (1964).
26. C. B. Moore and G. C. Pimental, *J. Chem. Phys.*, **41**, 3504, (1964).
27. D. E. Milligan and M. E. Jacox, *J. Chem. Phys.*, **47**, 278, (1967).
28. D. E. Milligan and M. E. Jacox, *J. Chem. Phys.*, **38**, 2627, (1963).
29. D. E. Milligan, M. E. Jacox, A. M. Bass, J. J. Comeford and D. E. Mann, *J. Chem. Phys.*, **42**, 3187, (1965).
30. R. D. Spratley, J. J. Turner and G. C. Pimentel, *J. Chem. Phys.*, **44**, 2063, (1966).
31. D. E. Milligan, M. E. Jacox and A. M. Bass, *J. Chem. Phys.*, **43**, 3149, (1965).
32. D. E. Milligan and M. E. Jacox, *J. Chem. Phys.*, **47**, 703, (1967).
33. D. E. Milligan and M. E. Jacox, *J. Chem. Phys.*, **48**, 2265, (1968).
34. M. E. Jacox and D. E. Milligan, *J. Chem. Phys.*, **47**, 1626, (1967).
35. M. E. Jacox, D. E. Milligan, N. G. Moll and W. E. Thompson, *J. Chem. Phys.*, **43**, 3734, (1965).
36. D. E. Milligan and M. E. Jacox, *J. Chem. Phys.*, **46**, 4562, (1967).

37. G. C. Pimentel, in *Formation and Trapping of Free Radicals*, eds. A. M. Bass and H. P. Broida, Academic Press, New York, 1960.
38. J. H. Current, personal communication.
39. L. Andrews and G. C. Pimentel, *J. Chem. Phys.*, **47**, 3637, (1967).
40. J. H. Current and J. K. Burdett, *J. Phys. Chem.*, in press.
41. A. Snelson, *J. Phys. Chem.*, **74**, 537, (1970).
42. J. M. Bassler, P. L. Timms and J. L. Margrave, *Inorg. Chem.*, **5**, 729, (1966).
43. D. E. Milligan and M. E. Jacox, *J. Chem. Phys.*, in press.
44. F. O. Rice and M. Freamo, *J. Am. Chem. Soc.*, **75**, 548, (1953).
45. K. B. Harvey and H. W. Brown, *J. Chim. Phys.*, **56**, 745, (1959).
46. R. Steudel, *Tetrahedron Letters*, **47**, 4699, (1967).
47. L. Y. Nelson and G. C. Pimentel, *J. Chem. Phys.*, **47**, 3671, (1967).
48. L. Y. Nelson and G. C. Pimentel, *Inorg. Chem.*, **6**, 1758, (1967).
49. P. N. Noble and G. C. Pimentel, *J. Chem. Phys.*, **49**, 3165, (1968).
50. J. S. Shirk and A. M. Bass, *J. Chem. Phys.*, **49**, 5156, (1968).
51. M. J. Linevsky, *J. Chem. Phys.*, **34**, 587, (1961).
52. W. L. S. Andrews and G. C. Pimentel, *J. Chem. Phys.*, **44**, 2361, (1961).
53. L. Andrews, *J. Chem. Phys.*, **50**, 4288, (1969).
54. L. Andrews, *J. Phys. Chem.*, **73**, 3922, (1969).
54a. L. Andrews, *J. Chem. Phys.*, **54**, 4935, (1971), and unpublished results.
55. L. Y. Tan and G. C. Pimentel, *J. Chem. Phys.*, **48**, 5202, (1968).
56. G. C. Carver and L. Andrews, *J. Chem. Phys.*, **50**, 4223, (1969).
57. L. Andrews, *J. Phys. Chem.*, **71**, 2761, (1967).
58. L. Andrews, *J. Chem. Phys.*, **48**, 972, (1968).
58a. E. E. Rogers, S. Abramowitz, D. E. Milligan and M. E. Jacox, *J. Chem. Phys.*, **52**, 2198, (1970).
58b. D. A. Hatzenbuhler and L. Andrews, unpublished results.
59. L. Andrews and T. G. Carver, *J. Chem. Phys.*, **49**, 896, (1968).
60. L. Andrews, *J. Chem. Phys.*, **48**, 979, (1968).
61. T. G. Carver and L. Andrews, *J. Chem. Phys.*, **50**, 4235, (1969).
62. D. W. Smith and L. Andrews, *J. Chem. Phys.*, **55**, 5295, (1971).
63. See textbooks on the subject; for example, E. B. Wilson, jr., J. C. Decius and P. C. Cross, *Molecular Vibrations*, McGraw-Hill Book Co., Inc., New York, 1955.
64. T. G. Carver and L. Andrews, *J. Chem. Phys.*, **50**, 5100, (1969).
65. D. E. Milligan and M. E. Jacox, *J. Chem. Phys.*, **47**, 5157, (1967).
66. M. E. Jacox and D. E. Milligan, *J. Chem. Phys.*, **49**, 3130, (1968).
67. N. G. Moll and W. E. Thompson, *J. Chem. Phys.*, **44**, 2684, (1966).
67a. R. A. Bernheim, H. W. Bernard, P. S. Wang, L. S. Wood and P. S. Skell, *J. Chem. Phys.*, **53**, 1280, (1970).
67b. G. Herzberg and J. W. C. Johns, *J. Chem. Phys.*, **54**, 2276, (1971).
68. D. E. Milligan and M. E. Jacox, *J. Chem. Phys.*, **41**, 3032, (1964); **51**, 277, (1969).
69. J. F. Ogilvie, *Spectrochim. Acta*, **23A**, 737, (1967).
70. J. S. Shirk and G. C. Pimentel, *J. Amer. Chem. Soc.*, **90**, 3349, (1968).
71. F. J. Adrian, E. L. Cochran and V. A. Bowers, *J. Chem. Phys.*, **36**, 1661, (1962).
72. R. D. Spratley and G. C. Pimentel, *J. Amer. Chem. Soc.*, **88**, 2394, (1966).
73. A. Arkell and I. Schwager, *J. Amer. Chem. Soc.*, **89**, 5999, (1967).
73a. L. Andrews and R. R. Smardzewski, *J. Chem. Phys.*, **58**, 15 Mar., (1973).
74. V. M. Khanna, R. Hauge, R. F. Curl and J. L. Margrave, *J. Chem. Phys.*, **47**, 5031, (1967).
75. D. E. Milligan and M. E. Jacox, *J. Chem. Phys.*, **49**, 4269, (1968).
75a. R. C. Spiker, jun. and L. Andrews, *J. Chem. Phys.*, **59**, 15 Sept., (1973).
75b. M. E. Jacox and D. E. Milligan, *J. Mol. Spectroscopy*, **43**, 148, (1972).

75c. L. Andrews and R. C. Spiker jun., *J. Chem. Phys.*, **59**, 15 Sept., (1973).

76. C. B. Colburn, *Endeavour*, **24**, 138, (1965).

77. M. D. Harmony and R. J. Myers, *J. Chem. Phys.*, **37**, 636, (1962).

78. J. J. Comeford and D. E. Mann, *Spectrochim. Acta*, **21**, 197, (1965).

79. M. E. Jacox and D. E. Milligan, *J. Chem. Phys.*, **46**, 184, (1967).

80. P. N. Noble and G. C. Pimentel, *Spectrochim. Acta*, **24A**, 797, (1968).

81. J. J. Turner and G. C. Pimentel, p. 101 in *Noble Gas Compounds*, ed. H. H. Hyman, University of Chicago Press, 1963.

82. J. J. Turner and G. C. Pimentel, *Science*, **140**, 974, (1963).

83. D. E. Milligan, M. E. Jacox and J. J. Comeford, *J. Chem. Phys.*, **44**, 4058, (1966).

84. L. Y. Tan and G. C. Pimentel, personal communication.

85. A. Snelson, personal communication.

86. M. E. Jacox and D. E. Milligan, *J. Chem. Phys.*, **53**, 2688, (1970).

87. D. E. Milligan and M. E. Jacox, personal communication.

88. J. I. Raymond and L. Andrews, *J. Phys. Chem.*, **75**, 3235, (1971).

89. D. E. Milligan and G. C. Pimentel, *J. Chem. Phys.*, **29**, 1405, (1958).

90. M. E. Jacox and D. E. Milligan, *J. Chem. Phys.*, **43**, 866, (1965).

91. A. Arkell, *J. Am. Chem. Soc.*, **87**, 4057, (1965).

92. P. N. Noble and G. C. Pimentel, *J. Chem. Phys.*, **44**, 3641, (1966).

93. D. E. Milligan, M. E. Jacox and W. A. Guillory, *J. Chem. Phys.*, **49**, 5330, (1968).

94. D. E. Milligan and M. E. Jacox, *J. Chem. Phys.*, **49**, 1938, (1968).

95. J. W. Hastie, R. Hauge and J. L. Margrave, *J. Phys. Chem.*, **72**, 4492, (1968).

96. L. Andrews and D. L. Frederick, *J. Amer. Chem. Soc.*, **92**, 775, (1970).

97. W. A. Guillory and C. E. Smith, *J. Chem. Phys.*, **53**, 1661, (1970).

98. L. Andrews and D. L. Frederick, *J. Phys. Chem.*, **73**, 2774, (1969).

99. J. F. Ogilvie, *Can. J. Chem.*, **46**, 2472, (1968).

100. M. E. Jacox and D. E. Milligan, *J. Chem. Phys.*, **48**, 4040, (1968).

101. J. F. Ogilvie and S. Cradock, *Chem. Commun.*, **364**, (1966).

102. D. E. Milligan and M. E. Jacox, *J. Chem. Phys.*, **44**, 2850, and **45**, 1387, (1966).

103. D. E. Milligan, *J. Chem. Phys.*, **35**, 1491, (1961).

104. D. E. Milligan and M. E. Jacox, *J. Chem. Phys.*, **39**, 712, (1963).

105. N. Acquista, L. Schoen and D. R. Lide, *J. Chem. Phys.*, **48**, 1534, (1968).

106. A. Arkell, R. R. Reinhard and L. P. Larson, *J. Am. Chem. Soc.*, **87**, 1016, (1965).

107. A. Arkell, *J. Phys. Chem.*, **73**, 3877, (1969).

108. M. M. Rochkind and G. C. Pimentel, *J. Chem. Phys.*, **46**, 4481, (1967).

109. W. G. Alcock and G. C. Pimentel, *J. Chem. Phys.*, **48**, 2373, (1968).

110. L. Y. Nelson and G. C. Pimentel, *Inorg. Chem.*, **1**, 1695, (1968).

111. A. K. Maltsev, R. G. Mikaelian, O. M. Nefedov, R. H. Hauge and J. L. Margrave, *Proc. Nat. Acad. Sci.* (*U.S.*) 1971 in press; see also *J. Phys. Chem.*, **75**, 3984, (1971).

112. D. W. Smith and L. Andrews, *J. Phys. Chem.*, **76**, 2718, (1972).

113. L. Andrews and D. W. Smith, *J. Chem. Phys.*, **53**, 2956, (1970).

114. D. W. Smith and L. Andrews, *J. Chem. Phys.*, **58**, 1 April, (1973).

115. D. W. Smith and L. Andrews, *J. Chem. Phys.*, **59**, Dec., (1973).

116. R. C. Spiker, jun., L. Andrews and C. Trindle, *J. Amer. Chem. Soc.*, **94**, 2401, (1972).

117. D. E. Milligan and M. E. Jacox, *J. Chem. Phys.*, **52**, 2594, (1970).

118. J. K. Burdett, L. Hodges, V. Danning and J. H. Current, *J. Phys. Chem.*, **74**, 4053, (1970).

119. R. J. Isabel and W. A. Guillory, *J. Chem. Phys.*, **55**, 1197, (1971).

120. R. C. Spiker, jun., and L. Andrews, *J. Chem. Phys.*, **58**, 15 Jan., (1973).

121. L. Andrews and J. I. Raymond, *J. Chem. Phys.*, **55**, 3078, (1971).

122. L. Andrews and J. I. Raymond, *J. Chem. Phys.*, **55**, 3087, (1971).

123. G. Mamantov, E. J. Vasini, M. C. Moulton, D. C. Vickroy, T. Maekawa, *J. Chem. Phys.*, **54,** 3419, (1971).
124. V. Bondybey, G. C. Pimentel and P. N. Noble, *J. Chem. Phys.*, **55,** 540, (1971).
125. G. R. Smith and W. A. Guillory, *J. Chem. Phys.*, **56,** 1423, (1972).

6 Infrared studies of vaporizing molecules trapped in low-temperature matrices

ALAN SNELSON

Contents

6.1 Introduction

The matrix isolation technique as developed by Becker and Pimentel[1] in 1956 provided a new tool for the study of highly reactive chemical species. Its adaptation to the study of vaporizing molecules* was not attempted until some four years later. The impetus for this latter development was largely due to a need which arose in the United States for good thermodynamic data which could be used in rocket propulsion performance calculations. For these calculations, it was necessary to characterize the high temperature species formed in the combustion process. A variety of techniques had demonstrated the existence of many stable polyatomic molecules in the rocket exhaust, but the determination of their molecular geometrics and energy levels presented many difficulties. Attempts to adapt IR spectroscopic techniques to obtain vibrational and rotational energy levels of these species were not too successful when applied under the high temperature conditions. Difficulties were encountered in designing optical cells which at the high temperatures were inert to the contained samples. Spectroscopic measurements were complicated by the high level of background IR radiation emanating from the heated sample cell. Interpretation of the spectrum, usually consisting of broad ill-defined bands resulting from transitions between highly populated vibrational and rotational energy levels, proved extremely difficult. Because of these problems, IR spectroscopic data for high temperature polyatomic molecules prior to 1960 were quite limited and often of questionable reliability.

The first successful application of the matrix isolation technique to the observation of infrared spectra of high temperature species was made by Linevsky in 1960.[2] Lithium fluoride was heated in a platinum Knudsen cell at about 900 °C and the resultant vapour species trapped in noble gas and nitrogen matrices. Absorption bands in the infrared attributable to monomeric lithium fluoride were identified. The possibilities of the then novel technique for the determination of IR spectra of high temperature polyatomic species were immediately recognized in several laboratories and studies initiated on species of particular interest in the rocket propulsion programme. From these early, and more recent studies, the following advantages of the method over conventional high temperature IR techniques have become apparent.

(i) Containment problems associated with high temperature vapours are greatly simplified. Only the Knudsen cell and furnace assembly must be unreactive to the vaporizing species. Since these components are generally quite small, exotic materials may be used in their construction if necessary, without the price becoming prohibitive.

* In the present chapter the term vaporizing molecules refers to those species which are introduced into the matrix by simultaneous condensation of a matrix gas and a molecular beam of the species.

(ii) Spectra can be obtained relatively easily for compounds which have vapour pressures in the range 10^{-4}–10^{-5} atm below 2800 K. Sufficient material to obtain the infrared spectra, usually between 10^{-5}–10^{-7} mol, can be isolated in the matrix in as little as a few minutes or over a period of several hours depending upon the reactivity of the vaporizing species. In principle, a sufficient amount of material may be trapped in the matrix for compounds whose vapour pressures lie below 10^{-5} atm by extending the period of time over which the matrix is formed. However, in practice, problems are usually encountered due to the matrix becoming non-transparent to the IR radiation.

(iii) Precise location of band centres is simplified since rotational structure is usually absent. The vibrational transitions appear as sharp features with band widths at half peak height of typically a few wave numbers. This behaviour is particularly valuable when several vibration transitions of a molecule are close together, or when several molecular species are present which have similar energy levels.

(iv) By virtue of the low temperature environment in which molecules are trapped only those transitions which originate from the molecular ground state are observed. This considerably simplifies interpretation of the spectrum.

(v) In systems containing several vapour species, it is often possible to distinguish absorption bands which are related to the same molecular species by comparing absorption band intensities under differing experimental conditions.

(vi) Bond angles can be calculated from matrix spectra when isotopic data are available. To obtain a reasonably small uncertainty in the calculated angles, frequencies must be measured to within $0 \cdot 1$ cm^{-1} or better.

To offset these advantages, there are some inherent difficulties associated with the approach:

(i) Suppression of rotational energy levels in matrix spectra prevents interatomic distances being deduced and prevents band contour shapes from being used to help in assigning frequencies to specific vibrational modes.

(ii) Relatively firm vibrational assignments for species isolated in matrices can only be made if spectra are available in which isotopic frequency shifts are observed. In those cases where isotopic studies are not possible, vibrational assignments have to be made solely on the basis of an assumed molecular model.

(iii) Frequently the matrix environment causes single frequencies to appear with multiplet structure, complicating interpretation of the spectrum. This so-called 'matrix splitting' can usually be detected by trapping the species under differing experimental conditions, and observing the form of the band envelopes. Often isotopic substitution is also useful in detecting 'matrix splitting' effects.

(iv) The magnitude of frequencies recorded in matrix spectra are usually different from the true gas phase values. In most cases these shifts are not too large and for the purpose of thermodynamic calculations their effects are usually not too serious.

Although the above difficulties are very real, there is no doubt that the advantages of the matrix isolation technique for the study of high temperature species methods far outweigh the disadvantages. With the greater availability of cryogenic refrigeration, the approach has now become relatively widely used to study IR spectra of high temperature polyatomic species. In the present chapter the achievements of the matrix isolation technique as applied to the study of vaporizing molecules will be presented.

6.2 Compounds containing group IA metals

6.2.1 Halides, LiF, Li_2F_2, LiCl, Li_2Cl_2, LiBr, Li_2Br_2, NaF and Na_2F_2

Prior to the adaptation of the matrix isolation technique to the study of high temperature vapours, little was known on the IR spectra of the alkali halide species. The vapour phase of the alkali halides consists largely of monomers and dimers[3,4] according to mass spectroscopic studies. Electron diffraction data[8] indicate the dimer to be planar and cyclic, with D_{2h} symmetry. High temperature IR spectra of the vapours over lithium fluoride[5,7] and lithium chloride[6,7] have been reported. Vibration frequencies for the monomer and dimers in both systems were recorded. For the dimers only two of the three possible IR active frequencies, assuming D_{2h} symmetry, were obtained. The unobserved frequencies were assumed to lie beyond the long wavelength limit of the spectrometer at 340 cm^{-1}.

Linevsky's[2] first investigation of the lithium fluoride system was essentially concerned with demonstrating the feasibility of using the matrix isolation technique to obtain infrared spectra of high temperature vapours and was not aimed at unravelling the spectrum of what proved to be a rather complex system. Matrices of Ar, Kr, Xe and N_2 were used and the frequencies for monomeric lithium fluoride in the various matrices reported. Two matrix phenomena immediately became apparent. The observed lithium fluoride frequencies in the different matrices were shifted from 7–14 per cent. to the red from the known gas phase value of 898 cm^{-1}, and the shape of the absorption band envelopes appeared matrix dependent, occurring as a singlet in Kr, Xe and N_2, and a doublet in Ar. The former effect was interpreted qualitatively in terms of different electrostatic interactions between the trapped molecule and the surrounding matrix, and the latter in terms of more than one trapping site in the matrix. Quantitatively it was demonstrated that a dipole-induced dipole type interaction between the trapped molecule and the rare gas matrix could account for a large fraction of the observed frequency shifts. For nitrogen a similar calculation was made and a dipole-quadrupole interaction was also considered. Again a large fraction of the observed frequency shift could be calculated (see Chapter 4). The differences between the calculated and observed frequency shifts were attributed to dispersion forces.

In a later study of the lithium fluoride system,[12] using matrices of Ne, Ar, O_2, CO and CH_4, it was shown that essentially all the observed noble gas frequency shifts could be calculated using the dipole-induced dipole model by assuming suitable orientation and distance parameters between the trapped molecule and the surrounding matrix. In the same investigation, the effect on the spectra of adding between 1–5 per cent. nitrogen to an argon matrix was investigated. Absorption band intensities were changed compared to those observed in the pure matrix, and although the frequencies were mostly unaltered, a few new weak absorption bands did appear. On annealing the 'doped' matrix, most of the features characteristic of the pure argon matrix were markedly reduced in intensity and new features appeared at frequencies rather close to the values obtained in a pure nitrogen matrix. This was interpreted in terms of the LiF molecule readily forming 'solvated' clusters with the available nitrogen in the matrix, resulting in a spectrum closely similar to that of lithium fluoride in pure nitrogen.

The first publications with the avowed intent of using the matrix isolation technique to investigate the spectra of some of the pure alkali halides appeared in 1963.[9,10] Matrices of Ar, Kr and Xe were used in both studies. Spectra were obtained for the vapour species over naturally occurring,[10] and isotopically enriched[9] lithium fluoride, lithium chloride[10] and sodium fluoride.[10] For lithium fluoride, intensity measurements and frequency shift data were used to assign two absorption bands to the B_{3u} and B_{2u} modes of an Li_2F_2 dimer with D_{2h} symmetry. Similar conclusions concerning the origin of pairs of bands in the lithium chloride and sodium fluoride systems were also made. In both investigations the long wavelength limit of the IR spectrometers was about 300 cm^{-1}, and the out-of-plane infrared active B_{1u} bending modes of the dimers were assumed to lie beyond this limit.

In later investigations, all three infrared active frequencies of the Li_2F_2,[13] Li_2Cl_2,[11,14] and Li_2Br_2[11,14] dimers, with D_{2h} symmetry, were reported. These are listed in Table 6.1. Application of the sum rule to the observed frequencies of the isotopically substituted halides $^6Li_2X_2$, $^6Li^7LiX_2$ and $^7Li_2X_2$, clearly demonstrated that the lowest frequency could be assigned to the B_{1u} mode. On the basis of theoretical ionic model calculations[19] on the Li_2X_2 dimers, the higher of the two remaining frequencies for each of the dimers was assigned to the B_{3u} mode. Further support for this assignment was deduced[5] from a consideration of the likely magnitudes of the F-matrix elements for the B_{3u} and B_{2u} modes. The most conclusive evidence in favour of this assignment in the lithium fluoride and sodium fluoride systems was obtained by Cyvin[17] from a normal co-ordinate analysis of Li_2F_2 and Na_2F_2.

Bond angle data for the dimers is limited, but may be approximated from the relationship $v(B_{2u})/v(B_{3u}) = \cot(\gamma/2)$ where γ is the X—Li—X angle. For

Table 6.1 Frequencies and assignments for the alkali halide vapour species

Species	Matrix	Observed frequencies, cm^{-1}	Point group	References
LiF^a	Ne	867	$C_{\infty v}$	9 through 15
Li$_2$F$_2$	Ne	641(B_{3u}) 553(B_{2u}) 287(B_{1u})	D_{2h}	
Li$_2$F$_2$	Ar	720·5, 497·5, 255·2($\Sigma +$), 245·7, 152·8(π)	$C_{\infty v}$	16
LiCl	Kr	569	$C_{\infty v}$	9 and 11
Li$_2$Cl$_2$	Kr	473(B_{3u}) 344(B_{2u}) 174(B_{1u})	D_{2h}	
LiBr	Kr	512	$C_{\infty v}$	
Li$_2$Br$_2$	Kr	421(B_{3u}) 298(B_{2u}) 150(B_{1u})	D_{2h}	11
LiI	Kr	433	$C_{\infty v}$	
NaF	Ne	515	$C_{\infty v}$	9 and 11
Na$_2$F$_2$	Ne	380(B_{3u}) 363(B_{2u}) (178)^b(B_{1u})	D_{2h}	
LiNaF$_2$	Ne	660, 321,^{c} 376(A_1), 589, 326(B_2) 238(B_1)	C_{2v}	17
LiAlF$_4$	Ne	902, 818, 650, 433, 244(A_1), 315(B_1), 562, 269(B_2)	C_{2v}?	18

a. All frequencies for lithium containing compounds are for the ^{7}Li isotope.
b. Estimated by comparison with Li$_2$F$_2$ and LiNaF$_2$.
c. Calculated from normal co-ordinate analysis.

the lithium fluoride, chloride and bromide dimers, angles of 99°, 108° and 110° were calculated. These compare well with the available electron diffraction data[18,21] for the chloride and bromide of 108 $\pm$ 4° and 110 $\pm$ 4°, respectively.

In the gas phase study[7] of Li$_2$F$_2$ and Li$_2$Cl$_2$, the B_{3u} and B_{2u} modes were observed at 640 cm^{-1} and 460 cm^{-1} for the former, and 460 cm^{-1} and 335 cm^{-1} for the latter. The agreement between these values and those listed in Table 6.1 is fairly good with the exception of the B_{2u} frequency of Li$_2$F$_2$, for which there is 93 cm^{-1} difference. However, the agreement with the gas phase matrix frequencies is more apparent than real. From a consideration of the order of magnitude of the frequency shifts for these species in the various inert gas matrices, the gas phase frequencies may be estimated at about 20 cm^{-1} and 30 cm^{-1} larger than the observed matrix values listed in Table 6.1 for Li$_2$F$_2$ and Li$_2$Cl$_2$, respectively.[15]

Although interpretation of the matrix isolation spectra of the alkali halides in terms of monomer and dimer species was relatively straightforward, absorption bands were observed in all investigations for which suitable assignments were not obvious. Attempts to assign some or all of these bands to cyclic trimers,[13,14] which according to mass spectral data exist in these systems, were not very convincing. Expected intensity ratios for the mixed isotopic species, and the corresponding frequency shifts, did not agree consistently with the chosen model. Other interpretations for these bands in terms of different trapping sites for the monomer[2] and the formation of agglomerates of the halide due to poor isolation[9] were also proffered, but not too convincingly. A satisfactory interpretation of most of these absorption bands in the lithium fluoride spectrum was made in a matrix isolation study[16] in which the previously unexplored spectral region from 200–120 cm^{-1} was examined. A total of five absorption bands were located which intensity ratio measurements suggested were related. From a careful study of isotope effects an assignment to an unsymmetrical linear dimer, Li—F—Li—F, was proposed. A normal co-ordinate analysis was made in an attempt to establish if some or all the bond lengths were equal, but for the three different models chosen equally good agreement between the observed and calculated frequencies was obtained. A suggestion had been made prior to this study[12] that some of the bands in the lithium fluoride spectrum might be due to a linear dimer being formed in the matrix during the trapping process. The existence of a gas phase stable linear dimer was thus not established, for the study did not completely obviate the possibility of a matrix stabilized species, Li_2F_2, since the efficiency of the trapping process in matrix studies performed up to this time had never been seriously tested.

At about the same time as the linear Li_2F_2 spectrum was being investigated, a study had been initiated aimed at adapting the matrix isolation technique to the determination of heats of vaporization of high temperature materials.[9] As part of this study it was necessary to establish experimental conditions under which effectively complete isolation of a reactive vaporizing material could be achieved. To this end, monomeric lithium fluoride vapour, obtained by superheating the saturated vapour at 1700 °C, was isolated in neon and argon matrices. The effectiveness of the isolation was judged by the appearance or non-appearance of dimer absorption bands. Using liquid helium as the coolant and neon as the matrix, dimer bands, although weak, were observed at estimated dilutions up to 2×10^5 : 1, but with argon, the dimer bands were completely suppressed at dilutions greater than 8×10^4 : 1. The applicability of the Lambert–Beer law to species trapped in the matrix under the good isolation conditions was tested for monomeric LiF and found to be obeyed with a precision of better than 1 per cent. (see also §3.7). After testing the procedure for determining heats of vaporization on aluminium fluoride with excellent results, the spectrum and heats of formation of the equilibrium vapour species over lithium fluoride were examined. Absorption bands assigned to the linear dimer were observed indicating that the species was vaporizing from the sample. From the measured heats of vaporization, thermodynamic arguments were used which showed that the absorption band assignment to a linear dimer, rather than a trimer, was definitely favoured. The concentration of the latter with respect to the monomer was estimated at less than 1 per cent.

To date the existence of a linear dimer has only been definitely established in the lithium fluoride system, but the existence of such dimers in other alkali halide systems seems very possible since absorption bands have been observed in various matrix studies which are difficult to assign to trimeric species. There is some evidence in experiments in which a large amount of lithium fluoride has been deposited in the matrix of absorption bands which can plausibly be assigned to a cyclic trimer.[22] Under similar conditions the same can probably be expected in other systems.

6.2.2 Mixed halides LiNaF$_2$ and LiAlF$_4$

The spectrum of the vapour species existing over mixtures of lithium fluoride and sodium fluoride at about 900 °C have been reported.[17] In addition to absorption bands which were obviously due to pure Li$_n$F$_n$ and Na$_n$F$_n$ species, several other new absorptions appeared. Five of these occurred with similar intensity as the Li$_n$F$_n$ and Na$_n$F$_n$ bands. Relative intensity measurements suggested they could all be assigned to the same molecular species. Isotope shifts were measured for ^{6}LiF and ^{7}LiF substituted species. From spectra obtained from vaporization of equimolar mixtures of ^{6}LiF, ^{7}LiF, and NaF, the presence of only one lithium atom in the molecule was confirmed. By analogy with the lithium fluoride and sodium fluoride systems, a planar rhombohedral mixed dimer, LiNaF$_2$, of C_{2v} symmetry was assumed. Calculated and observed frequency shifts were found to be consistent with such a model. A normal co-ordinate analysis, with a potential function containing six valence-type force constants was made and found to be capable of reproducing the observed frequencies of ^{7}LiNaF$_2$ exactly, and those of ^{6}LiNaF$_2$ to within less than 1 per cent. Comparison of the stretching force constants for the Li—F and Na—F bonds in the monomers and dimers shows the former to be about 2·4 times larger than the latter. This ratio is similar to the ratio of the non-bridge to bridge force constants in Al$_2$F$_6$, Al$_2$Cl$_6$ and B$_2$H$_6$ where values of 2·45, 2·4 and 1·9 are found.[23]

The only other mixed alkali halide for which matrix spectral data are available is lithium aluminium fluoride, LiAlF$_4$.[18] This is the major gas phase species which exists over equimolar mixtures of LiF + AlF$_3$ at about 1170 K.[23] In the spectral range 4000–200 cm^{-1}, eight absorption bands were observed which relative intensity measurements suggested were related. The molecular configuration of this species is not known. Of the various possibilities, a C_{2v} structure, having a planar LiF$_2$Al ring, with the remaining two fluorine atoms in a plane at right angles to the ring was suggested by analogy with Li$_2$F$_2$ and Al$_2$F$_6$.[24] For this symmetry, eleven IR active fundamental frequencies are expected. Some preliminary normal coordinate calculations have been made using frequencies for the ^{6}Li and ^{7}Li substituted species and the assignments shown in Table 6.1 derived. However, there are some problems with regard to the magnitude of the isotopic frequency shifts observed and calculated, and at the present time the C_{2v} assignment is questionable. Although a high temperature infrared gas phase study has been reported for LiAlF$_4$,[25] only two frequencies at approximately 780 cm^{-1} and

870 cm^{-1} were recorded. These probably correspond to the matrix frequencies of 902 cm^{-1} and 818 cm^{-1} given in Table 6.1.

6.2.3 Stable gas phase oxides LiO, Li$_2$O, Li$_2$O$_2$

Two infrared spectroscopic investigations of the vapour species over lithium oxide have been reported.[26,27] The first covered the spectral range 4000–250 cm^{-1} and included a mass spectroscopic and Knudsen effusion study of the vapour phase, while the second, some three years later, extended the spectroscopic observations out to 50 cm^{-1}. The mass spectroscopic study indicated a vapour phase over Li$_2$O at 1390–1640 K consisting of Li, O$_2$, Li$_2$O with the possibility of some LiO, and at higher temperatures, a small amount of Li$_2$O$_2$. Matrices containing lithium oxide vapour species were prepared using Knudsen cell temperatures in the range of 1625 K and 1760 K, and advantage was taken of the Li$_2$O$_2$ concentration-temperature dependence in analysing the spectra. Interpretation of the spectra was further aided by using isotopically enriched ^{6}Li and ^{7}Li samples.

Five sets of absorption bands appeared in the spectrum, these appeared as singlets when pure isotopic lithium oxide was vaporized, and as triplets when isotopic mixtures of the oxide were heated. This suggested the molecular species responsible for these bands contained two lithium atoms. Relative intensity measurements were used to demonstrate that three of the sets of absorptions were from the same molecular species, and likewise two from a second molecular species. Since the latter two absorption bands were the most intense at the lower effusion temperature, they were assigned to Li$_2$O. The absence of a third frequency which could be assigned to the same molecule suggested a symmetrical linear arrangement of the atom in agreement with the electric deflection data.[29] The calculated and observed isotopic shifts for the frequencies assigned to v_3 at about 1000 cm^{-1} (Table 6.2) for the ^{6}Li$_2$O and ^{7}Li$_2$O species were also consistent with a linear geometry. However, because of the relative insensitivity of this type of calculation with respect to the magnitude of an apex angle near 180°, a lower limit of 160° was suggested for the angle, based on the assigned experimental error of the frequency measurements. The isotopic shift ratio calculated for the v_2 frequencies lying near 100 cm^{-1}, assuming a linear model, was 1·043, compared to the experimental value of 1·054 ± 0·008. Although a slightly bent structure would give better agreement with the experimental ratio, a linear assignment was preferred based on experience with MgF$_2$,[30] which suggested that even in a weakly bent molecule the symmetrical stretching frequency should be easily observable. As noted above no frequency which could be assigned to v_1 of Li$_2$O was observed. A force constant analysis for Li$_2$O with $D_{\infty h}$ symmetry was made using a three term potential function, yielding stretching, stretching interaction, and bending force constants all in mdyn/Å of $f_r = 2·14$, $f_{rr} = -0·1$ and $f_\alpha = 0·014$.

Table 6.2 Frequencies and assignments for the alkali metal oxygen compounds

Species	Matrix	Observed frequencies, cm^{-1}	Point group	Bond angle	References
LiO	Kr	745	$C_{\infty v}$	180	26, 27
Li$_2$O	Kr	685^a(Σ_g^+) 112(π_u) 986·5(Σ_u^+)	$D_{\infty h}$	180	26, 27
Li$_2$O$_2$	Kr	241·5(B_{1u}) 325·4(B_{2u}) 522·0(B_{3u})	D_{2h}	116	26, 27
LiON	Ar	1352, 675, 333A'	C_s	100	34
LiO$_2$	Ar	1096·9(A_1) 698·8(A_1) 492·4(B_1)	C_{2v}	44^b	31
LiO$_2$Li	Ar	297·5(B_{1u}) 796·0(B_{2u}) 445·5(B_{3u})	D_{2h}	58^b	31
NaO$_2$	Ar	1080·0(A_1) 390·7(A_1) 332·8(B_1)	C_{2v}	38^b	35
NaO$_2$Na	Ar	239·0(B_{1u}) 524·5(B_{2u}) 254·0(B_{3u})	D_{2h}	52^b	35
NaOH	Ar	431(Σ^+) 337·0(π)	$C_{\infty v}$	180	41, 42
RbOH	Ar	354·4(Σ^+) 309·0(π)	$C_{\infty v}$	180	41, 42
CsOH	Ar	335·6(Σ^+) 302·4(π)	$C_{\infty v}$	180	41, 42
LiBO$_2$	Ar Kr	1976·0, 1094·0, 568·7, 470·6, 107·4(A'), 577·5(A'')	C_s	~90^c	44
CsBO$_2$	Ar Kr	1945·4, 1076·7, 576·3, 206·5, 60·0^a(A'), 581·0(A'')	C_s	~90	44
LiNO$_3$	Ar	1017(A_1') 823(A_2'') 1275(E') 1515(E')	D_{3h}		49a
NaNO$_3$	Ar	1023(A_1') 825(A_2'') 1283(E') 1484(E')	D_{3h}		49a
KNO$_3$	Ar	1031(A_1') 830(A_2'') 1291(E') 1462(E')	D_{3h}		49a
RbNO$_3$	Ar	1033(A_1') 830(A_2'') 1293(E') 1456(E')	D_{3h}		49a

a. Estimated.
b. $\angle$ (O—M—O).
c. $\angle$ MOB.

For the remaining three sets of absorption bands which appeared to be related via intensity measurements, an assignment was made to the species Li$_2$O$_2$. By analogy with the alkali halide dimers, D_{2h} symmetry was assumed. Observed and calculated isotopic shifts supported the assignment. The lowest frequency was assigned to the B_{1u} out-of-plane bending mode and the two higher frequencies to the B_{3u} and B_{2u} in-plane-bending modes. The angle O—Li—O was estimated at 116° using the relation $v_5(B_{2u})/v_6(B_{3u}) = \cot(\phi_1/2)$. A force constant analysis was made using as internal coordinates the bond distance r the O—Li—O angle ϕ_1 and the Li—O—Li angle ϕ_2. The interaction terms f_{rr}, $f_{r\phi}$ and $f_{\phi_1\phi_2}$ were assumed to be zero. The principal force constants f_r and f_ϕ were calculated at 0·56 mdyn/Å and 0·18 mdyn/Å, respectively. For ^{7}Li$_2$O$_2$, frequencies were calculated for the two unobserved A_g modes at 780 cm^{-1} and 212 cm^{-1}, and for the B_{1g} mode at 478 cm^{-1}.

Of the remaining absorption bands in the spectrum of the Li$_2$O vapours, isotope effects indicated the presence of species containing one lithium atom only. Observed ^{6}Li and ^{7}Li isotopic frequency shifts for the bands at 789 cm^{-1} and 745 cm^{-1} were in agreement with that expected for ^{6}LiO and ^{7}LiO. Comparison of these frequencies with those of ^{6}LiF (880 cm^{-1}) and ^{7}LiF

(833 cm^{-1}), suggest the assignment of the 789 cm^{-1} and 745 cm^{-1} bands in the lithium oxide spectrum to the monoxides to be reasonable.

Two other sets of absorption bands at 240 cm^{-1} and 700 cm^{-1} also appeared to result from a molecule containing one lithium atom. Intensity considerations suggested the same species was responsible for both sets of bands. Isotopic frequency shift data, and the appearance of only two frequencies was used to infer the existence of a linear or nearly linear symmetric species LiO_2. However in a later study[31] the assignment of the higher frequency at 700 cm^{-1} to LiO_2 was seriously challenged.

Finally, in the above investigations an attempt was made to estimate the LiO bond length, r, in the various lithium oxide species. Equations relating bond lengths to the force constants were used[32,33] to obtain: $LiO, r = 1.62 \text{ Å}$, $Li_2O, r = 1.59 \text{ Å}$ and $Li_2O_2, r = 1.90 \text{ Å}$.

6.2.4 Matrix stabilized oxides LiON, LiO_2, Li_2O_2, NaO_2 and Na_2O_2

The three oxide types discussed above are the only known stable gaseous oxides of the alkali metals under low pressure conditions. However, the existence of a new group of alkali metal oxides[31,34,35] which may be regarded as matrix stabilized has been established. The experimental technique used in these studies allows a molecular beam of alkali metal atoms to be trapped either in an inert gas matrix containing a small percentage of reactant, or in a matrix of pure reactant. The approach was first demonstrated by the identification of the species LiON,[34] formed by the interaction of lithium atoms with an argon matrix gas containing nitric oxide. Isotopes of lithium, oxygen and nitrogen were used in these experiments. Intensity measurements indicated that three new absorption bands were related to the same molecular species, and isotope effects suggested the presence of only one atom each of Li, O and N. A molecular configuration of LiON, as opposed to LiNO, was inferred qualitatively from the effect of isotopic substitution on the various frequencies. This arrangement of the atoms was supported by product rule calculations which also indicated a bent configuration. A normal co-ordinate analysis was made using a five constant potential function, the 17 observed frequencies, and plausible values for the LiO, LiN and ON bond lengths. The best agreement between the observed and calculated frequencies was obtained for the LiON configuration with an oxygen bond angle of $100 \pm 10°$. The principal stretching force constants for the LiO and NO bonds were calculated at 1.33 mdyn/Å and 7.97 mdyn/Å. Attempts to justify the structure of LiON in terms of classical valence rules or from simple molecular orbital theory were not too successful. An electrostatic view of the molecule in the form $Li^+(ON)^-$ was considered. It was assumed that the positive lithium ion would preferentially attach itself to the more negative atom of the NO group, presumably the oxygen. Although this approach was admitted to be somewhat naive, the relatively small value of the NO force constant in LiON at 7·97 mdyn/Å, compared to the value in NO at 15·4 mdyn/Å, was shown to be consistent with known trends in M-NO type molecules. It had been observed that as the electronegativity of M decreased, the NO force constant also decreased.[36] In the limit of M with zero electronegativity, the NO group may be regarded as essentially $(NO)^-$, and a force constant for the latter has been predicted at 5·6 mdyn/Å.

Two studies have been reported in which lithium[31] and sodium[35] atoms have been trapped in argon matrices doped with oxygen, and in pure oxygen matrices. The absorption band structure on isotopic substitution with ^6Li, ^7Li, $^{16}O_2$ and $^{18}O_2$, together with relative intensity measurements was used to identify the species LiO_2, NaO_2, LiO_2Li and NaO_2Na. Other absorption bands appeared in the spectra and were tentatively attributed to polymeric material.

For the species LiO_2 and NaO_2 a total of three absorption bands were observed for each. Oxygen isotope effects demonstrated that the highest frequency occurring at about 1090 cm^{-1} ($^{16}O_2$ substitution) in both molecules, could be attributed to an essentially O—O type vibration, and that both oxygen atoms were symmetrically equivalent. Lithium isotope effects were used to show that the two lower frequencies occurring at 698 cm^{-1} and 492 cm^{-1} (^7Li species) largely involved the lithium atom, and that only one lithium atom was present in the molecule. By analogy, frequencies at 390 cm^{-1} and 333 cm^{-1} in the NaO_2 spectra were assigned to similar vibrations of the sodium atom. From these data both molecules were assigned a triangular structure of C_{2v} symmetry. Consideration of the form the G-matrix elements on isotopic substitution, suggested assignment of the 698 cm^{-1} band to a symmetric, and the 492 cm^{-1} band to an asymmetric vibration of the lithium atom. Product rule calculations, and a normal co-ordinate analysis in terms of a 6 constant potential function were made utilizing the observed 18 frequencies for the various LiO_2 isotopic species. An O—O bond distance of 1·33 Å was assumed, taken from the X-ray diffraction data on solid NaO_2, and the Li—O bond distance and O—Li—O angle varied to obtain the best fit with the observed frequencies. Excellent agreement between the observed and calculated frequencies, were obtained for an Li—O bond distance of 1·77 Å and an O—Li—O angle of 44°. Calculations for linear and bent Li—O—O configurations were not consistent with the experimental data. Similar calculations for NaO_2, though less detailed since no alkali metal isotope frequencies were available, gave a Na—O bond distance of 2·07 Å, and an O—Na—O bond angle of 38·0°. In both LiO_2 and Na_2O the O—O force constant was calculated at about 5·5 mdyn/Å. This is very close to the value estimated for free O_2^-, at 5·6 mdyn/Å,[37] and suggests that the bonding in both the Li and Na compounds is electrostatic in nature, the molecule consisting of an alkali metal cation and a superoxide anion. In this type of structure, the electron from the alkali metal may be regarded as being transferred completely to one of the antibonding π orbitals in O_2, resulting, it has been argued,[38] in a decrease of the O—O force constant.

Isotopic substitution data for other absorption bands in the spectra indicate the presence of two equivalent alkali metal, and two equivalent oxygen atoms. The bands were assigned to LiO_2Li and NaO_2Na. Isotopic frequency shift calculations involving the three frequencies assigned to each molecule were consistent with both molecules having a planar rhombic structure of D_{2h} symmetry. In contrast to the Li_2O_2 dimers discussed earlier,[26,27] the two lithium cations in LiO_2Li were assumed to be bonded to a peroxide anion, containing an oxygen–oxygen bond. A detailed vibrational analysis was not possible on the bases of the three observed frequencies, but using the approximation involving the B_{3u} and B_{2u} frequencies,[30] O—M—O angles of 58° and 51·7° were obtained for the Li and Na species, respectively. Using these angles, and making some assumptions concerning the magnitude of interaction constants, LiO and NaO stretching force constants of 1·24 mdyn/Å and 1·03 mdyn/Å were estimated. The out-of-plane bending force constants calculated from the observed B_{1u} frequencies were 0·25 mdyn/Å and 0·32 mdyn/Å for the lithium and sodium compounds, respectively. The larger value or the latter, with respect to the former, is surprising since for the stretching frequencies the opposite trend appears to be true. In the planar molecules Li_2F_2, $NaLiF_2$ and Na_2F_2,[11] the out-of-plane bending force constant decreases in the order given.

Finally, with regard to the mode of formation of the LiO_2Li and NaO_2Na species, the observation was made that the existence of these was only evident in oxygen doped argon matrices and not at all in pure oxygen matrices, even though the species LiO_2 and NaO_2 resulted in either case. This behaviour was believed to be consistent with the formation of LiO_2Li by the addition of lithium atoms to LiO_2, since in a pure oxygen matrix, the possibility of a LiO_2 and Li encounter has a low probability compared to a Li and O_2 interaction.

6.2.5 Hydroxides NaOH, RbOH and CsOH

Mass spectroscopic data[39,40] have indicated a predominance of dimers in the saturated vapours of all the alkali metal hydroxides with the monomer present only as a minor species. Only one high temperature infrared gas phase study has been reported for any of these compounds, and although weak bands were observed their origin was quite uncertain. Matrix isolation studies[41,42] have been reported for CsOH, RbOH and NaOH and their deuterated derivatives. To observe the monomeric species a double oven technique was used to superheat the vapour. Experimentally the hydroxides were found to be extremely difficult to contain. Of a variety of materials tried, silver proved most successful but even this was attacked at the higher temperatures required in the superheating experiments. Interpretation of the spectrum was further complicated by the partial decomposition of the hydroxides to water and metal atoms. Attempts to observe the spectra of KOH and LiOH were not successful, the latter since no suitable Knudsen cell material could be found.

For the three hydroxides investigated, two frequencies were assigned to each, all falling in the range 225–400 cm^{-1}. Attempts to observe frequencies in the 3600 cm^{-1} and 2560 cm^{-1} region, corresponding to expected O—H and O—D stretches, respectively, were unsuccessful due to interference by H$_2$O and D$_2$O absorptions. Microwave data[43] for CsOH and RbOH, though not complete, were consistent with a linear structure for both these compounds and bond distances were determined. Using the latter data, and the observed matrix frequencies for the hydrogenated and deuterated species, it was shown that the lower of the two frequencies assigned to the Cs and Rb compounds corresponded to the bending modes. The bond length in NaOH was not known and assignment of the lower frequency to the bending mode was made by analogy with the Cs and Rb species. Assuming a linear structure for NaOH, bond lengths of 1·93 Å and 0·97 Å for the Na—O and O—H bonds, respectively, were found to satisfy the observed isotope shift of the v_2 bending mode.

From the assigned frequencies, the bending mode force constants were calculated at 0·053 mdyn/Å, 0·046 mdyn/Å and 0·047 mdyn/Å for NaOH, RbOH and CsOH, respectively, values which are considerably lower than those associated with normal covalent bonds. The suggestion was made that this trend may possibly be expected as a general feature of molecules which contain highly ionic bonds. Data obtained subsequently for the alkali metal metaborates[44] give some support to this speculation.

In an attempt to determine the metal-oxygen stretching force constants of the hydroxides, a two constant potential function was used. A force constant of 7·5 mdyn/Å was assumed for the O—H stretching mode, and quite similar values of 1·00 mdyn/Å, 1·05 mdyn/Å and 1·07 mdyn/Å were obtained for the Cs—O, Rb—O and Na—O stretching constants, respectively. A comparison of the calculated alkali metal-oxygen force constants in the hydroxides, with those in the analogous alkali metal fluorides show close similarity for Rb and Cs compounds, but in NaF the force constant is appreciably larger than in the hydroxide. It was concluded that the M—F force constants are probably somewhat larger than the M—OH values, but by how much, appears at present not certain.

6.2.6 Metaborates LiBO$_2$, NaBO$_2$, KBO$_2$, RbBO$_2$ and CsBO$_2$

Mass spectroscopic data[45] for the saturated vapour species over LiBO$_2$ and NaBO$_2$ at 1100–1200 K give a composition of about 90 per cent. monomer, 10 per cent. dimer and 0·1 per cent. trimer. Electron diffraction studies[46] on LiBO$_2$ and NaBO$_2$ vapours suggest the gaseous molecule has a structure

M—O—B=O, with a linear O—B=O group, and an M—O—B bond angle of about 90–105°. In a high temperature infrared gas phase study[47] on $LiBO_2$, $NaBO_2$ and $CsBO_2$, in the 2500–250cm^{-1} region, only two of the possible six infrared active vibration frequencies were reported for each species. These two frequencies, at about 1900 cm^{-1} and 600 cm^{-1}, were quite similar in all the compounds. The higher frequency was assigned to an O=B stretching mode and the lower to an O—B—O bending mode. No metal-oxygen stretching mode was reported. The observed spectra were rationalized by assuming an $M^+(O^- —B^+—O^-)$ type configuration in which the alkali metal cation was presumed to exert only a small perturbation on the BO_2^- anion, resulting in similar frequencies in the $O^- —B^+—O^-$ moiety.

In the matrix investigation[44] a double oven Knudsen cell was used to superheat the vapours by 200–400 °C, thus eliminating polymeric material. The spectral range from 3000–50 cm^{-1} was examined, and frequencies recorded for all the alkali metal metaborates. Isotope shifts due to ^6Li, ^7Li, ^{10}B and ^{11}B substitution were identified and recorded. In the frequency range 3000–500 cm^{-1} a total of five groups of absorption bands appeared for all the metaborates, and with closely similar frequencies. This was taken as evidence that these bands originated largely in the BO_2 moiety. Below 500 cm^{-1} two absorption bands were recorded for each of the species, $^7LiBO_2$, $^6LiBO_2$, $NaBO_2$ and KBO_2, but only one for $RbBO_2$ and $CsBO_2$. All these absorption bands showed a steady decrease in frequency as the mass of the alkali metal atom increased and were therefore assigned to vibrational modes involving the metal atom.

A vibrational assignment was made on the assumption of C_s symmetry, in accordance with the electron diffraction data,[46] and by comparison with the frequencies assignments in HBO_2,[48] and B_2O_3.[49] The M—O stretching and M—O—B bending modes were assigned to the two frequencies occurring below 500 cm^{-1}, the higher frequency to the stretching, and the lower to the bending mode. For the Rb and Cs metaborates the bending mode frequencies were not observed, presumably due to their low extinction coefficients, rather than the frequencies lying beyond the long wavelength limit of the investigation. Of the five frequencies above 500 cm^{-1}, the highest at about 1970 cm^{-1} was attributed to a B=O stretch. Two frequencies at about 570 cm^{-1} and 580 cm^{-1} were assigned to in-plane and out-of-phase bends, respectively, of O—B=O. Since the alkali metaborates with C_s symmetry have six infrared active frequencies, and seven were observed, it was found necessary to assign a band appearing at about 1190 cm^{-1} to an overtone of the O—B=O out-of-plane bend, in Fermi resonance with the B—O stretch at about 1090 cm^{-1}. In Table 6.2 the frequencies of the $^7Li^{11}BO_2$ and $Cs^{11}BO_2$ species only are given. It may be observed that the maximum differences between the frequencies assigned to the BO_2 groups in these two compounds are less than 2 per cent.

In the force constant analysis it was assumed that a valence force approximation was valid and that anharmonic and matrix shift corrections to the observed frequencies could be neglected. Bond distances for $B{=}O$ and $B{-}O$ at $1.20\,\text{Å}$ and $1.36\,\text{Å}$, were taken from the electron diffraction data[46] on $LiBO_2$ and $NaBO_2$, the same values being used for all the metaborates. From the same study[46] the $Li{-}O$ and $Na{-}O$ distances were reported at $1.82\,\text{Å}$ and $2.14\,\text{Å}$. It was observed that these values were intermediate between the bond distances in the corresponding chlorides and fluorides. This correlation was used to estimate values of $2.40\,\text{Å}$, $2.52\,\text{Å}$ and $2.63\,\text{Å}$ for the $M{-}O$ bond lengths in the potassium, rubidium and caesium metaborates, respectively. The $M{-}O{-}B$ bond angle was treated as a variable in the force constant calculation. Using six valence force field constants, one interaction constant, and an $M{-}O{-}B$ angle near $90°$, agreement between the observed and calculated frequencies to better than 0.3 per cent. was obtained. The principal force constants for the BO_2 group were very similar in all the metaborates, the following average values resulting: $F_{11}(B{=}O,\ \text{stretch}) = 13.6 \pm 1.2$ mdyn/Å, $F_{22}(B{-}O\ \text{stretch}) = 8.0 \pm 0.4$ mdyn/Å, $F_{33}(O{-}B{=}O,\ \text{out-of-plane bend}) = 0.4 \pm 0.01$ mdyn/Å, and $F_{44}(O{-}B{=}O,\ \text{in-plane bend}) = 0.35 \pm 0.3$ mdyn/Å. The $M{-}O$ stretching, and the $M{-}O{-}B$ bending force constants at 0.83 mdyn/Å and 0.095 mdyn/Å, respectively in $NaBO_2$, decreased steadily with increasing mass of the alkali metal atom. In this respect lithium metaborate does not fit the trend, values of 0.65 mdyn/Å and 0.044 mdyn/Å were obtained for the corresponding force constants. This situation, in which the force constant in the lithium compound is lower than the corresponding sodium compound, has also been observed for the out-of-plane bending modes in LiO_2Li and NaO_2Na.

6.2.7 Nitrates $LiNO_3$, $NaNO_3$, KNO_3, $RbNO_3$, and their dimers

There is little indication in the literature that the alkali-metal nitrates can be vaporized and it is generally believed that, with the exception of copper (II) nitrate, ionic nitrates decompose rather than vaporize. A matrix isolation study in Ar, CO_2 and CCl_4 has recently been reported[49a] of $LiNO_3$, $NaNO_3$, KNO_3 and $RbNO_3$ together with $TlNO_3$ and $Cu(NO_3)_2$. The IR spectra are simple and clearly indicate that the dominant species in each case is the monomer of the metal nitrate, although limited decomposition occurs with $LiNO_3$.

A comparison is made of the extent of the nitrate anion distortion for various chemical environments, by use of the splitting of the $v_3(e)$ degenerate vibrational modes. Smooth plots are found for the two v_3 components when plotted against cation polarizing power. This indicates that cation polarization of the anion is the dominant source of the anion distortion in the alkali-metal nitrate monomers. This is not so for $TlNO_3$ and $Cu(NO_3)_2$ since although the polarizing power of the Tl^+ ion is similar to that of K^+, the v_3

splitting is 1·5 times greater for the TlNO$_3$ monomer and Δv_3 for Cu(NO$_3$)$_2$ is 1·7 times as great for LiNO$_3$ despite comparable cation polarizing powers. This suggests that in the case of TlNO$_3$ and Cu(NO$_3$)$_2$ the covalent contribution to anion distortion is of the same order of magnitude as that from cation polarization of the anion.

Features indicative of dimer species, (MNO$_3$)$_2$, are also apparent for M = Li, Na, K and Rb; they are tentatively assigned cyclic structures. This particular study[49a] is noteworthy for one of the first reports of a Raman spectrum of a matrix-isolated vaporizing molecule (see Chapter 9). Raman bands were detected for the v_1 modes of the isolated potassium and copper (II) nitrates.

6.3 Compounds containing the group IIA metals

6.3.1 Halides BeF$_2$, BeCl$_2$, BeBr$_2$, BeI$_2$, MgF$_2$, MgCl$_2$, CaF$_2$, SrF$_2$ and BaF$_2$

High temperature infrared gas phase spectra have been obtained for the following alkaline earth halides; BeF$_2$, BeCl$_2$ and MgCl$_2$,[50] and the data interpreted in terms of species with symmetrical linear geometries. Subsequent electric deflection studies have demonstrated that many of the alkaline earth halides are non-linear.[51-54] Non-linearity has also been deduced from thermodynamic data.[55] Matrix isolation studies have been made on all the alkaline earth fluorides,[56-58] all the beryllium halides[59] and magnesium chloride.[60]

Mass spectroscopic data[61] on the beryllium halides indicates a gas phase containing both monomer and dimer. The relative amounts are uncertain, but the dimer probably does not exceed 5 mole per cent.[61] under low pressure conditions. The matrix spectra[56,59] in the spectral region 2000–200 cm^{-1} were characterized by a few strong, and several weak absorption bands. The former were assumed to be due to monomeric, and the latter to dimeric species. Two strong absorptions were recorded for each of the species BeF$_2$, BeCl$_2$ and BeBr$_2$, while for BeI$_2$ only one strong band was observed. The frequency assignments for these bands were made assuming $D_{\infty h}$ symmetry for all the beryllium halides. The higher frequency for each molecule was assigned to the asymmetric stretching mode, and the lower, to the bending mode. A force constant analysis, using a simple valence force field with no interaction terms, yielded a Be-halogen stretching constant of 5·15 mdyn/Å in BeF$_2$, decreasing with increasing mass of the halogen to 1·96 mdyn/Å in BeI$_2$. Bending constants of 0·12 mdyn/Å, 0·07 mdyn/Å and 0·06 mdyn/Å were obtained for BeF$_2$, BeCl$_2$ and BeBr$_2$, respectively. The BeI$_2$ bending constant was estimated at 0·04 mdyn/Å, and the bending frequency at 175 cm^{-1}. In Table 6.3, the calculated symmetric stretching frequencies are presented. For BeF$_2$, sufficient data were available to estimate[62] the interaction constant k_{12}, and the symmetric stretching frequency of BeF_2 was found to be approximately 100 cm^{-1} larger than the value listed in Table 6.3 for BeF_2 based on the two constant potential function. This result for BeF_2 suggests that a sizable error limit be attached to the symmetric stretching frequencies of all the beryllium halides, given in Table 6.3.

Table 6.3 Frequencies and assignment for the alkaline earth halides

Species	Matrix	Vibration frequencies, cm^{-1}	Point group	Bond angle	References
BeF_2	Ne	$(680^a)(\Sigma_g^+)$ $345(\pi)$ $1555(\Sigma_u^+)$	$D_{\infty h}$	180	56, 59
$BeCl_2$	Ne	$(390^a)(\Sigma_g^+)$ $250(\pi)$ $1135(\Sigma_u^+)$	$D_{\infty h}$	180	56, 59
$BeBr_2$	Ne	$(230^a)(\Sigma_g^+)$ $220(\pi)$ $1010(\Sigma_u^+)$	$D_{\infty h}$	180	59
BeI_2	Ne	$(160^a)(\Sigma_g^+)$ $(175)^b(\pi)$ $873(\Sigma_u^+)$	$D_{\infty h}$	180	59
$^{24}MgF_2$	Kr	$478{\cdot}04,\ 242{\cdot}2(A_1)$ $837{\cdot}41(B_1)$	C_{2v}	158	57, 58
$MgCl_2$		$(273)^a\Sigma_g^+$ $88(\pi)$ $590(\Sigma^+)$	$D_{\infty h}$	180	60
$^{40}CaF_2$	Kr	$484{\cdot}75,\ 163{\cdot}36(A_1)$ $553{\cdot}66(B_1)$	C_{2v}	140	58, 56
$^{86}SrF_2$	Kr	$441{\cdot}53,\ 82{\cdot}0(A_1)$ $443{\cdot}40(B_1)$	C_{2v}	108	58, 56
BaF_2	Kr	$413{\cdot}22,\ (64)^b(A_1)$ $389{\cdot}58(B_1)$	C_{2v}	100^b	58, 56

a. Calculated.
b. Estimated.

Comparison of the asymmetric stretching frequencies obtained for BeF_2 (1520 cm^{-1}) and $BeCl_2$ (1113 cm^{-1}) in a high temperature gas phase study,[50] with those reported in Table 6.3, show good agreement. However, the corresponding gas phase values for the bending frequencies at 826 cm^{-1} and 482 cm^{-1} are in obvious disagreement with the matrix study. In the matrix investigation, a weak band at 820 cm^{-1} was assigned to polymeric material, and this could be responsible for the reported gas phase band. An analogous polymeric band did not appear in the $BeCl_2$ spectrum in the 480 cm^{-1} region, and the source of this latter band is not clear. The correctness of the matrix assignment for the bending frequencies was inferred from the good agreement that resulted on comparison of the calculated entropies of BeF_2 and $BeCl_2$, based on the matrix frequencies, with those entropies derived from direct experimental determinations.

The results of investigations from two different sources[56,57,58] on the remaining alkaline earth fluorides have been reported. In the earliest of these,[56] the spectral range from 2000–250 cm^{-1} was investigated, but the measured frequency accuracy at ± 1 cm^{-1} was not sufficient to enable isotopic substitution effects to be used to determine molecular geometries. In the later investigations,[57,58] a higher frequency accuracy was available, $\pm 0{\cdot}05$ cm^{-1}, and the range investigated extended out to about 30 cm^{-1}. Frequency shifts due to the isotopes ^{24}Mg, ^{26}Mg, ^{40}Ca, ^{44}Ca, ^{86}Sr and ^{88}Sr were measured. For each of these fluorides three absorption bands were observed which relative intensity measurements suggested could be attributed to a single molecular species. Mass spectroscopic data indicates essentially no polymeric material in these systems, and hence the only molecular species present to which the bands could be assigned were MF_2 type molecules. The appearance of three bands for each species suggested molecules with C_{2v} symmetry. Using isotopic shift data, F—M—F bond angles of 158°, 140°, and 108° were obtained for MgF_2, CaF_2 and SrF_2, respectively. For BaF_2 only two frequencies were observed at approximately 400 cm^{-1}. These were assigned to BaF_2 assuming a bent structure. The absorption intensity difference between the two bands was used as the criterion for assigning the higher of the two observed frequencies to the symmetric stretching mode, an order which is reversed in the other alkaline earth fluorides. The bond angle in BaF_2 was estimated at 100° from an examination of the change of bond angle with the mass of the metal atom in the alkaline earth series. A four constant potential function was used to analyse the data. The M—F force constants were found to decrease with increasing mass of the metal atom from 2·81 mdyn/Å in MgF_2 to 1·59 mdyn/Å in BaF_2. A similar variation in the bending constants was also found; decreasing from 0·14 mdyn/Å in MgF_2, to 0·02 mdyn/Å in BaF_2. Both interaction constants were fairly small, f_{rr} ranging from 0·35 mdyn/Å to $-0·27$ mdyn/Å, and $f_{r\alpha}/r$ from, 0·03 mdyn/Å to 0 mdyn/Å. No systematic variation in these parameters with respect to the mass of the central metal atom was observed.

Electric deflection data[51–54] are consistent with bent configurations for Ca, Sr and Ba fluorides, while MgF_2 appears to be linear. The structures derived from IR spectroscopic data suggest all these fluorides are bent. The suggestion was made[57] that the apparent linearity of MgF_2 in the electric deflection experiments could result if its permanent electric dipole moment was small and below the resolving power of the electric deflection instrumentation. For a slightly bent molecule the suggested low dipole moment is not unreasonable. Further support for the non-linear configuration was inferred from the close agreement between the calculated, (81·8 eu), and observed,[64] (82·7 eu), entropy of MgF_2 at 1400 K.

Two theoretical approaches have been involved in justifying the structure of the alkaline earth halides. Ionic model calculations[65] predict the observed low bending frequencies of these species fairly well, but geometries are found to be quite sensitive to the chosen polarizability of the alkaline earth metal ion. No definitive predictions, other than the possibility of non-linearity, could be made. In the second approach, hybridization of the s, p and d orbitals of the alkaline earth metal has been considered.[60,67] For Ca, Sr and Ba the energies of the p and d orbitals are quite similar and contributions to the bonding from s-d type hybridization, resulting in non-linear geometries, appears to be justified. The non-linearity of MgF_2, and the sensitivity of the geometry with respect to the particular halogen[67] presents difficulties in this interpretation.

6.4 Compounds of the group IIIA elements

6.4.1 Oxides B_2O_3, B_2O_2, BO_2, Al_2O, Ga_2O, In_2O and Tl_2O

Of the known trivalent oxides of this group, only B_2O_3 exists in the gas phase as such.[68] Gas phase electron diffraction studies[69] favour a V-shaped geometry of C_{2v} symmetry, with an apex angle of about 95°. Electric deflection experiments indicate the molecule has a permanent electric dipole, which, though small, is consistent with the V-shaped structure. A high temperature IR gas phase study[70] has been interpreted as favouring the C_{2v} structure O=B—O—B=O, due to the presence of bands in the spectrum at about 2000 cm^{-1}, whilst in a second study[71] a single absorption band observed at about 1400 cm^{-1} was taken as implying a trigonal bi-pyramid arrangement of the atoms.

Two matrix isolation studies on the vapours over boric oxide have been reported.[73,74] Isotopic enrichment of the boric oxide samples with ^{10}B and ^{11}B was used in both studies. For unknown reasons cleaner and more well defined spectra were obtained in the later study.[74] Essentially the same spectra were obtained in both studies, and although there were a few minor differences in interpretation, these largely resulted from difficulties associated with the poorer spectra in the earlier study. The frequencies and assignments for B_2O_3 given in Table 6.4 are from the later study.[74]

Interpretation of the B_2O_3 spectra was made assuming C_{2v} symmetry, and that the O—B=O groups were linear. Such a molecule has nine fundamental frequencies of which 8 are infrared-active. In the matrix investigations in the range from 300–4000 cm^{-1}, six frequencies were observed which could reasonably be assigned to the B_2O_3 species; the only major species in the vapour phase.[68] Of the eight infrared-active frequencies, the B—O—B bend can be expected to lie below 300 cm^{-1}, the long wavelength limit of the investigation, leaving seven frequencies to be assigned. Assignment of the observed frequencies to the various symmetry species was made initially by analogy with those in known boron-oxygen compounds. Correlation between the symmetry species of the C_{2v} and $D_{\infty h}$ point group, combined with absorption band intensity observations suggested that the apex angle was considerably greater than the 95° deduced from electron diffraction data.[69] Features appearing in the spectra, when large amounts of boric oxide were trapped in the matrix were shown to be consistent with allowed overtone or combination bands based on the C_{2v} assignment. Evidence was presented which suggested that two of the infrared-active frequencies were probably identical. A valence force field analysis of the data with five principal force constants, five interaction terms and an apex angle of 150° was found to reproduce the observed frequencies, if the unobserved B—O—B bending frequency was assumed to lie between 260–172 cm^{-1}. Agreement between the calculated entropy of gaseous B_2O_3, based on this frequency assignment,

Table 6.4 Frequencies and assignments for the group IIIA oxides and halides

Species	Matrix	Vibration frequencies, cm^{-1}	Point group	Bond angle	References
$^{11}B_2O_3$	Ar	2060, 729, 518[172] or [200](A_1) 456$^a(A_2)$, 2060, 1239, 454(B_1) 477(B_2)	C_{2v}	150	73, 74
$^{11}B_2O_2$	Ar	1899 Σ^+	$D_{\infty h}$	180	73, 74
BO_2	Ar	1276(Σ_u^+)	$D_{\infty h}$	180	74
Al_2O	Ar	994(B_1) 714, 120$^a A_1$	C_{2v}	145	76, 77
Ga_2O	N_2	809·4(B_1)	C_{2v}	143	78
In_2O	N_2	722·4(B_1)	C_{2v}	135	78
Tl_2O	N_2	625·3(B_1)	C_{2v}	131	78
	N_2	623·2$^b(B_1)$	C_{2v}	77	84a
	N_2	622·0$^c(B_1)$	C_{2v}	95	84a
	Ar	643·3(B_1)	C_{2v}	89	84a
	Kr	634·6(B_1)	C_{2v}	104	84a
AlF_3	Ne	960, 252(E'), 284(A_2'') 660$^a(A_1')$	D_{3h}	120	85, 86
Al_2F_6	Ne	979, 340(B_{1u}), 600(B_{2u}), 805, 575, 300(B_{3u})	D_{2h}		85
AlF	Ne	785(Σ^+)	$C_{\infty v}$		85
TlF	Ar	441	$C_{\infty v}$		89a
Tl_2F_2	Ar	316(Σ_u^+) 256(π_u)	$D_{\infty h}$		89a
$TlCl$	Ar	261	$C_{\infty v}$		89a
Tl_2Cl_2	Ar	184(Σ_u^+) 168(π_u)	$D_{\infty h}$		89a
$TlBr$	Ar	179	$C_{\infty v}$		89a
TlI	Ar	143	$C_{\infty v}$		89a

a. Calculated values. [ ] Estimated values.
b. Data recorded with IR-7 spectrometer.
c. Data recorded with IR-11 spectrometer.

and the experimentally derived value, was satisfactory. Since the latter has a rather large experimental error, the comparison did not allow any firm conclusions to be made with regard to the reliability of the estimated B—O—B angle at 150°, or the magnitude of the estimated bending frequency at between 172–260 cm^{-1}.

In both the above studies,[73,74] spectra were observed of the vapour species over heated mixtures of B + B_2O_3. Mass spectroscopic data[68] show that the species B_2O_2 is formed under these conditions. Only one set of absorption bands assignable to this species was observed at 1955 cm^{-1} and 1899 cm^{-1} for the ^{10}B and ^{11}B substituted species, respectively. A linear symmetric structure O=B—B=O was assumed[74] with $D_{\infty h}$ symmetry. The observed frequency was assigned to the asymmetric stretching mode, and the only other IR active mode, the bending frequency, was assumed to lie below 300 cm^{-1}. Assuming values for the B—B stretching, and the B—B=O bending force constants, the IR inactive frequencies were calculated using a three constant potential function.

The only other boron-oxygen species for which there is matrix data[74] is BO_2. This was formed by the reaction; $B_2O_3 + ZnO = Zn + 2BO_2$. The gas phase electronic spectrum of this species has been observed and the three fundamental vibration frequencies of the symmetrical linear species deduced. The matrix spectra were obtained to help clarify certain details of the B_2O_3 spectra. Only the asymmetric stretching frequencies of $^{11}BO_2$ (1276 cm^{-1}) and $^{10}BO_2$ (1323 cm^{-1}) were observed. The gas phase value for $^{11}BO_2$ is reported at 1322 cm^{-1}. No matrix frequency corresponding to the gas phase bending mode at 464 cm^{-1} was observed. This may have been due to the band being overlapped by a B_2O_3 absorption in this region.

Matrix spectra for oxides of the type, M_2O, with M=Al, Ga, In and Tl have been reported.[76-78] Aluminium super oxide is the major gas phase species[79] when aluminium oxide and aluminium metal are heated at about 1600 K or when Al_2O_3 is vaporized at about 2600 K.[79] The first investigation[76] of this species covered the spectral range 2000–250 cm^{-1}. With $^{16}O_2$ and $^{18}O_2$ isotopic enrichment of the Al_2O_3, two sets of absorption bands at 994 cm^{-1} and 715 cm^{-1} (^{16}O), and 958 cm^{-1} and 700 cm^{-1} (^{18}O), were observed. No absorption bands corresponding to a mixed $^{16}O^{18}O$ species were recorded. The lower frequency for both isotopes occurred at a much lower absorption intensity than the higher frequency. The former was assigned to the symmetric stretching mode, and the latter to the asymmetric stretching mode of Al—O—Al, with C_{2v} symmetry. From the measured isotopic frequency shift of the asymmetric mode, the apex angle was calculated at 145 ± 5°. The unobserved bending frequency was estimated from electronic spectra attributed to Al_2O, at 240 cm^{-1}. In a later matrix study of the Al_2O spectrum,[77] extending out to 190 cm^{-1}, no frequency corresponding to the predicted bending mode was observed. Based on the proposed similarity of the Li_2O and Al_2O species,[81] the Al_2O bending frequency was estimated at 120 ± 30 cm^{-1}. For apex angles between 150–140°, the Al—O force constant F_r was determined at 5·25 ± 0·35 mdyn/Å, the interaction constant F_{rr} at 1·4 ± 0·5 mdyn/Å. The bending constant was estimated at 0·035 ± 0·003 mdyn/Å.

The oxides Ga_2O, In_2O and Tl_2O are the major vapour phase species over the heated solid oxides, M_2O_3.[82] Electron diffraction data,[83] and low resolution IR studies[84] on Ga_2O and In_2O indicate these molecules are bent. In the matrix isolation study,[78] ^{16}O and ^{18}O isotopic substitution was made. Only one intense absorption band was observed for each of the species in the 600–900 cm^{-1} range. This was assigned to the asymmetric stretching mode of a bent M—O—M species. Apex angles were calculated at 143 ± 5°, 135 ± 7° and 131 + 11° for the Ga, In and Tl oxides, respectively, in fairly good agreement with the electron diffraction values. The principal stretching force constants F_r were calculated assuming the interaction constant $F_{rr} = 0·1F_r$. Values of 3·05 ± 0·35 mdyn/Å, 2·65 ± 0·3 mdyn/Å and 2·12 ± 0·25 mdyn/Å were obtained for the Ga, In and Tl oxides, respectively. These values substituted in a bond length-force constant relationship,[33] resulted in estimated bond lengths in good agreement with the electron diffraction values.[83]

Thallous oxide has been more recently examined[84a] by vaporizing Tl_2O at 700–770 K and trapping in Ar, Kr, and N_2 at 10 K. From isotopic shifts for $Tl_2{}^{16}O$ and $Tl_2{}^{18}O$ it is calculated that the bond angle is near 90°: Ar 89°, Kr 104°, N_2 95°. Even though the uncertainty in these values is at least 15° it is still difficult to reconcile the difference between them and the previous[78] result. The Tl—O bond distance is estimated to be 2·63 Å and the Tl—Tl distance 2·63 Å, which is suggestive of strong metal–metal interaction in the Tl_2O molecule.

6.4.2 Halides AlF, AlF_3, Al_2F_6, TlF, Tl_2F_2, TlCl, Tl_2Cl_2, TlBr and TlI

The compounds AlF_3, Al_2F_6 and AlF were the first group IIIA halides to have been investigated using the matrix technique. High temperature gas phase spectra for AlF_3 have been recorded.[83a,84b] Three IR active frequencies were observed and assigned by analogy with BF_3. Two matrix studies have been reported[85,86] both obtaining similar results, but only one of these is in the open literature.[85] From mass spectral data[83] the vapour phase over aluminium fluoride is known to consist largely of monomer, together with a few per cent. of dimer. AlF_3 is expected to have D_{3h} symmetry by analogy with BF_3, three of the four fundamental frequencies being active in the IR. In the matrix spectra there were two groups of intense absorption bands, at 920–960 cm^{-1} and 244–284 cm^{-1}, and several much weaker absorptions in the range from 1000–200 cm^{-1}. The intense absorption bands were assigned to monomeric aluminium trifluoride. The high frequency bands at 920–960 cm^{-1} were assigned to a doubly degenerate E' mode. In neon, argon and krypton matrices, at least four well defined peaks appeared in this region, and it was assumed that these resulted from matrix effects on a single fundamental frequency. Different trapping sites, removal of degeneracy by the matrix environment, or a combination of both was suggested. In the intense low frequency region at 284–244 cm^{-1}, the spectra appeared differently in the three different matrices. In neon, there were two well defined absorption bands separated by 28 cm^{-1}, in argon, there were three, with separations of 11 cm^{-1} and 9 cm^{-1}, and in krypton, only one very intense band. By analogy with boron trifluoride, AlF_3 is expected to have one doubly degenerate, and one out-of-plane bending mode in this region, the former probably having the lower frequency. The features appearing in the matrix spectra between 288 cm^{-1} and 244 cm^{-1} were assigned to these two vibrational modes with the assumption that in the krypton matrix the two frequencies were accidentally degenerate. In order to assign the absorption bands in the argon and neon matrices to specific vibrational modes, use was made of an orientation effect which appears to exist for planar molecules in matrix environments. Investigation had shown[85] that for the planar molecules, BF_3 and Li_2F_2, variations in absorption band intensities with respect to the angle at which the analysing beam interacts with the matrix sample, could be simply interpreted if it was assumed that molecules preferentially orient with their planes parallel to the surface on which the matrix is deposited. Out-of-plane vibrational modes might be expected at lower absorption intensities with the analysing beam perpendicular to the matrix surface than when at some lower angle. This approach applied to AlF_3 indicated that the higher of the two frequencies occurring in the 284–244 cm^{-1} region could be assigned to the out-of-plane bending mode, and the lower, to a degenerate E' mode. Curiously the low frequency E' mode did not exhibit any matrix splitting analogous to that observed for the higher frequency E' mode. Comparison of the frequencies listed in Table 6.4 for AlF_3, with those obtained in the high temperature gas phase study at 940 cm^{-1} (E'), 263 cm^{-1} (E') and 297 cm^{-1} (A_2'') show good agreement. A simple valence force field analysis without interaction terms[87] gave $k_1 = 4.91$ mdyn/Å, $k_{\delta/l^2} = 0.185$ mdyn/Å and $k_{\Delta/l^2} = 0.324$ mdyn/Å. A value of 660 cm^{-1} was calculated for the IR inactive A' frequency. A similar force field analysis for the BF_3 molecule based on the three observed IR active frequencies results in a calculated A' value about 25 per cent. lower than the known value obtained from Raman spectra, and suggests that for AlF_3, the A' mode is probably closer to 800 cm^{-1} rather than 660 cm^{-1}.

Since aluminium fluoride vapour is known to contain a small amount of dimer, six weaker absorption bands observed in the matrix spectra[85] were assigned to Al_2F_6. That these absorption bands increased in intensity, compared to those assigned to AlF_3, under conditions of poor isolation lent some support to this conclusion.

The structure of Al_2F_6 is not known; Al_2Cl_2 and Al_2Br_6 have D_{2h} symmetry.[88,89] Assuming the same symmetry for Al_2F_6, the vibrational assignment given in Table 6.4 was made. A valence force field analysis, without interaction terms, gave rather poor agreement between observed and calculated frequencies.

In a more recent study[89a] all of the thallous halides have been examined in Ar and Kr at 4 K over the range 4000–33 cm^{-1}. In addition to the absorption frequencies of the TlX monomers (Table 6.4), the spectra of the Tl_2F_2 and Tl_2Cl_2 dimers have been observed. Only two bands were assignable to the dimers which is consistent with a linear symmetrical X—Tl—Tl—X geometry for these molecules. A comparison of the observed to calculated entropy of Tl_2F_2 and Tl_2Br_2 adds support to the spectral assignments.

6.5 Compounds of the group IVA elements

6.5.1 Carbon species C_3

The vaporization of carbon at 2500 K results in the species C_1, C_2, C_3, C_4, and C_5.[91] The approximate relative amounts are C_3 (60 per cent.), C_1 (30 per cent.), and C_2 (20 per cent.) with the C_4 and C_5 polymers minor constituents. Gas phase electronic spectra have been observed for C_1, C_2 and C_3, and energy levels determined for C_1 and C_2. Interpretation of the C_3 spectrum proved difficult, the molecule[92] was assigned a linear structure. The bending mode frequency was reported at 63 cm^{-1}.

The IR spectrum of C_3 has been observed in argon matrices[93,94] in the spectral range 3000–280 cm^{-1}. Under the best conditions of isolation, only one absorption band appeared at 2039 cm^{-1}. It was assigned to the $\sum_u^+$ mode of C_3. The non-appearance of any other absorption bands was interpreted as supporting a linear $D_{\infty h}$ structure for C_3. Using a graphite sample, containing 60 per cent. ^{13}C, six absorption bands were observed in the 2000 cm^{-1} region corresponding to the six possible C_3 isotopic species. For unknown reasons the expected intensity distribution based on the isotopic composition was not realized. A value of 1235 cm^{-1} was obtained for the $\sum_g^+$ mode of C_3 from matrix fluorescence spectra.[94] Coupling this with the two IR active frequencies, values were obtained for the stretching and interaction force constants of $k_r = 10\cdot34$ mdyn/Å and $k_{rr} = 0\cdot54$ mdyn/Å respectively. Using these force constants, the frequencies of the absorption bands attributed to the ^{13}C substituted species were calculated and found to agree with the observed values to within 1 cm^{-1}. The magnitude of the carbon–carbon force constant at $10\cdot34$ mdyn/Å in C_3 was observed to be similar to that found in ethylenic double bonds. This was taken as support for the theoretically predicted double bond character of the carbon–carbon linkages in C_3.[95,96]

In both the above matrix studies, and in a later one,[97] the effect of allowing diffusion of the trapped carbon species in the matrix was reported. Many new absorption bands appeared in the 2400–700 cm^{-1} region. These were assumed to be due to the formation of higher carbon polymers. Theoretically a linear configuration had been predicted for polymers up to C_{10}, with polymers containing an odd number of carbon atoms more stable relative to those with an even number.[95,96] Calculated stretching frequencies for polymers with equivalent bonds[98,99] were made, and used as a guide in interpreting the spectrum. Assignments of absorption bands to species ranging from C_4 to C_{10} were made. The theoretically predicted[95,96] structure for C_4, .C=C—C=C. was interpreted as being consistent with the observed spectra. Isotopic studies were not made, and in view of the complexity of the spectra, these assignments must be regarded as tentative.

6.5.2 Silicon carbide, SiC_2

The vaporization of silicon carbide yields a variety of silicon carbon polymers[100] not unlike those observed with pure graphite. The major species is Si (83 per cent.), together with SiC_2 (9 per cent.), Si_2C (8 per cent.). The total of all other polymers was put at probably less than 1 per cent. The electronic spectrum of SiC_2 has been observed in the gas phase,[101] and interpreted as favouring a linear unsymmetrical structure for the molecule. The matrix technique[102] has been used to observe the infrared, visible and near-ultra-violet spectra of the vapours over heated silicon carbide. From an analysis of the electronic spectra, the ground state vibration frequencies given in Table 6.5 were derived. In the IR region (280–4000 cm^{-1}), two strong bands

Table 6.5 Frequencies and assignment for the group IVA compounds

Species	Matrix	Vibration frequencies, cm^{-1}	Point group	Bond angle[b]	References
C_3	Ar	$1235(\Sigma_g^+)$ $63\cdot1(\pi u)$ $2038(\Sigma_u^+)$	$D_{\infty h}$	180	93, 94 and 97
C_2Si	Ne	$853(\Sigma^+)$ $300(\pi)$ $1742(\Sigma^+)$	$C_{\infty v}$	180	102
SiO	N_2	$1223\cdot9(\Sigma^+)$	$C_{\infty v}$		107, 108
Si_2O_2	N_2, Ne	$804\cdot7(B_{2u})$ $766\cdot3(B_{3u})$ $79\cdot7(B_{1u})$	D_{2h}	87	107, 108
Si_3O_3	N_2	$972\cdot6$, $631\cdot5$, $312(E')$	D_{3h}	100	107, 108
GeO	N_2	$973\cdot4(\Sigma^+)$	$C_{\infty v}$		113
Ge_2O_2	N_2	$667(B_{2u})$ $599(B_{3u})$	D_{2h}	83	113
Ge_3O_3	N_2	824, $440(E')$	D_{3h}	100	113
Ge_4O_4	N_2	$553\cdot3$, $490\cdot0$, $476\cdot0$, $457\cdot5$	C_{4v}?		113
PbO	N_2	$718\cdot4$	$C_{\infty v}$		113a
Pb_2O_2	N_2	$557\cdot4$, (B_{2u}), $467\cdot7(B_{3u})$	V_h	79	113a
Pb_4O_4	N_2	$474\cdot4(F_2)$ $374\cdot4(F_2)$	T_d	81	113a
SiF_2	Ne	$864\cdot6(B_1)$ $851\cdot5$ 345 (A_1)	C_{2v}	$100\cdot9$	119
GeF_2	Ne	685 $263(A_1)$ $655(B_1)$	C_{2v}	94	120
$GeCl_2$	Ar	$398(A')$ $373(B_1)$	C_{2v}	90–100[a]	121
$SnCl_2$	Ar	$354(A')$ $334(B_1)$ $120(A')$	C_{2v}	90–100[a]	121
$PbCl_2$	Ar	$297(A')$ $321(B_1)$	C_{2v}	90–100[a]	121

a. Estimated values.
b. Bond angles given for the cyclic oxides are for the $\angle$ OMO, where M = Si or Ge.

at 1751 cm^{-1} and 835 cm^{-1} were assigned to the C—C and Si—C stretching frequencies due to their close similarity with the values obtained from the electronic spectrum. No bending frequency was observed, possibly due to this mode having a low extinction coefficient, or because the matrix shift put it beyond the long wavelength limit of the investigation. The stretching force constants in SiC_2 were calculated on the basis of valence forces[87] without including interaction terms. This gave; $k(Si—C) = 7\cdot44$ mdyn/Å and $k(C—C) = 7\cdot98$ mdyn/Å. The force constant for the Si—C single bond is typically about $3\cdot1$ mdyn/Å. The considerably larger Si—C force constant in SiC_2 was taken as indicating the presence of a silicon-carbon double bond. That the C=C stretching force constant in C_3 was larger than in SiC_2 was interpreted in terms of some loss of stabilizing electron delocalization SiC_2 compared to C_3.

In addition to the two absorption bands assigned to SiC_2, many others appeared in the IR region from 2000–500 cm^{-1}. Their existence was justified in terms of higher silicon–carbon polymers, and an assignment of some of them to Si_2C_3 was made. In the absence of isotopic studies, the assignment must be regarded as tentative.

6.5.3 Oxides

Silicon oxides SiO, Si_2O_2, *and* Si_3O_3

The gas phase electronic spectrum of SiO is well characterized.[103] No spectral data are available for Si_2O_2 and Si_3O_3. SiO is known to be the major vapour species when SiO_2 is vaporized, together with a very small amount of SiO_2.[104] In heated SiO_2 and Si mixtures, the species Si_2O_2 is known to be formed.[105] The species SiO is also formed in largely monomeric state when solid SiO is heated.[106] Two matrix studies have been reported on the silicon oxide vapour species produced by heating SiO_2,[107,108] $SiO_2 + Si$[107,108] and SiO.[109] In both studies essentially the same results were obtained irrespective of the means of producing the silicon-oxide species. Even under the best conditions of isolation, many absorption bands appeared in the spectral region below 1300 cm^{-1}. This behaviour led to the suggestion that efficient isolation of SiO appeared to be impossible and that higher polymers were responsible for many of the absorption bands.

From intensity considerations, three different groups of absorption bands were characterized. The highest frequency group, at about 1224 cm^{-1}, was assigned to monomeric SiO by analogy with the gas phase value of 1220 cm^{-1}. Frequency shifts occurring in these bands due to the naturally occurring silicon isotopes ^{28}Si, ^{29}Si and ^{30}Si,[108,109] and for the ^{18}O substitution,[108] were in agreement with those expected for monomeric silicon oxide.

The second group of absorption bands at about 850 cm^{-1}, 766 cm^{-1}[108,109] and 79 cm^{-1}[109] were assigned to a cyclic dimer with D_{2h} symmetry. Using an oxide sample containing approximately 45 per cent. ^{18}O, the two higher frequency bands were shown to have the characteristic triplet structure and relative intensities expected of a D_{2h} molecule.[7] Oxygen isotope effects were not observed for the 79 cm^{-1} band, but growth rates on diffusion, strongly suggested its assignment to the same molecular species as was responsible for the two higher frequencies. Frequency shifts observed and calculated for the various isotopic species of the two high frequency modes were in agreement with those expected for the IR active B_{3u} and B_{2u} modes of an Si_2O_2 molecule with D_{2h} symmetry. Since the expected frequency ratios on isotopic substitution are identical for both the B_{3u} and B_{2u} modes, the data could not be used to decide which of the two observed frequencies corresponded to a given vibrational mode. In an attempt to make such an assignment, geometrical arguments were used.[108] It has been shown[20] that to a fairly good approximation, the ratio of the $B_{2u} : B_{3u}$ frequencies may be used to calculate the $O\widehat{S}iO$ angle. Assuming $v(B_{2u}) > v(B_{3u})$, the $O\widehat{S}iO$ was calculated at 87°, while if $v(B_{3u}) > v(B_{2u})$, the $O\widehat{S}iO = 93°$. The lower angle favours an Si—Si distance greater than the O—O distance, the converse being true for the larger angle. Assuming the bonding in Si_2O_2 is essentially covalent, and transannular repulsions are important, the larger covalent radius of silicon as opposed to oxygen[110] was taken as evidence in favour of the small $O\widehat{S}iO$ angle, and hence $v(B_{2u}) > v(B_{3u})$. Similar arguments applied to Li_2O_2, predict $v(B_{3u}) > v(B_{2u})$, in agreement with the original assignment and theoretical predictions for this molecule.[20] Assuming a three constant potential function for the in-plane vibration modes, and arbitrarily setting the OSiO and SiOSi bending constant equal to 0·075 mdyn/Å, the principal Si—O stretching constant was calculated at 3·55 mdyn/Å. This force constant was then used to estimate the SiO bond distance at 1·71 Å and an Si—O bond order of 1·27.[111] The latter value was taken as implying the presence of some π bonding in Si_2O_2.

An assignment for the third group of absorption bands at about 930 cm^{-1}, 630 cm^{-1} and 312 cm^{-1} was less certain. Only the higher frequency band appeared with sufficient absorption intensity to allow the effect of ^{18}O enrichment to be clearly observed.[108] The band exhibited a quartet structure. A tentative assignment for these frequencies to a cyclic trimer of D_{3h} symmetry was made. However the observed intensity variations in the isotopically substituted species did not agree with the predicted value. The three frequencies were assigned to the in-phase doubly degenerate E' modes of Si_3O_3, with the assumption that the A'', out-of-plane bending mode lay below 200 cm^{-1}.[108] In the region 200–33 cm^{-1} no absorption band appeared which could be clearly assigned to the A_2'' mode, but the observation was made that the rather broad dimer band at 79 cm^{-1} might possibly be caused by the overlapping of two different frequencies.[109] The $O\widehat{S}iO$ given in Table 6.5 for Si_3O_3, based on D_{3h} symmetry, and some simplifying assumptions with regard to the force constant, must be taken as tentative.

Germanium oxides GeO, Ge_2O_2, Ge_3O_3 *and* Ge_4O_4

The electronic spectrum of gaseous GeO is well characterized. No data are available for the other oxides. Mass spectroscopic studies[112] on the vaporization of germanium oxide have established GeO as the principal species, together with significant amounts of Ge_2O_2 and Ge_3O_3. A matrix isolation study of the system[113] has been made by the group of workers studying the silicon-oxygen species.[108] The same criteria as used in the silicon-oxygen study were used in making the assignments for the various absorption bands. The spectra in both studies were very similar except that more absorption bands appeared in the germanium-oxygen system on diffusion of the species in the matrix. Some of these latter features were tentatively ascribed to a tetramer Ge_4O_4, with C_{4v} symmetry. Frequency assignments to the B_{3u} and B_{2u} modes of the Ge_2O_2 dimers were made using similar geometrical arguments as were made for Si_2O_2. From a force constant analysis, analogous to that used in Si_2O_2, a GeO force constant of 2·98 mdyn/Å was derived, and a bond length of 1·87 Å estimated. The assignment of absorption bands given in Table 6.5 to a cyclic trimer, with D_{3h} symmetry, must be regarded as tentative since on isotopic substitution, expected absorption band intensity ratios were not realized.

Lead oxides PbO, Pb_2O_2 *and* Pb_4O_4

The matrix isolation of PbO or ^{18}O-enriched PbO vapours in N_2 and Ar at 15 K has led[113a] to the characterization of the three principal vapour species PbO, Pb_2O_2 and Pb_4O_4. A normal co-ordination analysis for Pb_2O_2 suggests that this molecule is a planar four-membered ring iso-structural with Si_2O_2, Ge_2O_2 and Sn_2O_2. The spectrum of Pb_4O_4 is interpreted in terms of the T_d structure adopted by the isoelectronic ion $Pb_4(OH)_4^{4+}$. Approximate molecular parameters for the species are derived.

6.5.4 Halides, SiF_2, GeF_2, $GeCl_2$, $SnCl_2$ and $PbCl_2$

Microwave[114] and electronic spectra,[115,116] and electron diffraction[117] and molecular beam electric deflection experiments,[54] have been used to demonstrate that all these halides are symmetric non-linear molecules. An IR gas-phase study[118] on SiF_2 was inconclusive, but indicated that the symmetric and asymmetric stretching frequencies were probably very similar. The matrix investigations[119,120] of the Si and Ge fluorides were made in the range 400–2000 cm^{-1} with the object of determining the two stretching frequencies. The bending frequency had previously been determined from electronic spectra.[115,118]

SiF_2 was obtained by passing SiF_4 over heated silicon and GeF_2 by simple vaporization of the solid difluoride. Due to some disproportionation of the GeF_2 on heating, a small amount of GeF_4 was formed. Absorption bands appearing in the 860 cm^{-1} and 675 cm^{-1} region were assigned to the two stretching modes of SiF_2 and GeF_2 respectively. Using the observed frequencies of the naturally occurring silicon and germanium isotopes, bond angles were calculated assuming that either of the two observed absorption bands for each of the species might be assigned to the antisymmetric stretching mode. For SiF_2 the higher frequency gave a bond angle of 97·5 $\pm$ 1°, compared to the gas phase microwave value[114] of 100·0°. The same calculation using the lower SiF_2 frequency resulted in an angle of 81 $\pm$ 2° in poor agreement with the microwave value. The GeF_2 bond angle is not known, but application of the same technique as in SiF_2 for calculating bond angles, resulted in angles of 94 $\pm$ 2° and 82 $\pm$ 3° for the lower and higher frequencies, respectively. A bond angle of <90° was considered very unlikely and the lower of the two frequencies was assigned to the asymmetric stretching mode. Using a three term potential function, to analyse the data, values of $k_1 = 4.90$ mdyn/Å and 4.07 mdyn/Å, $k_{\delta/l^2} = 0.456$ mdyn/Å and 0.319 mdyn/Å and $k_{12} = 0.18$ mdyn/Å and 0.22 mdyn/Å were obtained for the Si and Ge compounds, respectively.

In the matrix study of the germanium, tin and lead dichlorides,[121] in the range 200–2000 cm^{-1}, the MCl_2 species were generated by simple vaporization of the dichlorides. For each halide two absorption bands were observed. Growth rate studies and relative intensity measurements strongly implied that the two absorption features arose from the same molecular species in each case. They were assigned to the symmetric and asymmetric stretching modes of the dichlorides. The frequency accuracy in this investigation was ± 0.5 cm^{-1}, which did not permit bond angle calculations, based on isotope shifts to be made with sufficient accuracy to differentiate between the symmetric and asymmetric frequencies. For $SnCl_2$, the symmetric stretching frequency was known from the electronic spectrum,[116] and in this case it was clear that the lower of the two observed matrix frequencies corresponded to the asymmetric stretching mode. A similar definitive assignment for the two observed frequencies of $GeCl_2$ and $PbCl_2$ was not possible. Qualitatively the symmetric stretching mode is expected to occur at a lower absorption intensity than the asymmetric stretching mode. On this basis, and by analogy with $SnCl_2$, the lower of the two observed frequencies of $GeCl_2$ and $PbCl_2$ were assigned to the asymmetric stretching mode. The bending frequencies for all these molecules are expected to lie below 200 cm^{-1}, based on data from the gas phase electronic spectra.

6.6 Compounds of the group IIB metals

6.6.1 Halides ZnX_2, CdX_2 and HgX_2

Electron diffraction studies[122] and electric deflection[123] experiments on the group IIB halides are consistent with a linear symmetrical geometry, $D_{\infty h}$, being assigned to these species. The asymmetric stretching frequencies of all but the fluorides have been observed in high temperature IR absorption[124] and emission[125,126] studies. Raman studies[127,128] on all the solid and molten halides, with the exception of the fluorides, have been reported, and assignments made for the symmetric stretching modes. Bending frequencies have only been reported for $ZnCl_2$, $ZnBr_2$ (IR study),[129] and for $HgCl_2$, $HgBr_2$ and HgI_2 (Raman study).[130] These latter assignments have recently been questioned.[130]

The matrix method was used to obtain the IR spectra of $ZnCl_2$, $CdCl_2$ and $HgCl_2$ down to 250 cm^{-1}.[131] This study was later superseded by a more complete investigation of all the halides in the range of 800–33 cm^{-1}.[132,133] In all experiments, the spectra of the gaseous species in the saturated and unsaturated superheated vapour were observed. For the zinc halides isotopic enrichment of the samples was used. The spectra of all the halides showed many similarities with regard to the number of absorption bands, and their intensity behaviour with respect to isolation of the saturated and unsaturated vapour. The higher and the lower frequencies for each halide were assigned to asymmetric stretching, and the symmetric bending modes, respectively. The remaining absorption bands were assigned to polymeric material.

Table 6.6 Vibration frequencies (Kr matrices) and force constants of group IIB halides

| Species | Vibration frequencies,[a] cm^{-1} | | | k_1 mdyn/Å | k_{δ/l^2} mdyn/Å |
	v_1^b	v_2	v_3		
ZnF_2	600	150	758	4·04	0·080
$ZnCl_2$	351	102	508	2·55	0·052
$ZnBr_2$	217	74	404	2·20	0·038
ZnI_2	155	62	346	1·81	0·030
CdF_2	572	123	662	3·67	0·063
$CdCl_2$	327	88	419	2·24	0·050
$CdBr_2$	205	62	319	1·98	0·037
CdI_2	149	50[c]	269	1·67	0·029
HgF_2	588	171	641	3·87	0·139
$HgCl_2$	348	107	405	2·53	0·089
$HgBr_2$	219	73	294	2·27	0·070
HgI_2	158	63	237	1·87	0·065

a. All frequencies taken from references 132 and 133.
b. Calculated values.
c. Estimated by analogy with other halides.

The frequencies assigned to the group IIB halides are listed in Table 6.6. For unknown reasons, the v_2 bending frequency of CdI_2 was not observed. Its value was calculated using a bending force constant estimated from those of the other halides. Since only two of the three fundamental frequencies could be observed, the force constant analysis was made in terms of a two constant potential function: k_r, the MX stretching force constant and, k_{δ/r^2}, the XMX bending force constant. In those cases where the symmetric and asymmetric stretching frequencies were known,[125] the interaction constant k_{rr} was found to be less than 1 per cent. of k_r and its neglect is not likely to have a serious effect on the calculation of the infrared-inactive frequency v_1. In general, the previously reported bending frequencies[126,129] for these halides do not agree with those reported in Table 6.6. In the matrix spectra of $ZnCl_2$ and $ZnBr_2$ broad bands centred at about 300 cm^{-1} and 200 cm^{-1}, respectively, were assigned to polymeric species and it appears possible that these may correspond to the previously assigned gas phase bending frequencies.[124] As noted earlier, the previously reported bending frequencies for $HgCl_2$, $HgBr_2$ and HgI_2, obtained from the UV absorption spectra[129] have recently been questioned,[130] and cannot be regarded as seriously challenging the validity of the matrix assignments for these modes.

Inspection of the force constants given in Table 6.6 shows a decrease in magnitude occurs both in the stretching and bending constants when a halogen is exchanged by a heavier one, but a minimum is exhibited at cadmium when the metal atom is exchanged. It has been observed that this minimum also correlates with the minimum in the metal atom ionization potential, giving support to the importance of ionic character in the bonding of these molecules. Indeed, ionic model calculations[124] predict rather well the bending force constants for these species. That the bending force constants of the mercury halides are consistently higher than those of the corresponding zinc and cadmium halides has been attributed to increase in the covalent character of the mercury halogen bond and concomitant increased directional bond character.

In another matrix study on the ZnF_2[134] vapour species, the observed shifts in the asymmetric stretching frequencies due to the isotopes ^{64}Zn, ^{66}Zn, ^{67}Zn, ^{68}Zn and ^{70}Zn have been used to test the assumption of linearity of this species. Slightly better agreement between the observed and calculated frequencies for the various isotopes was obtained if a bond angle of 157° were assumed, rather than 180°. The significance of these conclusions at the present time must be considered somewhat questionable since extreme reliability must be placed on the accuracy of the frequency measurements, and anharmonic effects, which may or may not be important, were neglected.

6.7 Transition metal compounds

6.7.1 Fluorides TiF_2, TiF_3, CoF_2, NiF_2, and CuF_2; rare-earth trifluorides

There have been no gas phase IR spectroscopic studies reported for any of these halides. Molecular beam electric deflection studies indicate that the transition metal dihalides are linear,[135] though as has been noted,[134] for bond angles greater than 120°, the sensitivity of the method is probably insufficient to discriminate between linearity and nonlinearity.

In the matrix study on the titanium fluorides,[134] titanium trifluoride was vaporized to obtain TiF_3, and superheated TiF_3 vapour was passed over titanium metal to form TiF_2. The frequency range from 2000–400 cm^{-1} was examined. One strong absorption band appeared in the spectrum of both TiF_3 and TiF_2. From their well defined titanium isotope structure they were assigned to TiF_3 (800 cm^{-1}) and TiF_2 (750 cm^{-1}). Other absorption bands also appeared, and intensity changes on diffusion in the matrix, or under varying conditions of isolation, indicated that most of these could probably be assigned to polymeric material. However, in both the TiF_3 and TiF_2 spectra, one weak band was observed in each, at 768·5 cm^{-1} and 643 cm^{-1}, respectively, which was tentatively assigned to a non-polymeric species.

The strong absorption band at $\sim 750 \, \text{cm}^{-1}$ in the TiF_2 spectrum was assigned to the asymmetric stretching mode, and the observed isotopic frequencies were used to derive a bond angle of $130 \pm 5°$. For a bent molecule of C_{2v} symmetry, all three fundamental frequencies are IR active and the weak frequency of $643 \, \text{cm}^{-1}$ was assigned to the symmetric stretching mode. The bending frequency was assumed to lie below $400 \, \text{cm}^{-1}$. Two possible structures were considered for the trifluoride, planar symmetrical D_{3h}, and pyramidal, C_{3v}. The former would be expected to have one IR active frequency in the region examined and the latter two. Because of the uncertainty with regard to assigning both the $800 \, \text{cm}^{-1}$ and $768 \, \text{cm}^{-1}$ bands to the TiF_3 molecule, a firm assignment for the TiF_3 structure could not be made.

The matrix spectra of CoF_2,[136] NiF_2 [136,137] and CuF_2 [136] have been examined in the $2000-400 \, \text{cm}^{-1}$ region. In all cases one strong absorption band and several much weaker features were observed for each species. The weaker features, from intensity measurements under differing experimental conditions, were assigned to polymeric material. The intense absorption band in each spectrum was assigned to the asymmetric stretching mode of the dihalide. In an earlier investigation of NiF_2,[137] the agreement between the calculated and observed frequencies for the naturally occurring nickel isotopes was taken as evidence, within the frequency accuracy of the measurements ($0.3 \, \text{cm}^{-1}$) of a linear structure for the molecule. In the later investigation[136] in which frequency measurements were believed reproducible to $0.05 \, \text{cm}^{-1}$, the isotope data for both NiF_2 and CuF_2 were consistent with the molecules having a slightly bent structure. For a bent molecule, the symmetric stretching mode is IR active and might be expected to appear in the spectral region examined. The non-appearance of an absorption band which could be assigned to the symmetric mode in either spectrum presents a problem. Until additional evidence is obtained for these species, their assigned non-linearity must be regarded as tentative.

The trifluorides of lanthanum, cerium, praseodymium, neodymium, samarium, and europium volatilized at $1470-1600 \, \text{K}$ have recently[137a] been examined in Ar, Kr, and N_2 matrices at $21 \, \text{K}$. For LaF_3, CeF_3, SmF_3, and EuF_3 only the antisymmetric mode and the bending modes were observed and the molecules thus assigned a planar configuration. For PrF_3 the symmetric and antisymmetric stretching modes were observed together with the bending modes and the molecule accordingly assigned a pyramidal configuration. Force constants and entropies are calculated for all the molecules. The spectrum of NdF_3 cannot yet be satisfactorily interpreted.

6.7.2 Chlorides and bromides, $CrCl_2$, $MnCl_2$, $FeCl_2$, $CoCl_2$, $NiCl_2$ and $NiBr_2$

Mass spectroscopic studies[138] indicate the vapour over these halides consists of monomer and a small amount of dimer. High temperature infrared spectra[139,140] have revealed only the asymmetric stretching frequencies of the monomers, and unspecified dimer modes, but no definite conclusions on the molecular geometry. Within the limitations of their sensitivity, electric deflection experiments indicate a linear, $D_{\infty h}$ structure for the monomers.[135] Electronic spectra of the monomers have also been interpreted assuming the molecules are linear.[137,141]

Infrared matrix studies have been made on all the above halides in the 2000–250 cm^{-1} region[137,142,143] and on all but CrCl$_2$ in the 250–33 cm^{-1} region.[142] In two of the investigations, the dihalides were vaporized[137,142] directly into the matrix.[143] In the third,[143] in addition to the simple vaporization of the MX$_2$ species, the dihalides were formed directly by reaction of chlorine gas with the transition metal at high temperature. The frequencies derived from the different investigations and their assignments are in good agreement. In the spectra of all the halides, monomeric and polymeric absorption bands were differentiated by the absorption band intensity changes under varying experimental conditions. The asymmetric stretching and bending frequencies assigned are given in Table 6.7. Where possible, isotope effects were used to verify the linearity of the species. Within the frequency accuracy of the measurements[143] (0·5 cm^{-1}), only NiCl$_2$ and CrCl$_2$ indicated a possibility of non-linearity. However, the relatively low frequency

Table 6.7 Frequencies and assignments for the transition metal halides

Species	Matrix	Vibration frequencies, cm^{-1}	Point group	Bond angle	References
TiF$_2$	Ne	752·8(B_1) 643(A_1)	C_{2v}	130 ± 5	134
CoF$_2$	Ne	745·8	?	170 ± 10	136
NiF$_2$	Ne	800·7(B_1)	C_{2v}	100 ± 7	136, 137
CuF$_2$	Ne	766·5(B_1)	C_{2v}	166 ± 4	136
TiF$_3$	Ne	799·4, 768·5	D_{3h}, C_{3v}		134
CrCl$_2$	Ar	493·5(Σ_u^+)	$D_{\infty h}$	180	137, 142, 143
MnCl$_2$	Ar	476·8(Σ_u^+) 83(πu)	$D_{\infty h}$	180	137, 142, 143
FeCl$_2$	Ar	493·2(Σ_u^+) 88(πu)	$D_{\infty h}$	180	137, 142, 143
	Ar	498·2 496·1 494·0(Σ_u^+) 491·4 (Σ_g^+) 488·2	$D_{\infty h}$	180	143a
Fe$_2$Cl$_4$	Ar	438(B_{3u}) 325(B_{2u}) 249(B_{3u})	D_{2h}		143a
FeCl$_3$	Ar	464·8(E') 116(A_2'') 102(E')	D_{3h}		143a
Fe$_2$Cl$_6$	Ar	467·5(B_{1u}) 406(B_{3u}) 328·0(B_{2u}) 280·2(B_{3u}) 118·5(B_{1u}) 115·7(B_{3u}) 98·7(B_{2u}) 24·3(B_{1u})	D_{2h}	128 92	143a
CoCl$_2$	Ar	493·4(Σ_u^+) 94·5(π_u)	$D_{\infty h}$	180	137, 142, 143
NiCl$_2$	Ar	520·6(Σ_u^+) 365(Σ_g^+)ᵃ 85(π_u)	$D_{\infty h}$	180	137, 142, 143
NiBr$_2$	Ar	414·2(Σ_u^+) 69(π_u)	$D_{\infty h}$	180	137, 142, 143

a. Value derived from electronic spectra.

accuracy of the measurements did not allow firm conclusions to be made with regard to the molecular geometry. On the assumption of the linear geometry, the M—Cl stretching force constants were calculated and found to increase from 2·06 mdyn/Å to 2·53 mdyn/Å as the mass of the metal increased from Mn to Ni. Chromium chloride did not fit into this sequence, since the Cr—Cl constant at 2·14 mdyn/Å was slightly larger than that of the next member, manganese.

In one of the investigations[142] an attempt was made to assign some of the polymeric absorption bands to a dimer species. The latter was assumed to have a metal-halogen bridge structure of D_{2h} symmetry. Making some reasonable assumptions with regard to the magnitudes of the force constants for such a species, it was found possible to duplicate fairly well the observed frequencies assigned to the dimer.

A detailed study has recently been reported[143a] of the iron chloride system obtained as gaseous species from molecular effusion from a double-stage Knudsen cell operated over the range 440–1990 K, and trapped in Ar and N_2 at 5 K, the species being examined in the range 680–20 cm^{-1}. Fairly complete spectra are reported for the species $FeCl_2$ ($D_{\infty h}$), Fe_2Cl_4 (D_{2h}), $FeCl_3$ (D_{3h}), and Fe_2Cl_6 (D_{2h}), and the assignments supported by normal co-ordinate analysis and partially resolved isotopic splittings. An interesting comparison is given of the M—X stretching and bending force constants for the bridged D_{2h} molecules: Al_2F_6, Al_2Cl_6, B_2H_6, Nb_2Cl_{10}, Fe_2Cl_6. It is noted that bond stretching force constants of M—X in the outer MX_2 groups are nearly twice as large as in the M_2X_2 ring. Thermodynamic functions are calculated for all the molecules except Fe_2Cl_4. There are considerable discrepancies between the statistical values and thermochemical data for $FeCl_2$, which suggests that the sublimation and vapour pressure data for this compound are doubtful and require re-examination.

6.8 Group IVB, VB and VIB oxides

6.8.1 Monoxides of Ti, Zr, Hf, Ta, W, Th and U

No gas phase IR spectra have been obtained for any of these monoxides. This is not surprising since they exist only at extremely high temperatures. Electronic emission spectra[144] have been reported and interpreted for all except ThO. Because of the complexity of the spectra, there are some uncertainties attached to the reported ground state vibration frequencies. The monoxides are obtained by vaporization of the respective dioxides. The major vapour species over TiO_2, ZrO_2, TaO_2 and ThO_2, are known from mass spectral studies to be the monoxides and dioxides.[145] The vaporization of WO_2, or WO_3, produces a large number of vapour species, but at the higher temperatures the monoxide and dioxide are the major species.[145]

Table 6.8 Frequencies and assignments for some group IVB, VB, and VIB oxides

Species	Matrix	Vibration frequencies, cm^{-1}	Point group	Bond angle	References
TiO	Ne	$1005(\Sigma^+)$	$C_{\infty v}$		146
ZrO	Ar	$957(\Sigma^+)$	$C_{\infty v}$		146, 147
HfO	Ar	$959(\Sigma^+)$	$C_{\infty v}$		146, 147
TaO	Ar	$1020(\Sigma^+)$	$C_{\infty v}$		148
WO	Ne	$1055(\Sigma^+)$	$C_{\infty v}$		149
ThO	Ar	$880(\Sigma^+)$	$C_{\infty v}$		147
UO	Ar	$776 \cdot 0(\Sigma^+)$	$C_{\infty v}$		150a
ZrO$_2$	Ar	$818(B_1)\ 884(A_1)$	C_{2v}	109	147
TaO$_2$	Ar	$912(B_1)\ 971(A_1)$	C_{2v}		148
WO$_2$	Ne	$928(B_1)\ 992(A_1)$	C_{2v}		149
ThO$_2$	Ar	$734 \cdot 5(B_1)\ 786 \cdot 8\ 124^a(A_1)$	C_{2v}	106	147
UO$_2$	Ar	$874 \cdot 3(?)\ 81(A_1)$	C_{2v}		150a

a. Calculated.

The IR matrix spectra of TiO,[146] ZrO,[146,147] HfO,[146,147] TaO,[148] and ThO[147] were obtained by vaporization of the respective dioxides. In the hafnium oxide spectrum only one absorption band at about 960 cm^{-1} was recorded in the range 200–2000 cm^{-1}. Under reducing conditions, that is, vaporization from a mixture of Hf and HfO$_2$, the spectrum remained unchanged. This was taken as evidence for assigning the band to HfO rather than HfO$_2$. Spectra obtained from the vaporization of TiO$_2$, ZrO$_2$, TaO$_2$ and ThO$_2$ all contained several absorption bands. Assignment of absorption bands to the monomers were made on the basis of observed and calculated frequency shifts for the ^{16}O and ^{18}O substituted species, and from observation of expected intensity ratios based on the known isotopic composition of the oxide vaporized. For all these oxides, for which ground state vibrational frequencies had been deduced from electronic spectra, the matrix values were found to corroborate the electronic spectral assignments.

To obtain the matrix spectra of WO,[149] oxygen at low pressure was passed over tungsten at very high temperature and the various oxide species isolated. Comparison of spectra obtained under these conditions with those obtained by vaporization of WO$_2$ or WO$_3$ indicated that a band appearing at about 1050 cm^{-1} could probably be assigned to WO. Isotopic substitution of ^{18}O could not be used to verify the assignment due to the existence of absorption bands for W^{16}O$_x$ species which were always present in the region where W^{18}O was expected to appear. It was observed that in moving across the periodic table from HfO (974 cm^{-1}) to TaO (1020 cm^{-1}) the frequencies of the diatomic oxides increased, and a value of about 1050 cm^{-1} for WO was in keeping with this trend. Further support for the assignment was obtained from the gas phase electronic spectra of WO,[150] for which a value of $\Delta G''_{1/2} = 1054 \cdot 9$ cm^{-1} was tentatively derived for the vibration frequency of the ground state.

The matrix spectrum of UO has recently[150a] been obtained by volatilizing a mixture of UO$_2$ and U and trapping in Ar. By varying the O/U ratio of the condensed phase from 1·5 to 3·0 and comparing the observed spectra of the matrix-isolated vapours in equilibrium with the condensed phases it has been possible to assign a frequency of 776·0 cm^{-1} to U^{16}O. Similar experiments using ^{18}O-enriched uranium oxides yields a frequency of 736·2 cm^{-1} for U^{18}O.

6.8.2 Dioxides of Th, Zr, Ta, W and U

The dioxides of Th, Zr and Ta have been shown to be non-linear, with C_{2v} symmetry from electric deflection experiments.[151] In the matrix spectra reported for the monoxides of these metals, additional absorption bands were observed which isotopic studies with ^{16}O and ^{18}O suggested could be assigned to the dioxides. The most complete data was obtained for ThO$_2$.[147] Using 45 per cent. ^{18}O isotopic enrichment, two well defined sets of triplets were recorded at about 770 cm^{-1} and 710 cm^{-1}, respectively. The most intense of these triplets, at about 710 cm^{-1}, was assigned to the asymmetric stretching modes of Th^{16}O$_2$, Th^{16}O^{18}O and Th^{18}O$_2$. Isotopic frequency data was used to calculate a bond angle of 106°. The weaker set of triplets was assigned to the symmetric stretching frequencies of the three isotopic species. The bending mode was not observed, and was assumed to lie below 180 cm^{-1}, the long wavelength limit of the investigation. Using a three term potential function, and OThO = 106°, the stretching, bending and interaction constants were derived at $k_1 = 5 \cdot 11$ mdyn/Å, $k_{\delta/l^2} = 0 \cdot 067$ mdyn/Å and $k_{12} = 0 \cdot 44$ mdyn/Å, respectively. The bending frequency was calculated at 124 cm^{-1}.

In a similar study of ZrO$_2$ using oxygen isotopic enrichment,[147] only one clearly defined set of triplet absorption bands was observed. These were assigned to the asymmetric stretching modes of ZrO$_2$ and a bond angle of 109° calculated. Using zirconium isotopic enrichment, a tentative assignment for a second set of absorption bands to the symmetric stretching modes of ZrO$_2$ was made. Because of the uncertainty of the latter assignment, no force constant calculations were made.

For a Ta^{16}O$_2$,[148] the appearance of two absorption bands at about 912 cm^{-1} and 972 cm^{-1} was taken as evidence for the molecule's bent structure. Intensity variations on isotopic enrichment with ^{18}O were compatible with the assignment but due to the relatively low precision of the frequency measurements, a bond angle could not be calculated. By analogy with ZrO$_2$ and ThO$_2$ the lower of the two frequencies was assigned to the asymmetric stretching mode of TaO$_2$.

The IR matrix spectra of the tungsten oxides under a variety of vaporizing conditions always contained a large number of absorption bands in the 1100–280 cm^{-1} region, due to the species WO, WO_2, W_2O_6, W_3O_8, W_3O_9 and W_4O_{12}.[149] Even with ^{18}O enrichment, assignment of bands to specific molecular species was difficult. Two bands at about 990 cm^{-1} and 928 cm^{-1} were assigned to $W^{16}O_2$ on the basis of relative intensity changes and isotopic frequency shifts. By analogy with TaO_2, ZrO_2, ThO_2 the molecule was assumed bent, with the lower of the two frequencies being assigned to the asymmetric stretching mode. An attempt was made to assign absorption bands to other tungsten oxide species. However, these assignments were regarded as tentative due to the complexity of the spectra.

In the matrix spectra reported[150a] for UO additional bands were observed which could be assigned to UO_2. A stretching mode was observed at 874.3 cm^{-1} for $U^{16}O_2$, 854.6 cm^{-1} for $U^{16}O^{18}O$, and 826.6 cm^{-1} for $U^{18}O_2$ but a definitive assignment to either the symmetric or antisymmetric stretch of a bent UO_2 was not possible. The bending mode v_2 was observed at 81 cm^{-1} for $U^{16}O_2$ and 73 cm^{-1} for $U^{18}O_2$.

6.9 Conclusion

Over the period of ten years since its inception the matrix technique has been used to obtain the IR spectra of over one hundred high temperature vapour phase species. In many cases sufficient data were obtained for definitive vibrational and geometrical assignments to be made. There can be no doubt that without the matrix method most of these data would not be currently available. In the future a steady continuation of the high temperature matrix studies may be expected with the object obtaining both thermodynamic data and an improved understanding of chemical bonding. Recently[152] the adaptation of the matrix isolation technique to obtain Raman spectra has been demonstrated using a laser excitation source; these early studies are described in Chapter 9. In high temperature studies this should prove of great value in increasing the reliability of vibrational assignments.

6.10 References

1. E. D. Becker and G. C. Pimentel, *J. Chem. Phys.*, **25**, 244, (1956).
2. M. J. Linevsky, *J. Chem. Phys.*, **34**, 1956, (1961).
3. R. F. Porter and R. C. Schoonmaker, *J. Chem. Phys.*, **34**, 29, (1961).
4. M. Eisenstadt, G. M. Rothberg and P. Kusch, *J. Chem. Phys.*, **29**, 797, (1958).
5. G. L. Vidale, *J. Phys. Chem.*, **64**, 314, (1963).
6. W. Klemperer, W. G. Norris, A. Buchler and A. G. Emslie, *J. Chem. Phys.*, **33**, 1524, (1960).
7. W. Klemperer and W. G. Norris, *J. Chem. Phys.*, **34**, 1071, (1961).
8. S. H. Bauer, T. Ino and R. F. Porter, *J. Chem. Phys.*, **33**, 685, (1960).

9. A. Snelson and K. S. Pitzer, *J. Phys. Chem.*, **67**, 882, (1963).
10. M. J. Linevsky, *J. Chem. Phys.*, **38**, 658, (1963).
11. S. Schlick and O. Schnepp, *J. Chem. Phys.*, **41**, 463, (1964).
12. R. L. Redington, *J. Chem. Phys.*, **44**, 1238, (1966).
13. A. Snelson, *J. Chem. Phys.*, **46**, 3652, (1966).
14. M. Freiberg, A. Ron and O. Schnepp, *J. Phys. Chem.*, **72**, 3526, (1968).
15. A. Snelson, *J. Phys. Chem.*, **73**, 1919, (1969).
16. S. Abramowitz, N. Acquista and I. R. Levin, *J. Res. Nat. Bur. Stds.*, *Section A*, **72A**, 487, (1968).
17. S. J. Cyvin, B. N. Cyvin and A. Snelson, *J. Phys. Chem.*, **74**, 4338, (1970).
18. S. J. Cyvin, B. N. Cyvin and A. Snelson, *J. Phys. Chem.*, **75**, 2609, (1971).
19. J. Berkowitz, *J. Chem. Phys.*, **32**, 1522, (1960).
20. D. White, K. S. Seshadri, D. F. Dever, D. E. Mann and M. J. Linevsky, *J. Chem. Phys.*, **39**, 2463, (1963).
21. P. A. Akishin and N. G. Rambidi, *Z. Physik. Chem.*, **213**, 111, (1960).
22. R. R. R. Redington (private communication).
23. R. F. Porter and E. E. Zeller, *J. Chem. Phys.*, **33**, 858, (1960).
24. A. Snelson, *J. Phys. Chem.*, **71**, 3202, (1967).
25. L. D. McCory, R. C. Paule and J. L. Margrave, *J. Phys. Chem.*, **67**, 1083, (1963).
26. D. White, K. S. Seshadri, D. F. Dever, D. E. Mann and M. J. Linevsky, *J. Chem. Phys.*, **39**, 2463, (1963).
27. K. S. Seshadri, D. White and D. E. Mann, *J. Chem. Phys.*, **45**, 4697, (1966).
28. L. Andrews, *J. Chem. Phys.*, **50**, 4288, (1969).
29. A. Buchler, J. L. Stauffer, W. Klemperer and L. Wharton, *J. Chem. Phys.*, **39**, 2299, (1963).
30. D. E. Mann, G. V. Calder, K. S. Seshadri, D. White and M. J. Linevsky, *J. Chem. Phys.*, **46**, 1139, (1967).
31. L. Andrews, *J. Chem. Phys.*, **50**, 4288, (1969).
32. J. W. Linnett and M. F. Hoare, *Trans. Fara. Soc.*, **45**, 844, (1949).
33. D. R. Herschbach and V. W. Laurie, *J. Chem. Phys.*, **35**, 458, (1961).
34. W. L. S. Andrews and G. C. Pimentel, *J. Chem. Phys.*, **44**, 2361, (1966).
35. L. Andrews, *J. Phys. Chem.*, **73**, 3922, (1969).
36. R. D. Sprately, Ph.D. Thesis, University of California, 1965.
37. J. Rolfe, W. Holzer, W. F. Murphy and J. H. Bernstein, *J. Chem. Phys.*, **49**, 963, (1968).
38. R. D. Sprately and G. C. Pimentel, *J. Amer. Chem. Soc.*, **88**, 2394, (1966).
39. R. C. Schoonmaker and R. F. Porter, *J. Chem. Phys.*, **28**, 4554, (1958).
40. R. F. Porter and R. C. Schoonmaker, *J. Phys. Chem.*, **62**, 237, (1958).
41. N. Acquista, S. Abramowitz and D. R. Lide, *J. Chem. Phys.*, **49**, 780, (1968).
42. N. Acquista and S. Abramowitz, *J. Chem. Phys.*, **51**, 2911, (1969).
43. D. R. Lide and R. L. Kuczkowski, *J. Chem. Phys.*, **46**, 4768, (1967).
44. K. S. Seshadri, L. A. Nimon and D. White, *J. Mol. Spectroscopy*, **30**, 128, (1969).
45. A. Buchler and J. B. Berkowitz-Mattuck, *J. Chem. Phys.*, **39**, 268, (1963).
46. P. A. Akishin, L. N. Gorokhov and Y. S. Khodeev, *Zh. Strukt. Khim.*, **2**, 209, (1961).
47. A. Buchler and E. P. Marram, *J. Chem. Phys.*, **39**, 292, (1963).
48. D. White, D. E. Mann, P. N. Walsh and A. Sommer, *J. Chem. Phys.*, **32**, 488, (1960).
49. A. Sommer, D. White, M. J. Linevsky and D. E. Mann, *J. Chem. Phys.*, **38**, 87, (1963).
49a. D. Smith, D. W. James and J. P. Devlin, *J. Chem. Phys.*, **54**, 4437, (1971).
50. A. Buchler and W. Klemper, *J. Chem. Phys.*, **29**, 121, (1961).

51. L. Wharton, R. A. Berg and W. Klemperer, *J. Chem. Phys.*, **39**, 2023, (1963).
52. A. Buchler, J. L. Stauffer, W. Klemperer and L. Wharton, *J. Chem. Phys.*, **39**, 2299, (1963).
53. A. Buchler, J. L. Stauffer and W. Klemperer, *J. Chem. Soc.*, **40**, 3471, (1964).
54. A. Buchler, J. L. Stauffer and W. Klemperer, *J. Amer. Chem. Soc.*, **86**, 4544, (1964).
55. D. L. Hildenbrand and L. P. Theard, *J. Chem. Phys.*, **42**, 3230, (1965).
56. A. Snelson, *J. Phys. Chem.*, **70**, 3208, (1966).
57. D. E. Mann, G. V. Calder, K. S. Seshadri, D. White and M. J. Linevsky, *J. Chem. Phys.*, **46**, 1138, (1967).
58. V. Calder, D. E. Mann, K. S. Seshadri, M. Allavena and D. White, *J. Chem. Phys.*, **51**, 2093, (1969).
59. A. Snelson, *J. Phys. Chem.*, **72**, 250, (1968).
60. D. White *et al.*, unpublished work.
61. D. L. Hildenbrand, *J. Chem. Phys.*, **40**, 3438, (1964).
62. J. W. Linnet and M. F. Hoare, *Trans. Fara. Soc.*, **45**, 844, (1949).
63. M. Allavena, R. Rysnik, D. White, V. Calder and D. E. Mann, *J. Chem. Phys.*, **50**, 3399, (1969).
64. D. L. Hildenbrand, Aeronutronic (Div. of Philco Ford Rept. U-3183, Contract AF49(638)-1397, 30 June, 1965.
65. A. Buchler, W. Klemperer and A. G. Emslie, *J. Chem. Phys.*, **36**, 2499, (1962).
66. H. A. Skinner, *Trans. Fara. Soc.*, **45**, 20, (1949).
67. E. F. Hayes, *J. Phys. Chem.*, **70**, 3740, (1966).
68. M. G. Inghram, R. F. Porter and W. A. Chupka, *J. Chem. Phys.*, **25**, 498, (1956).
69. P. A. Akishin and V. P. Spiridenov, *Dokl. Akad. Nauk.*, **131**, 557, (1960).
70. D. White, P. Walsh and D. E. Mann, *J. Chem. Phys.*, **28**, 508, (1958).
71. E. W. Kaiser, J. S. Muenter and W. Klemperer, *J. Chem. Phys.*, **48**, 3339, (1968).
72. P. L. Hanst, V. H. Early and W. Klemperer, *J. Chem. Phys.*, **42**, 1097, (1965).
73. W. Weltner and J. R. W. Warn, *J. Chem. Phys.*, **37**, 292, (1962).
74. A. Sommer, D. White, M. J. Linevsky and D. E. Mann, *J. Chem. Phys.*, **38**, 87, (1963).
75. J. W. C. Johns, *Can. J. Phys.*, **39**, 1738, (1961).
76. M. J. Linevsky, D. White and D. E. Mann, *J. Chem. Phys.*, **41**, 542, (1964).
77. A. Snelson, *J. Phys. Chem.*, **74**, 2574, (1970).
78. A. J. Hinchcliffe and J. S. Ogden, *Chem. Comm.*, **1969**, 1053.
79. J. Drowart, G. E. DeMaria, R. P. Burns and M. G. Inghram, *J. Chem. Phys.*, **32**, 1366, (1960).
80. J. K. McDonald and K. K. Innes, *J. Mol. Spectroscopy*, **32**, 501, (1969).
81. A. Buchler, J. L. Stauffer, W. Klemperer and L. Wharton, *J. Chem. Phys.*, **34**, 587, (1961).
82. R. P. Burns, *J. Chem. Phys.*, **44**, 3307, (1966).
83. N. G. Rambidi and S. M. Tolmachev, *Zhur. Strukt. Khim.*, **9**, 363, (1968).
83a. A. Buchler, E. P. Marram and J. L. Stauffer, *J. Phys. Chem.*, **71**, 4139, (1967).
84. A. A. Maltsev and V. F. Shevel'kov, *Teplofiz, Vysok, Temp.*, **2**, 650, (1964).
84a. J. Brom, jr., T. Devore and H. F. Franzen, *J. Chem. Phys.*, **54**, 2742, (1971).
84b. L. D. McCory, R. C. Paule and J. L. Margrave, *J. Phys. Chem.*, **67**, 1086, (1963).
85. A. Snelson, *J. Phys. Chem.*, **71**, 3202, (1967).
86. M. J. Linevsky, Contract Report AF33(615–1150), November, 1964. General Electric Space Sciences Laboratory.
87. G. Herzberg, *Infrared and Raman Spectra of Polyatomic Molecules*, D. Von Nostrand Co., Inc., New York, 1949.
88. E. J. Rosenbaum, *J. Chem. Phys.*, **8**, 643, (1940).

89. W. Klemperer, *J. Chem. Phys.*, **24**, 353, (1956).
89a. J. M. Brom, jr., and H. F. Franzen, *J. Chem. Phys.*, **54**, 2874, (1971).
90. S. M. Naude and J. J. Hugo, *Can. J. Phys.*, **35**, 64, (1957).
91. J. Drowart, R. P. Burns, G. DeMaria and M. G. Inghram, *J. Chem. Phys.*, **31**, 1131, (1959).
92. L. Gausset, G. Herzberg, A. Lagerquist and B. Rosen, *Disc. Fara. Soc.*, **35**, 113, (1963).
93. W. Weltner, P. N. Walsh and C. L. Angell, *J. Chem. Phys.*, **40**, 1299, (1964).
94. W. Weltner and D. McLeod, *J. Chem. Phys.*, **40**, 1305, (1964).
95. K. S. Pitzer and E. Clementi, *J. Amer. Chem. Soc.*, **81**, 4477, (1959).
96. R. Hoffman, *Tetrahedron*, **22**, 521, (1966).
97. W. Weltner and D. McLeod, *J. Chem. Phys.*, **45**, 3096, (1966).
98. J. O. Halford, *J. Chem. Phys.*, **19**, 1375, (1951).
99. A. F. Pozubenkov, *Dokl. Akad. Nauk. SSSR*, **163**, 58, (1965).
100. J. Drowart, G. deMaria and M. G. Ingram, *J. Chem. Phys.*, **29**, 1015, (1958).
101. B. Kleman, *Astrophys. J.*, **123**, 162, (1956).
102. W. Weltner and D. McLeod, *J. Chem. Phys.*, **41**, 235, (1964).
103. W. Jevons, *Proc. Roy. Soc. London*, **106**, 174, (1924).
104. H. L. Schick, *Chem. Rev.*, **60**, 331, (1960).
105. R. F. Porter, W. A. Chupka and M. G. Inghram, *J. Chem. Phys.*, **23**, 216, (1955).
106. K. F. Zmbov and J. L. Margrave, *High Temp. Sci.* (in press 1969).
107. J. S. Anderson, J. S. Ogden and M. J. Ricks, *Chem. Commun.*, **1968**, 1585.
108. J. S. Anderson and J. S. Ogden, *J. Chem. Phys.*, **51**, 4189, (1969).
109. J. W. Hastie, R. Hauge and J. L. Margrave, *J. Phys. Chem.*, **73**, 1105, (1969).
110. L. Pauling, *The Nature of the Chemical Bond*, Cornell University Press, Ithaca, New York, 1959.
111. E. A. Robinson, *Can. J. Chem.*, **41**, 3021, (1963).
112. J. Drowart, F. Degreve, G. Verhaegen and R. Collins, *Trans. Fara. Soc.*, **61**, 1072, (1965).
113. J. S. Ogden and M. J. Ricks, *J. Chem. Phys.*, **52**, 352, (1970).
113a. J. S. Ogden and M. J. Ricks, *J. Chem. Phys.*, **56**, 1658, (1972).
114. V. M. Rao, R. F. Curl, P. L. Timms and J. L. Margrave, *J. Chem. Phys.*, **43**, 2557, (1965).
115. R. Hauge, V. M. Khanna and J. L. Margrave, *J. Mol. Spectroscopy*, **27**, 143, (1968).
116. D. Naegeli and H. B. Palmer, *J. Mol. Spectroscopy*, **21**, 325, (1966).
117. M. W. Lister and L. E. Sutton, *Trans. Fara. Soc.*, **37**, 406, (1941).
118. V. M. Khanna, R. Hauge, R. F. Curl and J. L. Margrave, *J. Chem. Phys.*, **47**, 5031 (1967).
119. J. W. Hastie, R. H. Hauge and J. L. Margrave, *J. Amer. Chem. Soc.*, **91**, 2536, (1969).
120. J. W. Hastie, R. H. Hauge and J. L. Margrave, *J. Phys. Chem.*, **72**, 4492, (1968).
121. L. Andrews and D. L. Frederick, *J. Amer. Chem. Soc.*, **92**, 775, (1970).
122. P. A. Akishin and V. P. Spiridenov, *Kristallografia*, **2**, 475, (1957).
123. A. Buchler, J. L. Stauffer and W. Klemperer, *J. Amer. Chem. Soc.*, **86**, 4564, (1964).
124. A. Buchler, W. Klemperer and A. G. Emslie, *J. Chem. Phys.*, **36**, 2499, (1962).
125. W. Klemperer, *J. Chem. Phys.*, **25**, 1066, (1956).
126. W. Klemperer and L. Lindeman, *J. Chem. Phys.*, **25**, 397, (1956).
127. D. E. Irish and T. F. Young, *J. Chem. Phys.*, **43**, 1765, (1965).
128. C. S. Venkateswaran, *Proc. Ind. Acad. Sci.*, **1A**, 850, (1935).
129. H. Sponer and E. Teller, *Rev. Mod. Phys.*, **13**, 75, (1941).
130. S. Bell, *J. Mol. Spectroscopy*, **23**, 98, (1967).

131. R. W. McNamee, jr., Ph.D. thesis, University of California, 1962.
132. A. Loewenschuss, A. Ron, and O. Schnepp, *J. Chem. Phys.*, **49**, 272, (1968).
133. A. Loewenschuss, A. Ron and O. Schnepp, *J. Chem. Phys.*, **50**, 2502, (1969).
134. J. W. Hastie, R. H. Hauge and J. L. Margrave, *J. Chem. Phys.*, **51**, 2648, (1969).
135. A. Buchler, L. J. Stauffer and W. Klemperer, *J. Chem. Phys.*, **40**, 3471, (1964).
136. J. W. Hastie, R. H. Hauge and J. L. Margrave, *High Temp. Sci.*, **1**, 76, (1969).
137. D. E. Milligan, M. E. Jacox and J. D. McKinley, *J. Chem. Phys.*, **42**, 902, (1965).
137a. R. D. Wesley and C. W. DeKock, *J. Chem. Phys.*, **55**, 3866, (1971).
138. R. C. Schoonmaker, A. H. Friedman and R. F. Porter, *J. Chem. Phys.*, **31**, 1586, (1959).
139. S. P. Randall, F. T. Green and J. L. Margrave, *J. Phys. Chem.*, **63**, 758, (1959).
140. G. E. Leroi, T. C. James, J. T. Hougen and W. Klemperer, *J. Chem. Phys.*, **36**, 2879, (1962).
141. J. T. Hougen, G. E. Leroi and T. C. James, *J. Chem. Phys.*, **34**, 1670, (1961).
142. K. R. Thompson and K. D. Carlson, *J. Chem. Phys.*, **49**, 4379, (1968).
143. M. E. Jacox and D. E. Milligan, *J. Chem. Phys.*, **51**, 4143, (1969).
143a. R. A. Frey, R. D. Werder and H. H. Gunthard, *J. Mol. Spectroscopy*, **35**, 260, (1970).
144. G. Herzberg, *Spectra of Diatomic Molecules*, Van Nostrand, 1957.
145. *International Symposium on High Temperature Technology, Oct. 1959*, McGraw-Hill, Inc., New York, 1959.
146. W. Weltner and D. McLeod, *J. Phys. Chem.*, **69**, 3488, (1965).
147 M. J. Linevsky, *Proceedings of the Third Meeting of the Interagency Chemical Rocket Propulsion Group on Thermochemistry, 1965*, 'Vol. I, Chemical Propulsion Information', p. 71, Agency, Silver Springs, Md., 1965.
148. W. Weltner and D. McLeod, *J. Chem. Phys.*, **42**, 882, (1965).
149. W. Weltner and D. McLeod, *J. Mol. Spectroscopy*, **17**, 276, (1965).
150. A. Gatterer and S. G. Krishnamurty, *Nature*, 169, 543, (1952).
150a. S. Abramowitz, N. Acquista and K. R. Thompson, *J. Phys. Chem.*, **75**, 2283, (1971).
151. M. Kaufman, J. Muenter and W. Klemperer, *J. Chem. Phys.*, **47**, 3365, (1967).
152. H. Classen and J. S. Shirk, *Proceedings of the Mid American Symposium on Spectroscopy Sponsored by the Society of Applied Spectroscopy, Chicago Section, 1970*.

7 Infrared and Raman studies on the vibrational spectra of impurities in ionic and covalent crystals

W. F. SHERMAN AND G. R. WILKINSON

Contents

7.1 Introduction

The physics and chemistry of solid state materials covers a wide, important, and still rapidly expanding field of study. Many of the interesting and useful experiments within this field have been carried out not only on pure materials, but also on impure materials which were produced specifically for investigating impurity activated effects. This article is concerned with one aspect of this subject; the impurity activated vibrational spectra of ionic and covalent crystals. Molecular crystals are dealt with elsewhere in this book.

A broad distinction will be drawn between *point*, or *atomic defects* which will be considered in paragraph 7.2, and *polyatomic defects*, which can be expected to show all the properties discussed for point defects, plus those extra effects resulting from their intrinsic vibrational and rotational motion and which will be dealt with in paragraph 7.3.

7.1.1 Sample preparation

Although the art of crystal growing is such that almost every single doped crystal requires some special individual treatment in order to produce satisfactory samples there are a few general points which can be made here. Most of the work reported will refer to single crystal samples grown from a suitably doped melt by one or other of the standard methods. In a few cases pure single crystal samples have been doped after growth by some form of inter-migration method, and a few examples of sintered polycrystalline material are given. For most experiments the minimum doping level consistent with making the desired measurements has been used in order to keep to a minimum the interaction between impurities, and this has typically been between 0·01 and 0·5 mole per cent., (i.e. between 100 and 5000 ppm). However, some impurities are so strongly infrared active that very small concentrations can be used. For example the v_3 vibration of NCO^- isolated in KBr gives an infrared absorption of about 10 per cent. for a crystal 1 cm thick containing $\frac{1}{3}$ ppm at 100 K. At the other end of the scale come the mixed crystals, where in some cases studies have covered the whole concentration range from 0 to 100 per cent.

7.1.2 Basic spectroscopic principles

7.1.2.1 Infrared

In the near infrared region conventional prism or grating spectrometers have been used, although usually these have had to be of a fairly high resolution type in order to do justice to the sharp spectral features that are frequently found. In some instances the instruments have been modified in various respects; for example for very low temperature work some experiments have required that the sample be mounted after the monochromator in order to reduce to the absolute minimum the amount of radiation falling upon it. Experiments which involve a directional dependence, (e.g. uniaxial stress, or electric field experiments), have required polarizers to be incorporated somewhere in the optical path and these have usually been of the 'stacked plate' type or 'line grid' type.

For far infrared work various types of instruments including grating monochromators, Michelson interferometers, and lamellar gratings have been used.

7.1.2.2 Raman

Relatively little Raman work has been done on vibrational spectra of impurities in solids because of the difficulty of obtaining a useable signal to noise ratio for the necessarily weak signals. However satisfactory laser-Raman spectrometers are becoming more common and more work of this sort should soon be available. A typical Raman spectrometer has an ionized argon laser operating with about 0·5 W in the 4880 Å line, followed by a high throughput double monochromator and pulse counting detector.

7.1.3 Other experimental variables

The recording of the spectrum of a particular defect-containing-crystal is the basic measurement that has to be considered, but many of the most informative experiments have involved the use of a further experimental variable. Some of these will be considered below.

7.1.3.1 Temperature

This, being probably the most easily and most commonly used variable, has been extensively employed in studies on defect activated vibrational spectra. The parameters which have been measured as a function of temperature are: peak absorbance frequency, half-band-width, integrated band area, peak absorbance, and band shape. Temperature has a special significance for this work, since the vibrational properties of the defect-containing crystal which are under investigation are just those properties which dictate the macroscopic thermal properties of the sample. It is for this reason that other techniques such as the temperature variation of the specific heat and thermal conductivity at low temperatures will be referred to at some places in this article. The interpretation of temperature dependent parameters will be dealt with later in appropriate places in the text.

7.1.3.2 Hydrostatic pressure

This is a very attractive variable for investigating those situations where the experimentally observable effects can be described in terms of the internuclear spacings, usually via the appropriate potential functions. By compressing the sample, the internuclear spacings are changed and in principle the best approximation to use for the potential functions can be found. However, fairly large pressures are required to obtain useful frequency shifts, which are typically of the order of 1 cm^{-1}/kbar (1 kbar $\equiv 10^9$ dyne cm^{-2} = 10^8 N m^{-2} ≈ 1000 atm).

Work in the near infrared up to 50 kbar has been reported for polyatomic impurity ions in alkali-halides, both on the 'internal' vibrations of the impurity and on the lattice side band structure.

7.1.3.3 Uniaxial stress

The most informative experiments are those on samples which can withstand sufficient uniaxial stress to allow accurate determinations of frequency shift rates as a function of applied stress and clear observation of the splitting patterns produced. The uniaxial stress experiment is technically easier to perform than the hydrostatic experiment, especially in experimentally difficult spectral regions such as the far-infrared, because no pressure transmitting medium is required, and no high pressure retaining windows are needed. The strength requirement is not as restricting as it may at first seem, because by working at very low temperatures many of the spectral features of interest in this work can be made very sharp indeed, ($\Delta v_{1/2} = 0.5$ cm^{-1} or less), and can have their band centres located to an accuracy of about 0.05 cm^{-1}. By using polarized radiation and recording two spectra at each stress value, one for the E vector of the radiation parallel to the stress axis and the other for the E vector perpendicular to it, very small splittings, much less than the half width of the band, can be quite accurately measured. Thus uniaxial stress experiments have been reported for such 'soft' materials as KI and KBr.

The extra information that can be derived from uniaxial stress experiments comes from the lowering of the symmetry of the crystal, and in particular the lowering of the symmetry of the crystal site occupied by the impurity. Clearly the orientation of the stress axis with respect to the crystal axis is a disposable experimental parameter which allows several different types of distortion to be made to the impurity occupied site in differently oriented uniaxial stress experiments. Specific examples illustrating this will be discussed later.

7.1.3.4 Electric field

Experiments of this sort are theoretically very attractive since they cause a lowering of the site symmetry similar to that discussed above for uniaxial stress, but more extreme in that the two directions along the field axis are not equivalent. However, only relatively small vibrational changes are caused by electric fields, and only for a few favourable examples have measurable effects been recorded before electrical breakdown has occurred within the sample.

7.1.3.5 Isotopic substitution

The substitution of a nucleus of different mass into a vibrating system, and the use of the observed changes in vibrational frequencies to infer information about the bonding within the system, is a well established technique which needs no general introduction here. One observation which may however be worth making at this point is that this technique, which changes a mass without changing any of the force constants, should be seen as complementary to the hydrostatic pressure experiment which changes the force constants without changing the masses. All the other effects discussed here are more complicated in that they change more of the parameters, or change the geometry of the system. In particular it is worth noting in this context that the complexity of the effects of temperature changes which are due to the phonon produced dynamic distortions of the system are usually not properly appreciated.

7.1.3.6 One defect in many different crystals

This type of experiment is clearly analogous to the common liquid phase experiment where a given solute is observed in a wide range of different solvents or the cryogenic studies where a given solute is isolated in a variety of solid molecular matrices. Since both the surrounding masses and the coupling potentials are changed when the given impurity is moved from one crystalline environment to the next, it is usually impossible to quantitatively determine all the parameter changes from the single piece of experimental information, the frequency shift. However a considerable amount of useful information has been collected by this sort of experiment and several examples, primarily for polyatomic impurities will be discussed later.

7.1.3.7 Interaction between impurities

Several different types of experiment can be considered under this heading but only the two general types will be introduced here.
(i) If a sufficient concentration of a given impurity is introduced into a crystal, effects arise due to the interaction between impurities which find themselves close to one another.
(ii) If a particular impurity within a particular crystal is studied as a function of the concentration of other different sorts of impurity within the same crystal, then effects will be observed due to the interactions between the different impurities, often due to the formation of specific complexes between them. One common example of this arises from the introduction of an impurity which is stable only in a state of charge different to that prevailing in the host crystal, when it is usually found that it can be introduced only if a further defect of a 'charge-compensating' type is allowed to accompany it.

7.1.3.8 Time

Only a few impurity activated spectra have been recorded systematically as a function of time, but, for example, certain metal-cyanide complexes within alkali halides dissociate at high temperatures and on quenching to a lower temperature the spectral features due to the complex can be seen to grow as a function of time. This particular method of monitoring the ability of certain impurities to migrate within crystals is a relatively new and interesting technique.

7.2 Point defects

7.2.1 Historical background

The vibrations associated with a point defect in a solid are clearly the defect-distorted vibrations of that solid. In principle a defect will affect all the vibrations of its host lattice, but it is usual to think only in terms of those modes which are appreciably changed by the presence of the defect, that is those modes which have relatively large amplitudes in the vicinity of the defect.

There has been a great deal of recent activity in the field of defect distorted lattice vibrations, but like many other branches of science it has identifiable roots in very early work. For example the result that a light particle in a long chain of coupled heavy particles gives rise to a non-propagating mode of vibration is contained work published in 1841. However the subject as it is today can be assumed to have started with the work of Lifshitz[1] in 1943, although this is to ignore the efforts of Mollwo[2] who in 1933 was trying to fit his *F*-centre frequencies to a theory based on a particle in a box, which in retrospect seems to have taken him fairly close to an extreme example of the light particle in a heavy lattice.

The theoretical work started by Lifshitz had to progress unaided by experimental results until 1960 when Schaefer[3] published his work on the *U*-centre in alkali-halides, although it did receive a stimulus in 1958 when Mossbauer[4] published his work on the recoilless resonant absorption or emission of γ-rays.

A fairly comprehensive review of both the theoretical and experimental work up to 1966 was given by Maradudin[5] and other useful collections of work on this subject are to be found in the *Proceedings of the International Conference on Lattice Dynamics* held at Copenhagen, edited by R. F. Wallis[6] (1963); *Phonons in Perfect Lattices and in Lattices with Point Imperfections*, edited by R. W. H. Stevenson[7] (1966); *Localized Excitations in Solids*, edited by R. F. Wallis[8] (1968); and *Localized Modes and Resonance States in Alkali Halides*, by M. V. Klein[9] (1968).

Schaefer's[3] work on the *U*-centre, a substitutionally isolated H$^-$ or D$^-$ ion, was the first, and for some time the only experimental information about defect distorted lattice vibrations. It is therefore not at all surprising that these *U*-centre results have been used to test the predictions of almost all the theoretical models which have so far been proposed. Much of the detail of these various theories can be found in references 5, 6, 7, 8 and 9 and therefore only a bare outline of them will be presented in the next subsection of this chapter. However in order to understand the infrared and Raman spectra associated with defects in solids it is necessary to say a little about the objectives and methods employed in some of these theoretical studies. From the spectroscopic standpoint the objective could be regarded as the acquiring of an understanding of all those changes which are produced in the spectra of solids when defects are introduced. This can be considered as a double problem: (a) what changes are produced by the defects in the vibrational properties of the solid, and (b) what infrared or Raman activity is possessed by these changed modes.

There are therefore two main types of effect to be looked for : (a) vibrational modes which are radically changed by the presence of the impurity, and these modes would almost certainly be expected to be either infrared or Raman active, and (b) vibrational modes which are not changed appreciably in form by the inclusion of the defect but which are given, either directly or in combination with other modes, a changed activity in the infrared or Raman spectrum of the solid.

From the theoretical standpoint the questions are:

(i) What mathematical model can be postulated to describe the defect distorted solid?

(ii) In which ways do the vibrational properties of this model differ from those of the equivalent model of the pure solid?

(iii) What experimental evidence could be expected to support the deductions?

7.2.2 Mathematical models

Small models—light mass in rigid lattice

Although it is of negligible direct interest to the lattice dynamical theorists, the first model to be considered is that of a single particle of mass m held centrally within an infinitely massive box by springs of force constant k. Figure 7.1 shows this in one dimension but the extension to three dimensions is obvious, and simply results in the single natural frequency becoming triply degenerate.

If such a model is to have any value at all for point defects it will only be for those cases where m is much less than the masses of the surrounding nuclei, and in fact we will see later that it is quite instructive to consider such a simple model in the case of the H^- local mode frequency. Note that this model predicts an isotopic frequency dependence of the form:

$$\frac{v_i}{v_j} = \sqrt{\frac{m_j}{m_i}}$$

(i.e. $v_{H^-}/v_{D^-} = 1 \cdot 414$ for the U-centre).

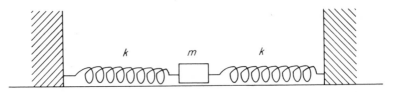

Figure 7.1 A single particle of mass m held centrally within an infinitely massive one dimensional box by two springs of force constant k, has a single vibrational frequency

$$v = \frac{1}{2\pi c}\sqrt{\frac{2k}{m}}$$

We could hope to improve this model by increasing its complexity and the next stage is that shown one-dimensionally in Figure 7.2.

For m appreciably less than m_1 the highest natural frequency is approximately given by the expression in the figure caption, which gives a value which is about 0·6 per cent. higher than the Figure 7.1 model for an H^-

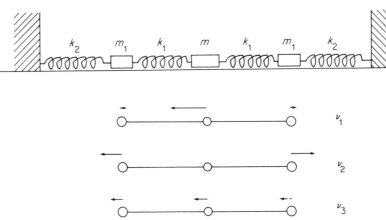

Figure 7.2 A one dimensional model in which the light impurity mass m is held centrally between two lattice masses m_1 which in turn are fixed to the walls of an infinitely massive box is shown, and the *eigen*vectors of the three bond stetching vibrations are indicated for $m \ll m_1$ and $k_1 \approx k_2$. The highest frequency

$$v_1 \approx \frac{1 + m/4m_1}{2\pi c}\sqrt{\frac{2k_1}{m}}$$

is only slightly greater than the Figure 7.1 frequency.

$$v_2 = \frac{1}{2\pi c}\sqrt{\frac{k_1 + k_2}{m}} \quad \text{and} \quad v_3 \approx \frac{1}{2\pi c}\sqrt{\frac{2k_2}{2m_1 + m}}$$

will be seen later to be closely analogous to the A_{1g} gap mode for KT/H^- and the low frequency resonance mode of U-centres respectively isolated in a potassium halide, and shows a v_{H^-}/v_{D^-} ratio predicted at 1·403 for potassium halides and 1·398 for sodium halides. The form of the highest frequency vibration is that of the antisymmetric stretching vibration of the $m_1 - m - m_1$ molecular species and it is interesting to note that a much lower frequency stretching vibration similar to the symmetric stretch is also predicted, at a frequency of

$$\frac{1}{2\pi c}\sqrt{\frac{k_1 + k_2}{m_1}}$$

which for $m_1 = 39$ and $k_1 \approx k_2$ is about one-sixth of v_{H^-}. A sharp feature in the lattice side band spectrum of H^- isolated in KI at 93·7 cm^{-1} (or about $v_{H^-}/4$) has been identified as due to a vibration which is a three-dimensional equivalent of the motion described above. In fact a slightly more realistic, three-dimensional model of the Figure 7.2 type would predict this band at about the correct frequency.

The third vibrational frequency of the Figure 7.2 model, when the three masses move together against the walls of the well, has a frequency of approximately

$$\frac{1}{2\pi c}\sqrt{\frac{2k_2}{2m_1 + m}}$$

and will later be seen to be similar in form and close in predicted frequency to the U-centre resonance mode.

This process of building up a sufficient region of the crystal around a defect to explain the phenomena of interest and regarding the rest of the lattice as rigid, or infinitely massive, has been employed by several authors, and somewhat similar methods have been developed to deal with disordered solids.[10] Methods like these are very good and convenient when highly localized modes of vibration are being considered but they are generally unhelpful or sometimes even misleading if they are applied to modes in which large volumes of the crystal participate. These modes are better considered in terms of the standard lattice dynamical arguments which have been developed over the years to deal with such modes.

7.2.2.2 Linear chains

The simplest mathematical model which shows many of the vibrational properties of a crystal lattice is the long monatomic chain, and the simplest type of defect is the isotopic one, which gives a mass change but no force constant change. As mentioned earlier the mathematics of this model were considered as early as 1841, although it was more than one hundred years later that Lifshitz[1] in 1943 first considered the value of such a model for investigating the vibrational properties of defects isolated in crystal lattices. In 1955 Montroll and Potts[11] considered the case of localized modes occurring in a one-dimensional monatomic lattice in which the mass and coupling force constants of one atom was changed, and in 1956 Mayer, Montroll, Potts and Walsh[12] considered the case of a one-dimensional diatomic lattice containing a single isotopic impurity. Bjork[13] in 1957 considered a diatomic chain containing a single defect which had both mass and force constant different from those of the replaced atom, and by this time the model had become sufficiently refined to be able to predict under what range of mass and/or force constant change, localized modes above the optic band, or within the gap between acoustic and optic bands, could be expected. Remember that there was still no experimental evidence available at this time.

7.2.2.3 Lattice dynamics of large three-dimensional crystals

After Schaefer's[3] U-centre results became known in 1960 a greater interest in this type of work was generated and the more realistic, although more complex, three-dimensional lattices were analysed and the existence of resonance modes was predicted.[14,15,16] The rapid strides taken by the theory since that time have been in the direction of using more realistic three-dimensional models with changed force constants. An increasing use has been made of group theoretical arguments to reduce the complexity of the problems encountered in the evaluation of numerical solutions of these more complicated mathematical models. References 5, 6, 7, 8 and 9 trace this development; which has now reached the stage that given sufficient information to allow a reasonable estimate of the force constant changes to be made, the theory can now be used to predict with reasonable reliability the existence and approximate frequencies of localized modes, gap modes and resonance modes for a given impurity in a given lattice. It can also be used with some confidence to predict the far-infrared absorption of defect activated modes, the probable lattice side band structure on localized modes (see Figure 7.12), and the Raman spectrum which is to be expected. Although not of direct interest to the work discussed here, it should be noted that these defect distorted lattice dynamical arguments have also been used very successfully to explain the observed experimental results in a number of related fields, for example, defect produced anomalies in neutron scattering, low temperature specific heat, low temperature thermal conductivity, and low temperature electrical conductivity.

7.2.2.4 Selection rules for defect activated lattice bands

The infrared or Raman activity of the defect distorted lattice vibrations received little attention until 1963 when Dawber and Elliott[17] gave the first detailed discussion of the IR activity expected for a defect-distorted lattice vibration for the case of a charged impurity within a lattice composed of neutral atoms. In 1964 Loudon[18] published a set of selection rules for point-defect activated lattice bands for face-centred cubic, diamond, and zinc-blende structures. He showed that all phonons can be defect activated for face-centred cubic and diamond lattices, but that some selection rules do occur for rock salt, calcium fluoride and zinc blende lattices. Thus his tables allow selection rules to be easily determined for phonon processes associated with substitutional impurities in a number of important lattices.

7.2.3 Definitions of lattice dynamical terms

Using the long chain model only, the following four figures are presented to illustrate the terms used above and discussed for specific examples later.

Figure 7.3 shows a section from a long, pure, monatomic chain and indicates the type of motion described in general by the terms transverse acoustic and longitudinal acoustic.

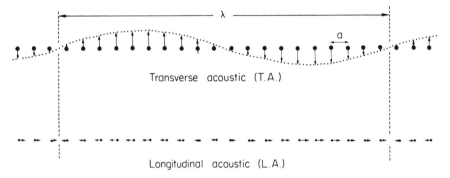

Transverse acoustic (T.A.)

Longitudinal acoustic (L.A.)

Figure 7.3 A section of a long monatomic chain illustrating the type of motion described by the terms transverse acoustic (TA) and longitudinal acoustic (LA)

Since the vibrational disturbances travel along the chain at velocities which change as a function of the wavelength of the amplitude envelope λ, non-linear dispersion curves can be drawn for the phonon energy as a function of $1/\lambda$. This follows from the direct proportionality of $1/\lambda$ to the pseudo-momentum vector of a quantized vibrational motion. The $1/\lambda$ axis only has to extend up to $1/a$ since any attempt to draw the displacements for a wavelength less than a simply results in displacements which are completely described by a wavelength longer than a (i.e. this is the zone boundary value). Figure 7.4a shows the dispersion curves for the phonons pictorially displayed in Figure 7.3, and Figure 7.4b shows the density of vibrational states of this system $g(v)$ as a function of the energy of the quantized vibrations, and illustrates the relationship between $g(v)$ and the reciprocal of the slope of the dispersion curves.

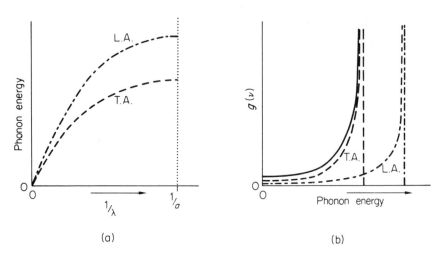

Figure 7.4 The phonon dispersion curves for longitudinal and transverse acoustic modes of vibration showing the variation of phonon energy kv_{vib} as a function of the reciprocal phonon wavelength. Also shown is the density of vibrational states $g(v)$ which represents the number of modes of vibration between v and $v + dv$

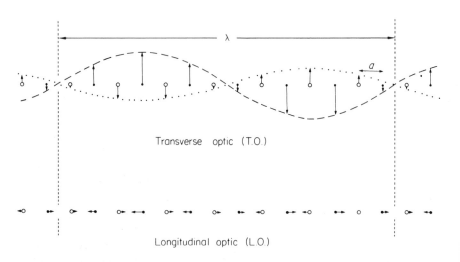

Figure 7.5 Representation of transverse optic and longitudinal optic modes on part of a long diatomic chain. Note that two amplitude envelopes in anti-phase are required to show an optic mode. The different masses of the two atoms leads to the two envelopes having different peak amplitudes. A diatomic chain also has acoustic type vibrational modes as shown in Figure 7.3. For a diatomic chain the shortest permitted wavelength λ is $2a$, where a is the interatomic separation.

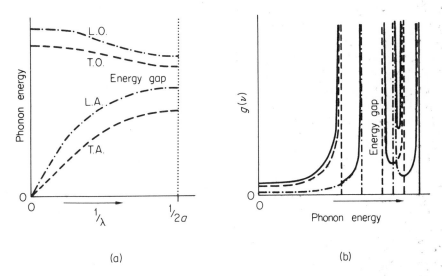

(a)   (b)

Figure 7.6 The phonon dispersion curves for a diatomic lattice are shown in (a), and the corresponding density of vibrational states $g(v)$ as a function of frequency in (b). The separation between the zone centre transverse and longitudinal optic frequencies depend upon the effective dynamic charge of the atoms. An energy gap between the acoustic and optic modes only exists if the mass ratio of the two types of atoms in the lattice is sufficiently different from unity. In the alkali halides energy gaps occur in NaI, NaBr, KI and KBr but not in NaCl, KCl, RbCl, RbBr or RbI

Figure 7.5 indicates the form of the optic modes for a diatomic chain, which can also vibrate in the acoustic modes illustrated in Figure 7.3, and Figure 7.6 shows the form of the dispersion curves and density of states curves for this type of chain.

Figure 7.7 illustrates for an impurity-containing diatomic chain those impurity induced vibrational effects which were mentioned earlier. Only transverse modes are shown, and completely analogous longitudinal modes could be envisaged; but for non-propagating modes in a three-dimensional lattice, these terms tend to lose their significance.

7.2.3.1 Localized mode

The effect of a light or tightly bound impurity giving rise to a mode of vibration which has a frequency higher than any of those of the pure system is shown in Figure 7.7a. This '*localized mode*' can be seen to have a form similar to a long wavelength optical mode centred on the impurity and having displacement amplitudes which decrease at least exponentially with increasing distance from the impurity.

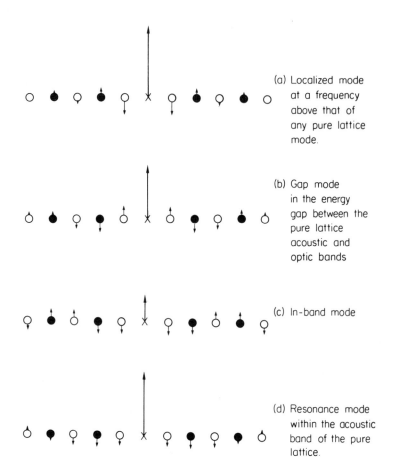

(a) Localized mode at a frequency above that of any pure lattice mode.

(b) Gap mode in the energy gap between the pure lattice acoustic and optic bands

(c) In-band mode

(d) Resonance mode within the acoustic band of the pure lattice.

Figure 7.7 An illustration of the atomic displacements for a single impurity in a long diatomic chain showing the form of vibrations described by the terms (a) 'localized mode' at a frequency greater than any pure lattice mode of vibration, (b) 'gap mode' which occurs in the energy gap between the pure lattice acoustic and optical branches, (c) 'in-band mode' in which the impurity plays a fairly normal part in a mode which is only slightly distorted from the pure lattice mode, (d) 'resonance mode' which lies within the low frequency tail of the acoustic band and in which the impurity shows a relatively large amplitude of vibration (after Sievers and Renk and Weber)

7.2.3.2 Gap mode

Impurities of suitable mass and/or force constant change can also produce normal modes of vibration of a localized type at frequencies between those of the various branches of the pure lattice system. These are termed *gap modes* and Figure 7.7b shows an example of such a mode falling in a gap between the acoustic and optic branches. Not surprisingly each half of this vibration looks part way between the shortest wavelength acoustic and optic modes (which define the edges of the gap), but there is a change of phase in passing the very large amplitude of the impurity, re-establishing the symmetry of the vibration about the impurity. The stationary centre of mass requirement of a normal mode coupled with the movement of the nearest neighbours in-phase with the impurity in a gap mode, rather than in anti-phase as in the localized mode, results in the gap modes being less highly localized than the super-optic local modes.

7.2.3.3 Resonance mode

As well as playing a dominant part in super optic localized modes and gap modes, an impurity also has to play some part in the whole range of lattice vibrations of the impurity doped system. Although a very small percentage of impurity could not be expected to produce any very radical change in the general pattern of the vibrations of the host lattice, it could cause marked effects in the IR or Raman spectrum of a crystal by changing the activity of some modes from that which they would have had in the pure host crystal. In particular it could render active, either directly or in combination with some other mode, some modes whose pure lattice counterparts were strongly forbidden. An effect of this type would generally be termed an *in-band mode* (see Figure 7.7c), but if such a vibration involved a relatively large impurity participation (i.e. large changes in impurity to nearest neighbour distance during the vibration), which in turn is most likely in regions in which the pure lattice shows a low density of vibrational states, then relatively sharp spectral features can be found and these are referred to as *resonance modes*. Figure 7.7d shows a resonance mode within the acoustic band of the diatomic chain, and apart from the large out of phase movement of the impurity, and the suppressed movement of its near neighbours, the vibration can be seen to closely resemble a long wavelength acoustic mode.

A figure similar to Figure 7.7 was presented by Sievers[19] after calculations by Renk[20] and Weber[21] on the eigenvectors for IR active impurity modes in a diatomic chain. For a mathematical definition of these impurity-lattice-dynamical terms, section II of part I of Maradudin's review article[5] is recommended.

Figure 7.8 shows schematically the density of vibrational states of a diatomic lattice containing impurities of different masses so that the diagram contains examples of the various types of mode discussed above.

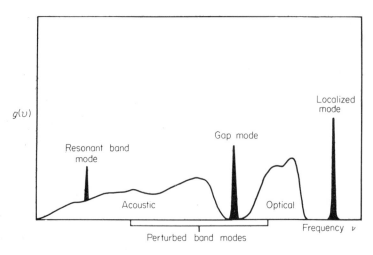

Figure 7.8 Schematic representation of the density of vibrational states of a diatomic lattice containing impurities of different masses showing (a) a localized mode, (b) a gap mode, (c) perturbed band modes, and (d) a resonant band mode

7.2.4 Specific examples

The general experimental and theoretical approaches to the subject of the vibrations of defects in solids which have just been outlined will now be illustrated with specific examples. Since the U-centre was historically the first system to be experimentally examined as an example of point-defect-produced lattice vibrations, it will be considered first here also. Not only was the U-centre the first system to yield experimental results, it was for two or three years the *only* system to yield experimental results, and this was at a time when the theoretical side of the subject was beginning to expand rapidly. For these reasons the U-centre story embodies almost a complete history of the development of the subject.

7.2.5 The U-centre

When an H^- or D^- ion substitutionally replaces the anion of a metal-halide crystal the resulting defect centre is called a U-centre. Although the ultra-violet absorption associated with hydride impurities in alkali-halides had been studied as early as 1930,[22] it was not until 1960 that Schaefer[3] published the first information about the infrared absorption by such systems.

7.2.5.1 U-Centre local mode

The results of Schaefer,[3] shown in Table 7.1 together with some other later results on U-centres, showed a localized vibrational mode at about 2·5 times the highest pure lattice vibrational frequency. The effect of replacing the

H^- by D^- was to reduce the local mode frequencies by a factor of about 1·39 which suggests that about 95 per cent. of the vibrational energy is carried by the defect ion itself. These two facts led Smart, Price and Wilkinson[23,24] to investigate the possibility of explaining these vibrations in terms of an H^- ion contained in an infinitely massive Born–Mayer[25] potential well.

Table 7.1 The frequency (in cm^{-1}) of the localized mode of H^- in alkali halides (U-centres). The upper figure is for H^- and the lower for D^-. The ratio of the two is very close to $\sqrt{2}$

	F	Cl	Br	I	
Na	858	565	498	427	H^-
	607·5	408	361	318	D^-
K	726	496·5	447	384	H^-
		357·5	319·7	279	D^-
Rb	703·1	476	425	360	H^-
		339			D^-
Cs		424	367		H^-
^6Li	1031				H^-
^7Li	1027				

This is the three-dimensional form of the very simple model shown in Figure 7.1 and discussed earlier; and it predicted very closely the frequencies at which the vibrations were found to occur. The force constants which were calculated to be acting between each of the lattice ions and the H^- ion were appreciably different from those coupling the host lattice ions to one another, and in order to bring the calculated and observed frequencies into complete agreement, systematic lattice distortions had to be included. Nearest neighbour expansions of about 0·19 Å were required in the fluoride lattices, 0·02 Å contractions in the chloride lattices, 0·11 Å contractions in the bromide lattices, and 0·22 Å contractions in the iodide lattices, as shown in Figure 7.9. These distortions can be seen to be quite reasonable and in keeping with the finding of Dick and Das[26] about the probable distortions to be expected in the vicinity of an impurity ion substituted into an alkali-halide.

The further implications of these calculated distortions, which cause changes in the force constants between the nearby host lattice ions through the anharmonic terms in the relevent potential terms, were not considered at that time. However, the points that do arise quite naturally from the above treatment are important ones for most of what follows and are therefore

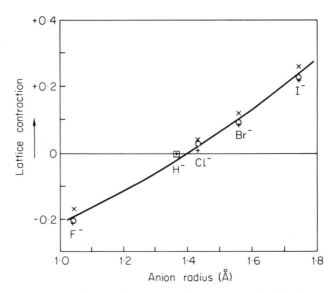

Figure 7.9 The localized lattice distortions in alkali halides caused by the substitution of an H⁻ ion into the fluorides, chlorides, bromides and iodides of sodium ($\times$), potassium ($\odot$) and rubidium ($+$) plotted as a function of anion radius (Smart)

worth restating as follows:

(i) For defect produced localized vibrational modes well above the highest pure lattice frequency it is a good approximation to assume that the defect vibrates within a rigid lattice.

(ii) For defects which can be assigned 'Born–Mayer' type parameters, isolated within alkali-halides it is possible to obtain good values for the force constants coupling the defect to its surrounding lattice ions, and these force constants must be expected to be in general appreciably different from those which coupled the replaced host lattice ion.

(iii) Distortion of the host lattice in the vicinity of the defect can be expected to be appreciable, at least at the nearest neighbour level, and this distortion must be expected to change, through anharmonic terms, the force constants coupling the motion of these near neighbour ions to the rest of the lattice.

(iv) A very large part (typically about 95 per cent.) of the restoring force constant operating on a defect during a local mode vibration comes from interaction with the nearest neighbour ions.

Turning now to the lattice dynamics approach to the U-centre problem, it will be appreciated that the considerable amount of theoretical work which had been done since 1943[1] on the problem of defect distorted vibrations, had reached a fair degree of sophistication before Schaefer's[3] experimental results became available. The techniques developed were essentially concerned with the mathematical approximations which could be employed to get useful numerical solutions to the equations describing the vibrations of an infinite, or at least very extensive lattice, containing a single point defect. In the earliest attempts to fit such theories to the localized U-centre vibrations the isotopic substitution approximation was used. That is, the H^- ion was regarded as an isotope of the halide ion it replaces with a mass change, but with no change of force constants and no distortion of the surrounding lattice. Not surprisingly this did not result in very accurate predictions of U-centre localized mode frequencies, because, as discussed earlier, the important H^- to nearest neighbour force constant is in general quite different from the corresponding pure lattice value. When a changed force constant was included in the lattice dynamical models, the mathematical difficulties increase considerably and simplifying assumptions had to be included at some stage of the numerical solution. In the earliest attempts to include changed force constants the mathematical approximations employed reduced the solutions for the localized mode to effectively that which results from the simple Figure 7.1 model, or later, the Figure 7.2 model. However, the value of the lattice-dynamical approach to this problem lies in its ability to provide an understanding of those effects which occur in the region covered by the host-lattice vibrational frequencies, rather than in its ability to closely predict the frequencies of high frequency localized modes.

Table 7.2 The frequency (in cm^{-1}) of the localized mode of H^- in the alkaline earth fluorides for the infrared active F_2 components of $v = 0 \to 1$, $0 \to 2$ and $0 \to 3$ transitions.

	CaF_2		SrF_2		BaF_2	
$v - v'$	100 K	300 K	100 K	300 K	100 K	300 K
$0 \to 1$	964·6	956·2	892·1	884·5	802·3	795·3
$0 \to 2$	1917·5	1901·6	1773·3	1759·4	1593·2	1580·5
$0 \to 3$	2911 2822		2691·5 2602·7	2664·6 2591·1	2421·9 2326·3	2403·6 2318·6
$0 \to 2$	1384	1373	1275	1266	1144·7	1133

7.2.5.2 U-Centre band modes

Immediately following Schaefer's[3] discovery of the local mode absorption due to the H^- ion isolated in alkali-halides, virtually all interest in this system was centred on this mode of vibration. However, as the lattice dynamicists began predicting other types of defect activated effects, so the experimentalists sought evidence of these other effects. In 1963 Fritz[27] first reported the existence of lattice side bands on the U-centre local mode (these will be discussed in the next subsection) and in 1964 Sievers[28] observed some in-band absorption with the KI/H^- system.

Figure 7.10 shows the absorption of a KI/H^- crystal at 7 K, in the $45 \rightarrow 100$ cm^{-1} region, and the calculated absorption coefficient for this same region as evaluated by Timusk, Woll and Gethins.[29]

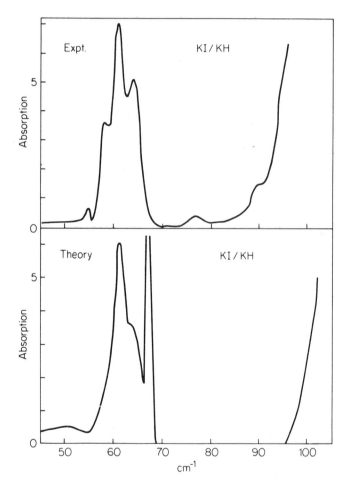

Figure 7.10 Comparison of the far infrared 'in-band' absorption below 70 cm^{-1} due to hydride ions in KI (Sievers) with the calculated absorption coefficient (Timusk, Woll and Gethins)

The calculations[29] involved the evaluation of the absorption coefficient, $\alpha_{(\omega)}$, as a function of ω and the expression used was:

$$\alpha_{(\omega)} = \frac{(n_\infty^2 + 2)^2}{9n_\omega} \frac{4\pi}{C} \frac{e^2}{\mu} \frac{N}{V} I_m\{\bar{G}_{RxRx}\}$$

This expression is similar to that suggested by Dawber and Elliott,[17] and is essentially the same as that used by Benedek and Nardelli,[30] or Patnaik and Mahanty.[31]

This expression for $\alpha_{(\omega)}$ contains the expected constants (i.e. refractive index, electronic charge, velocity of light, etc.) and the term $\bar{G}_{RxRx}$, which is a diagonal element of the Green's function matrix $\bar{G}(\omega^2)$ for the impurity containing crystal. The matrix was taken in the representation of eigenmodes of the pure lattice, so that the element required corresponded to the reststrahlen frequency, i.e. the TO mode at zero wavevector. Only modes polarized in the x direction were considered and $\bar{G}$ was evaluated using the T-matrix defined by

$$T = \Gamma(1 + G\Gamma)^{-1} \quad \text{and} \quad \bar{G} = G - GTG$$

where Γ is the matrix of force constant changes and G is the Green's function matrix of the pure crystal. This allowed the $\bar{G}_{RxRx}$ element to be expressed in terms of the reststrahlen frequency and the corresponding T matrix element.

$$T_{RxRx} = \frac{1}{N} \sum_{\substack{L \ L' \\ \pm 1 \ \pm 1 \\ \alpha \ \alpha'}} \varepsilon_\alpha^{\pm 1}(Rx)\varepsilon_{\alpha'}^{\pm 1}(Rx)T_{L, \pm 1, \alpha, L', \pm 1, \alpha'}$$

where L is the position of the ion, ± 1 refers to the sign of the ionic charge and α is a cartesian co-ordinate, and the polarization vectors $\varepsilon_\alpha^{\pm 1}(Rx)$ were given by

$$\varepsilon_\alpha^{\pm 1}(Rx) = (\pm 1)\sqrt{\frac{\mu}{M(\pm)}} \delta_{\alpha x}$$

where $M(\pm)$ is the mass of the appropriate ion and ± 1 refers to its charge.

By making full use of the cubic symmetry, the matrix inverted for the evaluation of T_{RxRx} was reduced to a 3 times 3 matrix corresponding to T_{1u} symmetry. Two changed force constant values were required, H^- to nearest neighbour (obtained from the local mode frequency) and nearest neighbour to fourth nearest neighbour (obtained from the A_{1g} mode shown by the lattice side band spectrum discussed later). The shell model of Dolling et al.[32] was used together with the summing technique of Gilat and Raubenheimer[33] to cover an effective 32 000 points within the Brillouin zone, in the evaluation of the six elements in G.

The above brief description of the mathematical steps involved in obtaining a calculated curve such as Figure 7.10b has been included to give some idea of the computational labour involved in testing this type of theory by numerically evaluating a function such as $\alpha_{(\omega)}$ over a range suitable for comparison with experiment.

The peak in the experimental curve at about 61 cm^{-1} is identified as due to the resonance mode which the theory predicted to be about 6 cm^{-1} higher. At 64 cm^{-1} the theoretical curve shows a feature whose frequency is determined by a Van Hove singularity (VHS) in a high symmetry direction. Since the VHS has been directly determined by neutron diffraction,[32] it is encouraging that the experimental curve also shows a peak at this frequency. A VHS in an off-symmetry direction is responsible for the frequency of the 61 cm^{-1} theoretical peak but since this frequency was not directly determined by the neutron work, but has to rely on the accuracy of the shell model calculation, it seems reasonable to believe that this VHS really exists at about 58 cm^{-1} where the experimental curve shows an otherwise unexplained feature. In order to try to find supporting information for their assignment, Timusk et al.[29] measured the far IR absorption of KI/D^{-} and detected a slight shift to lower frequency for the 61 cm^{-1} band with the other two features unaffected. This does indeed help to substantiate their assignment and this point will be taken up again later. The absorption in this region is due to the presence of a resonance mode, but the shape of the absorption band is very dependent on the density of states in that frequency region, and any Van Hove singularities for vibrations of appropriate symmetry can be expected to be clearly shown.

If Figure 7.7 is considered for the case of the U-centre, and it is remembered that a very large part of the restoring force on the impurity comes from its interaction with its nearest neighbours, it will be appreciated that the type of motion depicted for the resonance mode (Figure 7.7d) would have a frequency quite close to that of the local mode (Figure 7.7a). However, the distortion of the lattice around the H^{-} ion and the resulting anharmonic lowering of the force constants between the H^{-} ion's nearest neighbours and their nearest neighbours might reasonably allow the phase reversal motion of the H^{-} to be shared by its nearest neighbours. The resultant motion might therefore be expected to look more like that shown in Figure 7.11a.

Figure 7.11 (a) Shows the probable form of the U-centre resonance mode and should be compared with Figure 7.7d and v_3 in Figure 7.2. (b) is shown because it allows a rough value of the frequency to be very easily obtained although it is not as realistic a model as (a)

Figure 7.11b is included because, although not as realistic as Figure 7.11a, it does allow a rough value for the frequency to be very easily obtained. Thus, for KCl/H^-, KBr/H^- and KI/H^- if the two potassium ions in line with the H^- motion move with it as a rigid mass in the resonance mode they will see a restoring force in the three-dimensional case which is two to three times greater than that seen by the H^- when it moved essentially by itself in the local mode; the resonance mode frequency v_R would be given approximately in terms of the local mode frequency, v_L, by

$$v_R \simeq v_L \sqrt{\frac{2 \times 1}{1 \times 79}}$$

For KI/H^-

$$v_R = 384 \sqrt{\frac{2}{79}} = 61 \text{ cm}^{-1}$$

This chance agreement with the number discussed above should not be seen as giving any great significance to this relationship for the U-centre especially since, if the factor two is still used for the force constant ratio, it predicts a value of only 71 cm^{-1} for KBr which Timusk et al.[29] give as 89 cm^{-1}. (Using the more appropriate factor 3 for the force constant ratio in the more rigid bromide lattice gives a value of 87 cm^{-1}.) No other experimental values for far IR absorption are yet available for resonance modes due to U-centres (although several lattice side band spectra are available), but it is interesting to note that comparable scaling of local mode frequencies for sodium bromide and sodium iodide predict gap modes rather than resonant modes, and Figure 7.11a does begin to look something like the gap mode of Figure 7.7b and would look even more like it if the wavelength of the lattice acoustic wave were further shortened.

One further point about this suggested relationship between v_R and v_L is that it would suggest that the effect of replacing H^- by D^- would be to lower the resonant mode frequency by about $v_R/4M$, where M is the atomic mass (in a.m.u.) of the nearest neighbour metallic ions. That is about 0.38 cm^{-1} for the resonance mode in KI to be compared with the 'slight lowering' observed by Timusk et al.[29] which they interpreted as being due to different anharmonic contributions in the two cases.

7.2.5.3 U-centre lattice side band structure

As mentioned earlier, Fritz[27] reported as early as 1963 that the absorption by the localized mode of the U-centre was not a single absorption peak, but was always accompanied by sidebands which appeared to be sum-and-difference combinations with lattice vibrational modes. Figure 7.12 illustrates

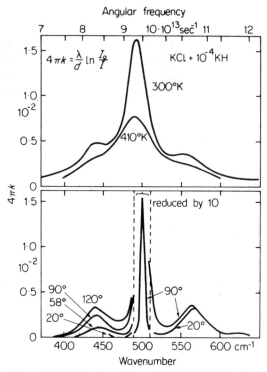

Figure 7.12 IR absorption of KCl/H⁻ showing that the U-centre local mode is not a single absorption peak but is accompanied by sidebands which involve sum and difference combinations with lattice vibrational modes (Fritz)

the spectra reported by Fritz[27] for the KCl/H⁻ system, Figure 7.13a shows the lattice summation band on the localized mode of the U-centre in KI at 6 K as reported by Fritz et al.[34] and Figure 7.13b shows the similar spectrum for KBr/H⁻ as reported by Timusk and Klein.[35] Fritz, Gross and Bauerle[36] have published lattice sideband spectra for U-centres in NaBr, KCl, KBr, KI, RbCl and RbBr.

Theoretical treatments of this sideband structure can be made which are essentially similar to that outlined earlier for the resonance mode and fair agreement between calculated and observed spectra can again be obtained. The qualitative results which emerge from the theoretical treatment of these bands are:

(i) In the lattice sidebands on the U-centre local mode it is those perturbed lattice modes which possess A_{1g}, E_g or F_{2g} symmetry which are expected to appear (i.e. only these types of modes have IR active F_{1u} sublevels when in binary combination with the F_{1u} local mode).

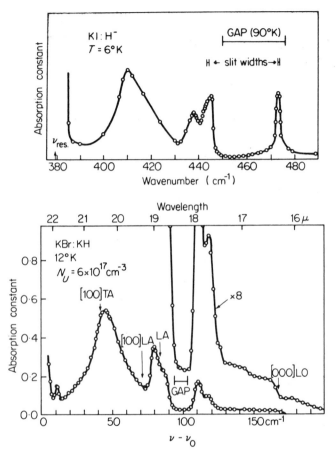

Figure 7.13 (a) The summation sideband structure on the high
energy side of the localized hydride mode in KI at 6 K (Fritz,
Gerlach and Gross). The position of the localized mode is at
ν_{res}, and (b) the corresponding spectrum due to H$^-$ in KBr
(Timusk and Klein)

(ii) The degree of perturbation of these modes by anharmonic relaxation
of the force constant around the H$^-$ ion is of great importance in
determining the relative activity of the modes in the sideband structure,
and hence dictates the shape of the band. The very existence of the in-gap
mode in KI/H$^-$ depends upon this relaxation.

(iii) Two different mechanisms, anharmonic coupling and second order
dipole moment coupling, could render these combination bands active
but the rapid decrease of sideband intensity with increasing separation
from the local mode frequency suggests that for this system the contribution of the anharmonic coupling is dominant.

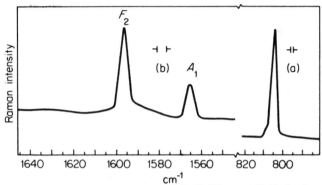

Figure 7.14 (a) Raman spectra of BaF_2/H^- at 6 K F_2 fundamental, and (b) second harmonics of A_1 and F_2 symmetry.
(The E symmetry second harmonic which is predicted to occur
at 1617 cm^{-1}) (Harrington, Harley and Walker)

7.2.5.4 U-Centre Raman spectra

Although very little Raman work on U-centres has been published, such
studies are quite feasible as is shown in Figure 7.14, which shows the Raman
spectrum of BaF_2/H^- not only for the fundamental, but also for the first
overtone levels.[37] However, the IR measurements are undoubtedly more
straightforward as is indicated by Figure 7.15, which shows the IR absorption

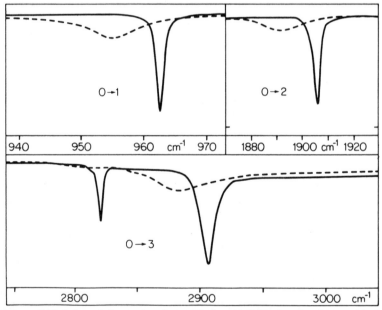

Figure 7.15 Infrared absorption due to H^- ions in CaF_2 showing the $v = 0 \to 1$,
$0 \to 2$ and $0 \to 3$ transitions in samples of different concentration and thickness
at 300 K (dotted) and 100 K (Hirst)

of the similar system CaF_2/H^- for the fundamental, first overtone and second overtone regions.[38]

7.2.6 Effects of temperature on U-centre spectra

7.2.6.1 Local mode

Measurements on the frequency, half bandwidth, and integrated intensity of many of the U-centre local modes have been made over wide temperature ranges, in some cases from 4 K to 400 K.

Figure 7.16 shows the data of Fritz *et al.*[36] and Mirlin and Reshina[39] for the half width of the KCl/H^- and KCl/D^- local mode bands as a function

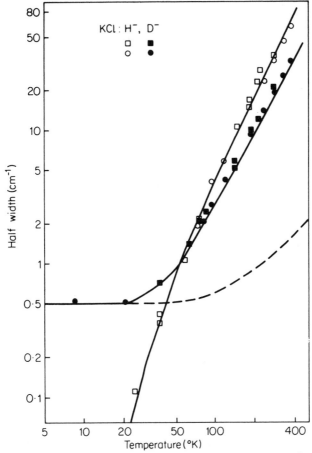

Figure 7.16 The half widths of the H^- and D^- local mode absorption in KCl as a function of temperature (Fritz, ref. 36). Note the high temperature ($\alpha \propto T^2$) region where the H^- band is approximately twice as broad as the D^- band and the low temperature region where the H^- band half width is significantly narrower than that of the D^- band

of temperature; the log–log plot used shows the region over which a T^2 dependence can be fitted and also brings out the low temperature region very clearly. At temperatures above about 100 K the H⁻ peak is seen to be roughly twice as broad as the D⁻ band, but below this temperature the H⁻ band sharpens more quickly than the D⁻ one, and both bands become equal to 1 cm⁻¹ in width at 50 K after which the H⁻ peak continues sharpening down to less than 0·1 cm⁻¹ at 10 K, whereas the D⁻ residual width seems to be about 0·5 cm⁻¹.

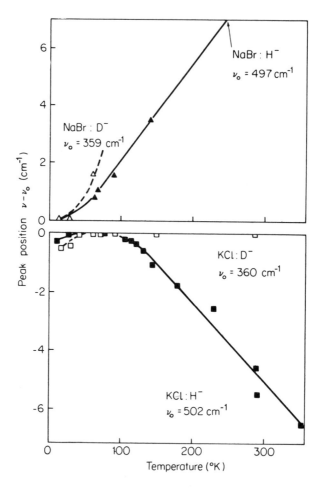

Figure 7.17 The frequency shift $v - v_0$ of the maximum absorption of local mode as a function of temperature for KCl/H⁻, KCl/D⁻, NaBr/H⁻ and NaBr/D⁻. The complexity of temperature induced shifts is evident from the qualitatively different forms of the curves

The low temperature residual widths are explained in terms of a two phonon decay process for the D^-, and a three phonon decay process for H^-, giving a longer lifetime and therefore a narrower half width to the H^-. Whereas in the high temperature T^2 region a fourth-order phonon scattering process proposed by Elliott[40] which depends on the square of the local mode amplitude would explain the factor of two by which the H^- band width exceeds that of the D^- band.

Changes in the U-centre peak frequency as a function of temperature have been found difficult to interpret quantitatively, because although most of the frequencies are found to decrease with increasing temperature (see, for example, KCl/H^- in Figure 7.17), some are found to remain essentially constant (e.g. KCl/D^-) and some are found to increase (e.g. $NaBr/H^-$). Experiments on mixed crystals, and experiments as a function of applied pressure (see later), suggest that thermal lattice expansion should always be accompanied by U-centre frequency decrease.

There is not general agreement in the literature about the changes with temperature of the total band area of the U-centre local mode absorption. This is due, at least in part, to the presence of the lattice side bands which extend for over $100 \, cm^{-1}$ each side of the main band at room temperature and make the construction of a true base line against which to measure the band rather difficult. In particular this would have been true for the earlier work on these systems when often only relatively small spectral ranges were scanned to record the central peak. Figure 7.12 showed the temperature effects on the KCl/H^- main band with lattice side bands, and it will be appreciated that a short scan through the central region only could easily result in a false baseline being taken and a drastically low value being obtained for the higher temperature band areas.

7.2.6.2 Lattice side bands

Figure 7.12 showed the temperature dependence of the lattice side bands for the KCl/H^- systems, which behave in a way quite typical of the other side bands which have been studied as a function of temperature. One point worth emphasizing here is the decrease in separation between the main band and the maximum in the absorbance of the lattice difference band, as the temperature is lowered. This is to be expected, as the higher energy lattice vibrations are preferentially frozen out at the lower temperatures. Thus only at the higher temperatures, when these bands are relatively uninformative, is the difference band a reasonably undistorted measure of the relevant lattice modes, and the summation band is the one to be used for most purposes. This point is emphasized because some authors have made misleading comments about differences observed between peak frequency positions as measured in the sum and difference parts of the lattice side band structure. Only for very sharp features could these be expected to give the same values.

7.2.6.3 Far-infrared bands

For defects in ionic crystals the strength of the host lattice absorption at any but the lowest temperature tends to make temperature dependence studies difficult in the far-infrared region. Some work will be discussed later, on bands at very low frequencies, but no work seems yet to have been published on the temperature dependence of the U-centre resonance band.

7.2.7 Effect of hydrostatic pressure on U-centre spectra

Only one attempt has been made to apply hydrostatic pressures to the infrared spectra of U-centres and that was by Jacobson.[41] He used a modified type of Drickamer high pressure optical cell working at 90 K to investigate the pressure dependence of some of the local mode frequencies. Within the 0–10 kbar pressure range available to him, he found frequency increases which were noticeably curved for NaCl/H$^-$ and KBr/H$^-$, but which were apparently linear with pressure for KCl/H$^-$ and RbCl/H$^-$ in its high pressure, caesium-chloride-type structure. He found that these shifts were quite in keeping with modified Smart-type[23] calculations on the increased restoring force to be expected within the compressed lattices.

7.2.8 Effect of uniaxial stress on U-centre spectra

Several authors have managed to perform uniaxial stress experiments on defect-activated vibrations in alkali halides and one such experiment on the U-centre, reported by Fritz *et al.*,[34] will now be considered. Figure 7.18 shows the splittings and linear shift rates with uniaxial pressure applied along a $\langle 100 \rangle$ direction of both the local mode, and in-gap lattice side band, features of the KI/H$^-$ system. The $\langle 100 \rangle$ direction was chosen because the largest splittings could be obtained this way and the experiment was carried out at liquid helium temperature in order to have the absorption bands as sharp as possible. In fact two separate experiments were performed, one on a relatively weakly doped crystal to measure the effects on the local mode itself, and the other with a more strongly doped sample to investigate the lattice side band. The local mode is seen to split into two components with the E parallel to P component at the higher energy. This is essentially as expected except that the E perpendicular to P component may have been expected to shift slightly negative, but this is just a further indication that the local elastic constants, very close to the H$^-$ ion, are different to those of the bulk crystal. The in-gap lattice side band splits by apparently exactly the same amount as the local mode but both side band components have an added increase in frequency with applied pressure, and it is this added increase which is the only stress induced effect on the vibration responsible for the in-gap mode.

These results, supported by analogous measurements under $\langle 110 \rangle$ stress, show that the 93·7 cm^{-1} side band mode is of symmetry A_{1g}. (Either of the

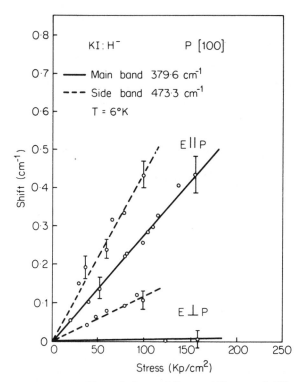

Figure 7.18 The splitting and linear shift rates dv/dP due to uniaxial pressure applied along a $\langle 100 \rangle$ direction for both the local mode and 'in-gap' lattice side band feature for hydride ions in KI (Fritz, Gerlach and Gross). Note that the sideband and main localized mode show identical splittings but the sideband components show an extra, linear, increase in frequency with pressure

other two possible symmetries E_g or F_{2g} would have caused further splittings.) This agrees with a calculation by Gethins et al.[42] which predicts an in-gap mode of this symmetry to result from the relaxation of the force constants in the vicinity of the H⁻ ion, and is consistent with its non-appearance in the far IR absorption spectrum of Figure 7.10.

Hayes and Macdonald[43] have made uniaxial stress measurements on U-centre modes in CaF_2, where they have observed the expected splittings in the local mode absorption and recorded linear shift rates with pressure up to the strength of their crystals. They also recorded absorption in the overtone region under uniaxial stress, where they found transitions becoming weakly allowed to states which were A_1 and E under the full T^d symmetry, but to which transitions became allowed from the A_1 ground state as a mixing took place of the three states, A_1, E and F_2 of the overtone, as the T_d symmetry was destroyed.

7.2.9 Effect of electric field on U-centre spectra

Hayes and Macdonald[43] used electric fields to investigate the CaF_2/H^- localized mode, but even at the maximum field that their crystals would tolerate (about 10^5 V/cm) the splittings could not be properly resolved. In so far as these results could be interpreted, they were in agreement with their uniaxial stress results mentioned above.

7.2.10 Effect of isotopic substitution on U-centre spectra

This technique has been extensively exploited by H^- to D^- substitution in local mode studies on U-centre spectra (but not a single T^- result has been produced although in some lattices this would bring the local mode quite close to the top of the optic band). Little effect is to be expected from the H^- to D^- change on the in-band modes, although one reference to this was given earlier. The only really sharp feature in the U-centre spectra which is not a super optic local mode is the in-gap lattice side-band feature in the KI/H^- system. This has been assigned as an A_{1g} mode in which the H^- is stationary and its mass unimportant. However it would probably be worth investigating this mode in the KI/D^- system both for support for the assignment, and to investigate any second order effects which though relatively small could possibly be detected on this very sharp band.

7.2.11 Comparison of spectra for U-centres in many crystals

Although the U-centre has been studied in many crystals (Tables 7.1, 7.2) few authors have tried to exploit the similarities and differences which can be seen when these spectra are compared. Smart[23] used this approach for some of his work as discussed earlier (Figure 7.9), and as shown in his plot of reduced half width of the local mode as a function of reduced temperature (Figure 7.19).

7.2.12 Interaction between U-centres and other impurities

From among the many different experiments which could be treated under this heading, only one, on the U-centre in mixed KCl, RbCl crystals will be considered. Various workers have performed experiments with mixed crystals which also contain U-centres and these will generally be considered later together with other multi-component (polyatomic) systems. One example, however, is given now since it has a bearing on several points which have been under discussion. Figure 7.20 shows the effect of various concentrations of Rb^+ on the absorption of the U-centre in KCl.[27] At low concentrations of Rb^+ the main peak broadens and shifts slightly to lower frequency, as the lattice spacing becomes slightly less well defined but, on average,

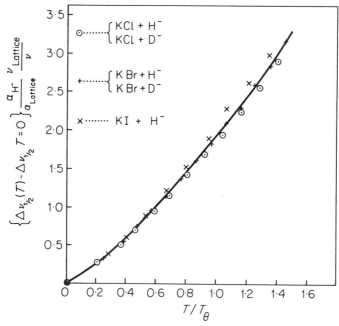

Figure 7.19 The reduced half width of the U-centre local mode absorption for KCl/H⁻, KCl/D⁻, KBr/H⁻, KBr/D⁻, KI/H⁻ plotted against the reduced temperature T/T_θ (Smart[23]). Over this temperature range the experimental points fall within experimental error on a single line for many different alkali halides

larger; two new bands, α, β due to H⁻ ions having one nearest neighbour Rb⁺ grow linearly with Rb⁺ concentration. Notice that the high frequency band β is only about one half of the intensity of low frequency band α; β represents vibrations which can be regarded as the H⁻ oscillating between a K⁺ and a Rb⁺ whereas α is the doubly degenerate vibrational perpendicular to this. Bands ε and δ have intensities which vary with the square of the Rb⁺ concentration and are thus probably due to H⁻ ions having two of their nearest neighbours as Rb⁺ and therefore a third line in this group, f, might be expected at high frequencies almost twice as far above the main band as β. This band would be expected to be weak since the relative intensities of $f : \varepsilon : \delta$ could be expected to be $1 : 6 : 8$ on the simplest argument. (If vibration towards a Rb⁺ is designated $+$ and vibration at right angles to a Rb⁺ is designated $-$, then statistically $(+\,+):(-\,-):(+\,-)$ occur in the ratio $1 : 6 : 8$; with the $(-\,-)$ made up out of a 2 and a 4 of different character which might have split or broadened this band.) See later when this same system is considered in terms of the reduced symmetry caused by the Rb⁺ ions.

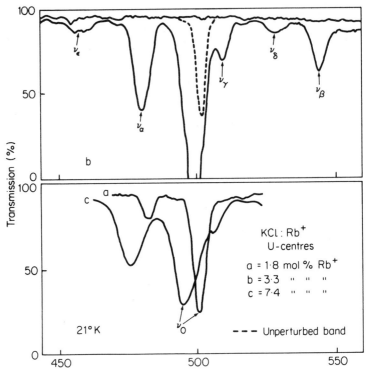

Figure 7.20 The effect of various concentrations of Rb^+ on the KCl/H^- absorption spectrum in the vicinity of the local mode. Note that of the two bands (α and β) which grow linearly with rubidium concentration the intensity of one (α) is about twice the intensity of the other (β). The intensities of ε and δ increase with the square of the Rb^+ concentration. Another line at about twice the separation of β from the main line is predicted. The frequency of the localized $RbCl/H^-$ mode is $475\ cm^{-1}$ (Fritz)

7.2.13 Other point defects in alkali halides

The U-centre was considered in considerable detail partly because it is the system which has been most intensively studied, and partly because it has been possible to introduce in this one context most of the ideas which are discussed in this chapter. However, many other point defect systems have been investigated, and some points of interest, noticeably the resonance mode, are more clearly illustrated by these other systems. Therefore before proceeding to discuss polyatomic impurities we will briefly consider some of these other point defects. Tables 7.3 and 7.4 list most of the other point defect systems which have been investigated by IR or Raman spectroscopy. Table 7.3 refers to strongly ionic crystals and Table 7.4 to crystals which are normally regarded as semiconductors.

Table 7.3 A list of point defects studied by IR or Raman spectroscopy in alkali halides

	Cl	Br	I
Na	K^+, Li^+, Cu^+, Ag^+ F^-, Br^-, I^-	Ag^+	Ag^+
K	Ag^+, Li^+ H_i^-	Ag^+, $^6Li^+$, $^7Li^+$ H_i^-	Na^+, Cs^+, Tl^+, Ag^+ H_i^- Cl^-, Br^-, F^-
Rb	H_i^-		
Cs	K^+, Rb^+	Li^+, Na^+, K^+, Rb^+, Tl^+	Li^+, Na^+, K^+, Rb^+, Tl^+

Table 7.4 A list of point defects in semiconductors whose vibrational spectra have been investigated by IR or Raman spectroscopy

C	N, B
Si	^{10}B, ^{11}B, ^{12}C, ^{13}C, ^{14}C
Ge	Si
GaAs	P, Al, ^6_iLi, ^7_iLi, ^6Li, ^7Li, Si_{Ga}, Si_{As}
InSb	Al
GaSb	Al, P
ZnTe	Li, Al
CdSe	S
CdTe	Li, Si, Al

Apart from the localized mode of the interstitial H^- ion, the vibrations referred to in Table 7.3 are either gap modes or resonance modes within the acoustic band. The bands which have received the greatest attention have been the low frequency resonance modes. Bearing in mind the fact that by far the largest contribution to the restoring force on an impurity ion comes from its nearest neighbours it will be appreciated that a low frequency resonance mode of the type analogous to the one-dimensional example of Figure 7.7d, can only exist for impurities for which the nearest-neighbour force-constant to impurity mass (k^*/m^*) ratio is considerably less than the equivalent ratio for the ion it replaces. (As $v \propto \sqrt{k/m}$, and an optic type of mode has to fall by a factor of say 5 to bring it low down in the acoustic region, clearly k^*/m^* will need to be down by a factor of about 25 with respect to k/m). Thus when a heavy, compact silver ion replaces the bigger, lighter potassium ion in KCl, KBr or KI a low frequency resonance mode results. When however the relatively light lithium ion replaces the potassium ion the situation is less obvious, but in fact in KBr a resonance mode is found because of a very big reduction in the force constant. Takeno[44] calculated this reduction in force constant to be by a factor of 200.

7.2.14 The Li$^+$ defect in alkali halides

Figure 7.21 shows the far infrared absorption of KBr/Li$^+$ as measured by Sievers and Takeno[45] for both the isotopes of Li$^+$. Note the large isotopic

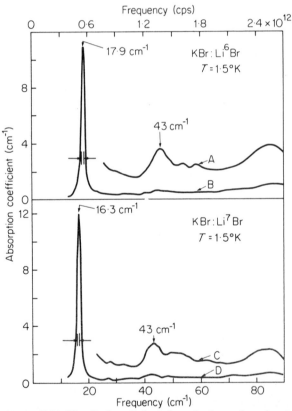

Figure 7.21 The 'in-band' far-infrared absorption due to lithium ions in potassium bromide (Sievers and Takeno). Note the frequency shift of the resonance mode between ^{6}Li and ^{7}Li. The observed frequency ratio for the two isotopes (1·098) is greater than the square root of the masses $(\frac{7}{6})^{1/2} =$ 1·080 suggesting a highly anharmonic potential well

percentage frequency shift of the sharp resonance band which gives a ratio of 1·098 for v_{6Li}/v_{7Li} which is actually larger than the ratio $(M_{Li^7}/M_{Li^6})^{1/2}$ which equals 1·080. This is an indication of the strongly overharmonic nature of the well in which the Li^+ ion is held. Many different types of experiment have been performed on this system; both spectroscopic, (isotopic[45,46] mixed crystals,[47] uniaxial pressure[48]) and non-spectroscopic, (low temperature dielectric constant,[49] thermal capacity,[50] specific heat[51] and thermal conductivity.)[52,51]

Much of the recent interest in this KBr/Li^+ system has centred on the discussion about whether or not the centre of mass of the lithium ion is central with respect to the supporting lattice. It was to investigate the symmetry of the Li^+ occupied site and thus to gain information about the position of the ion within the lattice that Nolt and Sievers[48] performed the uniaxial pressure experiments on the resonance mode. Figure 7.22 shows their

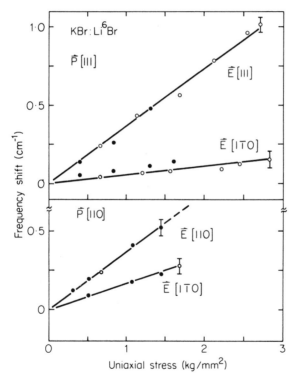

Figure 7.22 The frequency shift and splitting of the KBr/Li^+ line induced by uniaxial pressure along the principal crystal axes. The notation $P\langle 111\rangle$ refers to the stress direction and $E\langle 110\rangle$ to the direction of the electric vector of the polarized infrared radiation (Nolt and Sievers)

results for pressure along two of the three principle crystallographic directions, $\langle 110 \rangle$, $\langle 111 \rangle$. These results are consistent only with a T_d or O_h symmetry for the defect, (which could be resolved by an electric field experiment but not a uniaxial stress experiment), but since there are no plausible T_d models for this system it seems very likely that the lithium ion is in fact centrally held within a symmetrically distorted KBr lattice site.

The KCl/Li$^+$ system on the other hand is almost certainly an example in which the defect sits off-centre in its lattice site. This is discussed for example by Nolt and Sievers,[48] but since most of the evidence is non-spectroscopic it will not be dealt with here.

7.2.15 Mixed alkali halide systems

The Li$^+$ centre discussed above is a special case of the mixed alkali halide type, of which several examples have been investigated. For example Figure 7.23 shows the far IR absorption of the KI/Cl$^-$ system.[53]

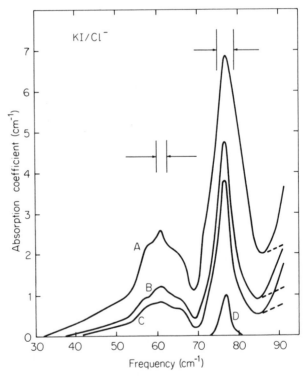

Figure 7.23 The far-infrared absorption of KI containing substitutional Cl$^-$ ions showing the gap mode at 77 cm^{-1} between the acoustic and optical branches for different impurity concentrations (Sievers *et al.*)

7.2.16 The Cu$^+$ defect in alkali halides

Figure 7.24 shows one of the few temperature dependent studies to have been published on low energy resonance modes of point defects in alkali-halides. This is work on the NaCl/Cu$^+$ system by Weber and Siebert.[54]

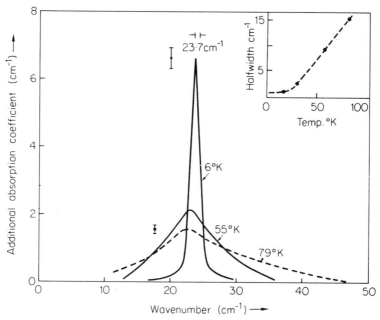

Figure 7.24 The far-infrared absorption due to in band resonance of Cu$^+$ in NaCl at 23·7 cm^{-1} is shown for three different temperatures (Weber and Nette). The inset is a plot of the half width of this resonance versus temperature

7.2.17 The Ag$^+$ defect in alkali halides

Figure 7.25 shows the observed far IR absorption of KBr/Ag$^+$ compared with two different calculations of this absorption,[55] and Figure 7.26 shows the effects of uniaxial pressure on the resonance mode absorption of the KI/Ag$^+$ system.[48]

7.2.18 Point defects in semi-conductors

Table 7.4 indicates some of the semi-conductor crystals whose point defect activated vibrational spectra have been studied. It should be noted that parts of the impurity-activated lattice vibrational spectrum of naturally occurring diamonds were studied as long ago as 1934[56] and studies on radiation damaged crystals of diamond, silicon and germanium all pre-date the references in Table 7.4. However the value of data from well-specified

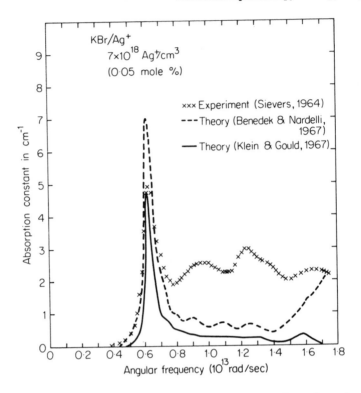

Figure 7.25 Comparison of the observed far IR absorption of
KBr/Ag$^+$ with two different theoretical calculations (ref. 55)

point defect systems is that it can be much more readily compared with
theoretical models; and so only when the methods required for controlled
crystal preparation become available was the interplay of theory and
experiment able to further our understanding of this subject.

References 5, 6, 7, 8 and 57 contain discussions of many of the systems in
Table 7.4 and therefore only three examples will be briefly considered below.
The lattice dynamical arguments are essentially the same as for the systems
considered earlier, and localized and resonance modes are again observed.
Calculations on the activity to be expected in the infrared for the defect
distorted lattice vibrations is less straight-forward for this type of crystal, but
the absence of strong one-phonon absorption makes the experimental
detection and measurements easier than for the ionic solids. However
problems due to conflicting absorption do arise if large concentrations of
impurity are substituted into semi-conductors, since this invariably leads to
an increase in the free carrier absorption.

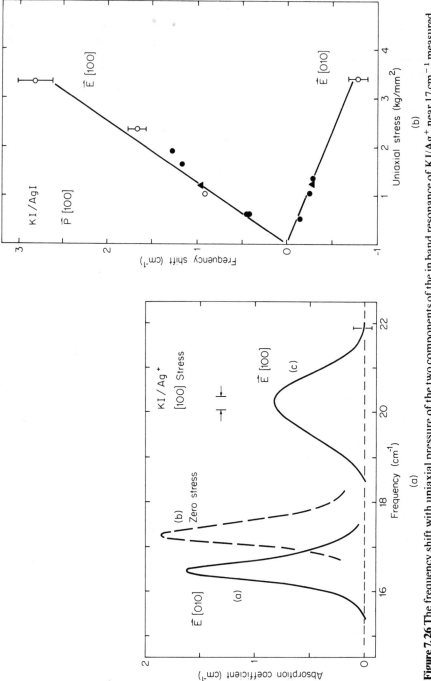

Figure 7.26 The frequency shift with uniaxial pressure of the two components of the in band resonance of KI/Ag$^+$ near 17 cm^{-1} measured with the electric vector of the infrared radiation parallel [100] and perpendicular [010] to the applied uniaxial stress (Nolt and Sievers)

7.2.18.1 Aluminium in the indium antimonide lattice

Figure 7.27 shows the IR absorption which was predicted by Dawber and Elliot[17] as likely to result from some group III and group V elements substitutionally isolated in silicon; and Figure 7.28 shows the observed IR absorption of the InSb/Al local mode both in its fundamental and also in its

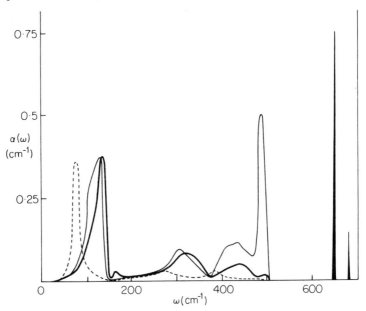

Figure 7.27 The calculated infrared absorption due to 10^{19} cm^{-3} atoms of charged group III and group V elements in silicon. The full curve is for ^{10}B and ^{11}B in natural abundance, the light curve represents the phosphorus doped sample and the dashed line is for antimony (Dawber and Elliot)

first overtone region as measured by Goodwin and Smith.[58] Aluminium atoms replace the indium atoms in the InSb lattice and high concentrations are readily attained as AlSb is soluble in InSb in all concentrations. The In and Sb atoms have almost the same mass (approximate atomic weights 115 and 122 respectively). Consequently replacement of indium by aluminium of mass 27 leads to a highly localized mode above the highest optic branch frequency of InSb. For this band, the isotopic substitution approximation predicts a frequency which is quite close to that which is observed, showing that there is only a small force constant change in this case. The first overtone has been observed to have a strength one sixtieth of that of the fundamental.

7.2.18.2 Beryllium in the cadmium telluride lattice

This is another system which has a local mode which has been studied[59] in its higher harmonics as well as at its fundamental frequency. Figure 7.29

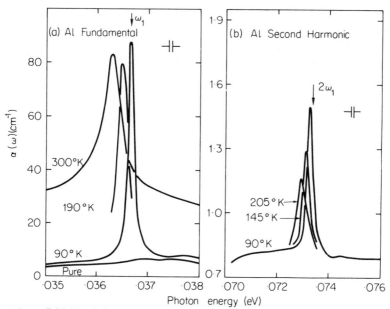

Figure 7.28 The infrared absorption due to Al in InSb showing the fundamental v_1 and second harmonic $2v_1$ of the localized Al mode at different temperatures (Goodwin and Smith)

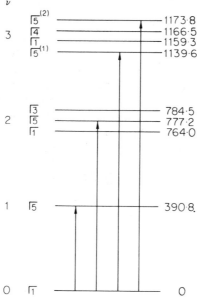

Figure 7.29 The energy level diagram for localized vibrational modes of Be in CdTe. The observed infrared transitions are indicated. The positions of the other levels have been calculated (Hayes)

shows the observed and calculated frequencies for this system, and this energy level diagram is qualitatively very similar to that for CaF_2/H^- (see reference 43) because of the similar site symmetries in the two cases.

7.2.18.3 Aluminium and phosphorus in the gallium antimonide lattice

Aluminium impurities in GaSb replace Ga atoms, and P impurities replace Sb atoms. Since the valence structures are similar in all cases, these impurities do not create difficult free carrier problems, and localized modes have been observed[59] at 317 cm^{-1} for Al and 324 cm^{-1} for P. At 77 K these localized mode absorptions have half widths of about 2 cm^{-1}.

7.3 Polyatomic defects

7.3.1 Introduction

Most of this section is devoted to the discussion of polyatomic impurity ions isolated in alkali halide lattices. As well as the translational degrees of freedom possessed by the point defects discussed earlier, these polyatomic impurities also possess vibrational and rotational degrees of freedom. The basic question to be answered is what happens to these degrees of freedom when the impurity becomes isolated within a crystal. Since the internal vibrational frequencies of all the impurity ions discussed are above the highest pure crystal vibrational frequencies of the supporting lattices we can extend the earlier discussion of localized modes to see that the lattice participation in these vibrations is bound to be very small. These internal modes will therefore be treated as properties of the impurity which are perturbed by its crystalline environment. The rotation of the impurity would generally be expected to be almost completely curtailed, but in certain very interesting cases we shall see that the barriers to rotation are in fact surprisingly low. Thus the rotational degrees of freedom of the 'free' polyatomic impurity ion will need to be considered in each case individually to see what contribution they can be expected to make to the dynamical behaviour of the impurity distorted lattice. The translational degrees of freedom of the polyatomic impurity, like those of the point defects considered earlier, are incorporated into the lattice vibrational spectrum of the host crystal with similar sorts of effect to be looked for, but with the added complications that the symmetry is further reduced.

7.3.2 Internal modes of vibration

Maslakowez,[60] in 1928, published the first IR spectrum of a molecular impurity ion isolated in an alkali halide lattice, but very little further work appeared until the late 1950s.[61,62,63]

Table 7.5 lists the various polyatomic ions which have been studied as substitutional impurities in alkali halides and Figure 7.30 shows the approxi-

Table 7.5 Molecular ions whose internal modes
of vibration have been studied in alkali halides
using IR or Raman spectroscopy

Diatomic

$D_{\infty h}$	O_2^-, S_2^-, Se_2^-
$C_{\infty v}$	$CN^-, OH^-, SH^-, SeH^-, TeH^-$

Triatomic

$D_{\infty h}$	$N_3^-, FHF^-, BO_2^-, S_3^-$
$C_{\infty v}$	NCO^-, NCS^-
C_{2v}	NO_2^-, NH_2^-, ND_2^-

Tetratomic

D_{3h}	NO_3^-, CO_3^{2-}
C_{3v}	ClO_3^-
C_{2v}	HCO_2^-

Pentatomic

T_d	$NH_4^+, ND_4^+, BH_4^-, BD_4^+, SO_4^{2-}$
C_{3v}	$NH_3D^+, ND_3H^+, BD_3H^-, BH_3D^-$
C_{2v}	$NH_2D_2^+, BH_2D_2^-$

mate size of these ions and the alkali halide ions they have been found to replace. Most of the ions will be seen to be singly charged, and these are generally found to replace suitably sized alkali halide ions in concentrations of up to one mole per cent. The doubly charged ions are only found to form good solid solutions if there is a suitable charge compensation mechanism provided, and this in general would be expected to lower the symmetry of the system. Only one positive ion, the ammonium ion, will be seen listed in Table 7.5 and this ion is unusual also in that it exhibits a tendency to hydrogen bond to its nearest neighbour halide ions. Data are also available for the PH_4^+ ion isolated in alkali halides (Sherman and Smulovitch unpublished).

The crystals most commonly used as hosts for these ions are the chlorides, bromides and iodides, of sodium, potassium, rubidium and caesium, although several of these ions have also been isolated in the halides of thalium and ammonium, and the hydroxide ion can also be made to replace the fluoride ion. With the exception of TlI which is hexagonal, these crystals are found to form lattices with either sodium chloride, or caesium chloride, structures, as indicated in Figure 7.31.

All of the impurity ions have been studied in at least four different lattices, and many of them have been studied in twelve. Frequency shifts which result from isolating an ion in different lattices have been accurately measured, and

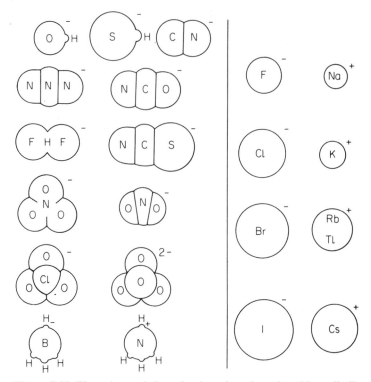

Figure 7.30 The polyatomic ions that have been introduced into alkali halides as substitutional impurities. The relative sizes shown are on approximately the same scale as those of the host lattice ions that they replace

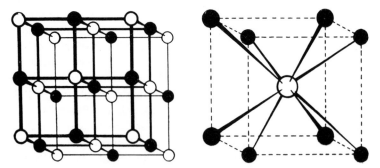

Figure 7.31 The 27 ions which form the unit cell for the NaCl structure and the 9 ions which form the unit cell for the CsCl structure. The chloride, bromide and iodides of sodium, potassium and rubidium all have the NaCl structure at normal temperatures and pressures, whereas the chloride, bromide and iodide of caesium and the chloride and bromide of thallium have a CsCl structure. By applying pressures of up to about 5 kbars for the rubidium halides, 20 kbar for the potassium halides, these crystals transform to the CsCl structure

although the over-all spread of values is usually not more than 3 per cent. of the mean frequency,* the sharpness of the absorption lines allows measurements of considerable accuracy to be made. The relatively small shifts which result from changing one impurity ion from lattice to lattice have caused most of the interpretations of these shifts, to be given in terms of a 'free ion' frequency which is perturbed to a different extent by the different host lattices. Before proceeding to discuss the various ions individually, the lattice perturbation of the 'free ion' frequencies, will be considered.

7.3.3 Frequency perturbation expressions

Not only ought the frequency shifts to be small if they are to be regarded as due to perturbations, but also it is necessary to be able to regard the normal modes of vibration of the perturbed molecular ion as being closely similar to those of the 'free ion'. This has been shown to be the case for small polyatomic impurity ions isolated in alkali halides by a study of the frequency shifts which result from isotopic substitution.

If the environmentally produced perturbation of a solute molecule is expressed in terms of the interaction energy, U, arising from the presence of the environment, then the frequency shift of its ith normal mode can be calculated as follows. The total potential function governing the vibration, V, is obtained from the 'free ion' potential function, V_F, and the interaction energy, U, by simple addition.

$$V = V_F + U$$

$$V_F = a_{11}S_1^2 + a_{22}S_2^2 + \ldots + a_{111}S_1^3 + a_{222}S_2^3 + \ldots \qquad (7.1)$$
$$+ a_{122}S_1S_2^2 + \ldots$$

$$= \sum a_{ii}S_i^2 + \sum_{ijk} a_{ijk}S_iS_jS_k + \text{higher order terms} \qquad (7.2)$$

Where the S_i are normal co-ordinates and the a are potential constants.

The molecule will be distorted by the environment and the new equilibrium configuration will be given in terms of distortion along the normal co-ordinates by equating all first derivatives of the total potential to zero.

$$\frac{\partial V}{\partial S_1} = \frac{\partial V}{\partial S_2} \cdots \frac{\partial V}{\partial S_i} = 0 \qquad (7.3)$$

Since the distortions will only be small, otherwise this approach is not justified, it is only necessary to consider the largest, harmonic terms in obtaining solutions to the equation (7.3). These equations therefore reduce to a series of very simple expressions for the new equilibrium configuration which is given in terms of $(S_i)_e$, the new equilibrium values of the normal co-ordinates with respect to the free ions configuration.

---

* BH_4^- is an exception to this, having about 5·5 per cent. NaCl to RbI shift.

$$2a_{11}(S_1)_e + \frac{\partial U}{\partial S_1} = 2a_{22}(S_2)_e + \frac{\partial U}{\partial S_2} = 2a_{ii}(S_i)_e + \frac{\partial U}{\partial S_i} = 0 \qquad (7.4)$$

that is

$$(S_i)_e = -\frac{1}{2a_{ii}}\frac{\partial U}{\partial S_i} \qquad (7.5)$$

If the force constant for the ith mode is now evaluated at this new equilibrium configuration $(k_i)_e$ then the environmentally produced frequency shift will follow directly.

$$(k_i)_e = \frac{\partial^2 V}{\partial S_i^2} = 2a_{ii} + 6a_{iii}(S_i)_e + \sum_r 2a_{iir}(S_r)_e + \frac{\partial^2 U}{\partial S_i^2}$$

$$= 2a_{ii} - \frac{3a_{iii}}{a_{ii}}\frac{\partial U}{\partial S_i} - \sum_r \frac{a_{iir}}{a_{rr}}\frac{\partial U}{\partial S_r} + \frac{\partial^2 U}{\partial S_i^2} \qquad (7.6)$$

$$\frac{\Delta v_i}{v_i} \approx \frac{\Delta k_i}{2k_i} = \frac{1}{2k_i}\left\{\frac{\partial^2 U}{\partial S_i^2} - \frac{3a_{iii}}{a_{ii}}\frac{\partial U}{\partial S_i} - \sum_r \frac{a_{iir}}{a_{rr}}\frac{\partial U}{\partial S_r}\right\} \qquad (7.7)$$

Equation (7.7) gives a general expression for the environmentally produced frequency shift which for many of the cases of interest is found to simplify down to a fairly tractable expression, because the symmetry of the ions and the lattices render many of the individual terms in equation (7.7) equal to zero. For example a diatomic ion adequately described by a Morse potential of the form $V_F = D(1 - e^{-\beta\xi})^2$ reduces equation (7.7) to the form

$$\frac{\Delta v}{v} = \frac{1}{2k}\left\{\frac{\partial^2 U}{\partial \xi^2} + 3\beta\frac{\partial U}{\partial \xi}\right\} \qquad (7.8)$$

Where $\xi = r - r_e$, the change in internuclear spacing.

Expressions like those in equations (7.7) and (7.8) have been derived by several authors for the environmental perturbation of the vibrational frequencies of a solute; for example, Benson and Drickamer,[64] Pullin,[65] Bryant and Turrell,[66] Field and Sherman,[67] have used arguments similar to those above; whereas Buckingham[68] has obtained essentially the same result using a perturbation theory argument (see §4.2.1).

The second-derivative term in equation (7.7) or (7.8) will be seen to represent an additional, external, or environmental force constant, whereas the first-derivative term is the change in internal force constant caused by the changed internuclear spacings within a molecule which is represented by an anharmonic potential function.

In every known case where all the terms in equation (7.8), or its equivalent, have been evaluated, it has been found that force multiplied by anharmonicity (i.e. first derivative) terms have dominated the frequency shift calculation. (See for example references 65, 66, 67 and 69.) This was recognized as early as 1938 by Bauer and Magat[67] who make this point about their equation which can be obtained from equation (7.7) by using the appropriate external potential function U. It is therefore most unfortunate that these terms are often ignored by users of the much quoted Kirkwood-Bauer-Magat equation. The general dominance of the first derivative term can be seen to follow from the general applicability of potential functions of the Morse type. Only a very unusual form of external potential would be capable of supplying a meaningful contribution to the frequency shift by way of the second derivative term without causing an over-riding effect through its first derivative distorting the anharmonic solute molecule.

It is interesting to note this dominance of the first derivative term in the wider context of the group-frequency against bond-length plots which have been made for several groups.[65,70,71] These plots suggest that even when the internuclear spacings within a group of interest are changed by chemical bonding effects within different parent molecules, it is still a fair approximation to regard the nuclei as moving within a single, well defined, Morse-like potential system, but considered to vibrate about different equilibrium spacings by low curvature 'extra-group' effects.

Thus Figure 7.32, which is a pictorial presentation of the argument which leads to equation (7.8), can be considered not only as a description of an environmentally perturbed diatomic molecular ion; but also as a fair description of a specific molecular group which is subjected to a perturbation U on being chemically bound into a larger molecule. Since the frequency perturbation described by equation (7.8) is due to the change in curvature from $k = (\partial^2 V/\partial r^2)_{r=r_f}$ to $k + \Delta k = [\partial^2 (V_F + U)/\partial r^2]_{r=r_e}$ the relative contributions to the frequency shift from the terms containing $(\partial U/\partial r)$, dictating the value of r_e, and $(\partial^2 U/\partial r^2)_{r=r_e}$ the curvature of U, can be visually assessed from Figure 7.32b which shows the variation of the components of $(\partial^2 V/\partial r^2)$ with r, drawn approximately to scale for CN^- in NaCl.

Equation (7.7) or (7.8) can be used in a variety of ways depending on which of the parameters appearing in these equations are sufficiently well known to be used with confidence in conjunction with the available experimental data. Only in one study on polyatomic defects in alkali halides so far, by Field and Sherman,[67] was there sufficient information adequately to test the validity of this approach, and even in this case certain reservations, to be discussed later, had to be made. Other authors to apply this approach to polyatomic defects in solids (e.g. Bryant and Turrell,[66] Cundill[72]) assumed the applicability of the method and used it to evaluate unknown anharmonicity terms which appeared in their equations.

Since frequency shifts have been given considerable prominence in almost all papers dealing with the spectra of polyatomic impurities in solids, the

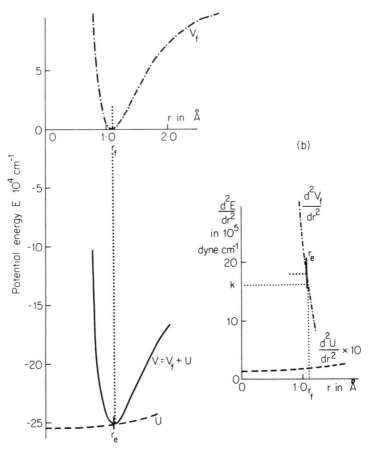

Figure 7.32 The potential energy of CN^- isolated in NaCl (cf. equation 7.8). Note that d^2U/dr^2 has been plotted multiplied by 10, and that the change of d^2V_f/dr^2 when evaluated at $r = r_e$ instead of $r = r_f$ is about ten times larger than the value of d^2U/dr^2

investigation by Field and Sherman[67] into the CN^- ion substitutionally isolated in various alkali halides will be considered first in order to compare with the environmentally induced frequency shifts of all other molecular impurities.

7.3.4 CN^- frequency shifts in alkali halides

The arguments used to derive equations (7.7) and (7.8) made no reference to the source or form of the perturbing potential, U. Therefore these equations can be expected to be of wide general applicability if the terms appearing in them can be evaluated to a usable accuracy. In seeking to test these equations however the use of a diatomic substitutional impurity ion isolated within an alkali halide appears to have certain distinct advantages over other systems. These are:

(i) The observation of two absorption bands, the fundamental and the first overtone, are sufficient to completely define the Morse potential and establish a good value for the only anharmonic potential constant required in equation (7.8).

(ii) The absorption bands at low temperatures are quite sharp and their positions can be measured to considerable accuracy.

(iii) Since the supporting lattice is relatively rigid and well defined it should be possible to give a good mathematical description of it.

(iv) The perturbing potential, U, which must be explicitly calculated for each case should be adequately described by the much investigated Born–Meyer[25] theory of ionic crystals.

By-passing, for the present, the difficulties involved in the determination of the correct orientation to be used for the CN^- ion and the precise form used for the potential U, consider now the result of comparing observed and calculated frequency shifts for the CN^- in various alkali halides at various 'quasi-hydrostatic' pressures. Figure 7.33 shows the observed frequencies of the fundamental CN^- stretching vibration when the ion is isolated in the chlorides, bromides and iodides of sodium, potassium, rubidium and caesium at pressures up to 50 kbar at a temperature of 100 K, plotted versus the corresponding frequency shift calculated using equation (7.8). Since the potassium and rubidium halides each change from a sodium chloride structured lattice to a caesium chloride structured lattice within this pressure range, this figure includes data on the CN^- frequency for the ion isolated at various pressures in eighteen different environments. Over 250 experimentally recorded frequencies are effectively represented on this graph extending over a range of about 70 cm^{-1}, and the line $v = 2038 + \Delta v$ drawn on this graph passes within ± 4 cm^{-1} of 90 per cent. of these points. The extrapolation to zero Δv, cuts the frequency axis at the 'free ion' frequency of 2038 cm^{-1}, after an extrapolation which is only about one quarter of the length of the experimentally investigated range. A value for the reliability of this number could be obtained from the standard deviation to be associated with the intercept of the least mean squares line through these points. It would be about ± 0.3 cm^{-1}, and a similar number would result from assessing the reliability of the mean of 250 values of $v_{obs} - \Delta v_{cal}$. However although this fairly high accuracy can be defined mathematically for this particular graph there are good reasons for believing that equally valid assumptions in the performance of the calculations of the Δv could produce another equally good looking graph but giving a 'free ion' frequency different to the one given above by up to about ± 3 cm^{-1}. This point is made now in order to avoid any confusion which might arise when this result is employed later to assist in obtaining approximate 'free ion' frequencies for other ions.

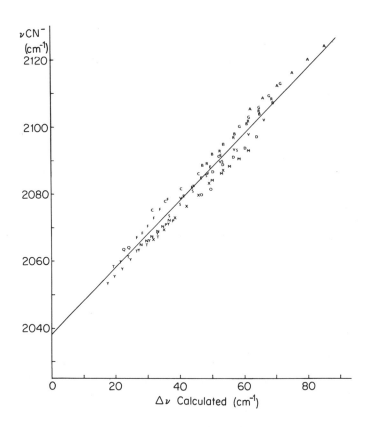

Figure 7.33 The observed fundamental CN⁻ stretching
vibration frequency for this ion isolated in various alkali
halides at 90 K and pressures up to 50 kbar plotted
versus the equivalent frequency shift calculated from
equation (7.8) (Field and Sherman). Since the potassium
and rubidium halides change structure in the pressure
range, each of the halides is denoted in the graph by two
types of symbols. The sodium chloride, bromide and
iodide data is plotted A, B and C respectively with the
equivalent potassium halide data plotted D and G, E and
M, F and N; the rubidium halide data is plotted O and R,
P and S, Q and T and the caesium halide data plotted
V, X and Y. The best line of unit slope through all these
points is shown to be $v_{obs} = 2038 + \Delta v_{calc}$. Thus 2038
cm⁻¹ is the estimated 'free space' vibrational frequency
of the CN⁻ ion

Although the numerical parameters used in the calculation of the Δv for Figure 7.33 were those selected before the calculations were started,[67] the degree of lattice distortion and the direction of orientation of the CN^- ion axis, was changed in the light of preliminary calculations.

The repulsive term in the Born–Mayer potential was found to dominate in the calculations of the frequency shifts with about 85 per cent. of each calculated shift with respect to the free ion frequency coming from the repulsive force shortening the anharmonic CN bond. Only when lattice distortions, of about the magnitude predicted by Dick and Das,[26] had been included in the calculations did it prove possible to obtain a reasonable overlap of data from the different environments as shown in Figure 7.33. This extreme sensitivity to very small changes in the position of the nearest neighbour ions is a clear warning: (a) against the inclusion of small effects to improve an apparent calculated-to-observed frequency fit, since quite acceptable changes in proposed nearest neighbour position even within a relatively well ordered solid will completely swamp such effects, (b) against very detailed calculations of frequency shifts for less well ordered systems, (e.g. liquids) unless there is quite precise information available about the geometry of the environment closely surrounding the relevant group.

7.3.5 Comparison of O_2^- and CN^- frequency shifts

When considered together with the CN^- calculations described above, the O_2 frequency shift study also has important implications for much of the polyatomic ion work described later.

The Raman spectrum of O_2^- isolated in several different alkali halides has been studied by Holzer et al.,[73] and the frequencies compared with the CN^- frequencies of Field and Sherman[67] in order to get a 'free ion' frequency. This is a very interesting study, because although within a common cation group of alkali halides the two frequencies, $v(O_2^-)$ and $v(CN^-)$ show similar shifts, within a common anion group $v(O_2^-)$ undergoes very little change, whereas $v(CN^-)$ shows quite large frequency changes. Therefore when the two frequencies are plotted against one another, as in Figure 7.34, a two-dimensional array of points is obtained which shows fairly linear common-cation lines. These lines tend to converge and cut the $v(CN^-) = 2038 \, cm^{-1}$ line within quite a small range of $v(O_2^-)$ frequencies, which is presumed to include the 'free ion' frequency of the O_2^- ion.

Figure 7.34 although similar to that shown by Holzer et al.[73] has been redrawn using 300 K values for $v(CN^-)$. This makes a slight difference because the temperature induced frequency shifts for CN^- in alkali halides are quite large in the chloride lattices, (up to $5 \, cm^{-1}$ for $300 \rightarrow 100$ K), smaller in the bromides and very small in the iodides. Thus using 300 K values for both ions in Figure 7.34 has slightly changed the slopes of common cation lines from those reported in ref. 73. Care has been taken to draw the

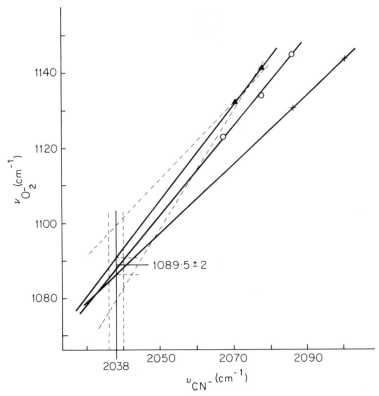

Figure 7.34 Shows the comparison of the O_2^- vibrational frequencies measured by Raman spectroscopy with CN^- frequencies in the same alkali halide host lattice. The points plotted correspond to the frequencies for the two ions in one particular alkali halide at 300 K. The lines are common cation lines with sodium halides plotted +, potassium halides O, and rubidium halides ▲. Since these common cation lines converge and intercept the $\nu(CN) = 2038$ cm^{-1} line in a small range of values given by 1089.5 ± 2 cm^{-1} Holzer *et al.* have deduced this is the 'freespace' O_2^- vibrational frequency. The above figure has been redrawn using 300 K data for both ions, and the dashed lines indicate the range of rubidium lines which would be covered by frequency changes in the range ± 1 cm^{-1} in the rubidium chloride and bromide points

lines exactly through the two sodium and two rubidium points but the dashed lines show the range of lines which would be covered by frequency changes in the range ± 1 cm^{-1} on the values of the rubidium chloride and bromide points. Clearly much less uncertainty is associated with extrapolating the potassium line once the assumption is made that it should in fact be a straight line. It would obviously be very helpful to have the points for sodium iodide and rubidium iodide and also to have a fourth line from the caesium halides. However, Figure 7.34 shows that if the three available lines are to converge upon a point on the $v(CN^-) = 2038 \text{ cm}^{-1}$ line, then this point must have $v(O_2^-)$ in the range $1089 \cdot 5 \pm 2 \text{ cm}^{-1}$. If the lines are to converge to a point having a $v(CN^-)$ co-ordinate in the range $2038 \pm 2 \text{ cm}^{-1}$ then somewhat greater uncertainty must be associated with the $v(O_2^-)$, (say $1089 \pm 5 \text{ cm}^{-1}$). Without any information about the CN^- 'free ion' frequency Figure 7.25 would suggest a range of possible values for both 'free ion' frequencies. This would not be a very well defined range particularly in view of the near parallelism of the potassium and rubidium lines. These ranges would seem to be given by vCN^- (free ion) $= 2033 \pm 10 \text{ cm}^{-1}$ and vO_2^- (free ion) $= 1084 \pm 10 \text{ cm}^{-1}$.

This careful examination of the $v(CN^-)$ versus $v(O_2^-)$ plot of Holzer et al.[73] is necessary if their technique is to be of more general use. There is in fact sufficient data available for most of the ions which have been isolated in alkali halides for this technique to be applied.

7.3.6 Intercomparison of O_2^-, CN^- and N_3^- frequencies

Figure 7.35 shows the comparison of the $v_3(N_3^-)$ frequencies with the $v(CN^-)$ frequencies for the two ions in each of nine different alkali halide lattices at 300 K. Although the common cation lines are fairly straight they are also nearly coincident and therefore no assistance in locating the 'free ion' frequencies is to be obtained from the convergence of these lines. However even in Figure 7.35 a reasonable value for $v_3(N_3^-)$ 'free ion' frequency can be obtained *if* the free ion frequency for $v(CN^-)$ is reliable. Figure 7.36 shows the $v_3(N_3^-)$ data plotted versus the $v(O_2^-)$ frequencies and here again a reasonable degree of convergence helps to remove the extreme dependence on the $v(CN^-)$ 'free ion' frequency exhibited by Figure 7.35. The reliability of this treatment clearly depends on the justifiability of the extrapolation, which cannot ever be expected to be exactly linear, but which is probably fairly straight for two ions which show comparable importance to the equivalent terms in the two expressions for the frequency shifts, (i.e. similar terms in equation (7.7) show a similar relative importance for the two ions.

A similar treatment, but using the pressure variable in place of the common cation line is less suspect as regards the extrapolation, particularly since sufficient points can be plotted along each line to indicate its curvature. Unfortunately there is no pressure dependence data for the O_2^- frequency and therefore Figure 7.37 has been drawn using the CN^- data of Field and Sherman[67] and the N_3^- data of Cundill.[64] Although the eighteen individual lines have not been drawn on Figure 7.37 it will be seen that such convergence and crossing of lines as would have taken place in this example suggest an appreciably higher 'free ion' frequency for the CN^- than that calculated in reference 67. Sixteen out of the eighteen lines, would cut the $v(CN^-) = 2038 \text{ cm}^{-1}$ line in a range of values given by $1970 \pm 14 \text{ cm}^{-1}$, and if these

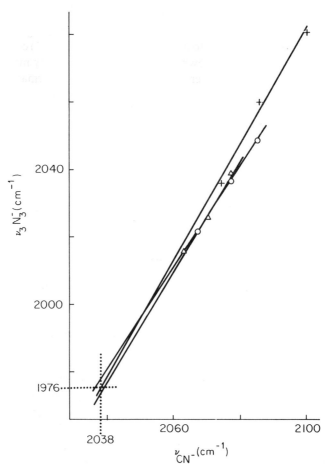

Figure 7.35 A comparison of the $v_3(N_3^-)$ frequencies with the $v(CN^-)$ frequencies for the two ions in each of nine different alkali halide lattice at 300 K. The points for the chloride, bromide and iodide of sodium are plotted $+$, the equivalent potassium halides are plotted O, and the rubidium halides are plotted $\triangle$. Although the common cation lines are fairly straight, they are nearly coincident and parallel. Therefore the convergence shown in Figure 7.34 is missing and the $v_3(N_3^-)$ free ion frequency can only be assessed in terms of the free CN^- frequency of 2038 cm^{-1}. The estimated free space $v_3(N_3^-)$ frequency is 1976 $\pm$ 4 cm^{-1}

points of intersection are regarded as providing independent estimations of the N_3^- 'free ion' v_3 frequency, then averaging these gives a best value of about 1970 $\pm$ 3 cm^{-1}. This value for the $N_3^- v_3$ 'free ion' frequency is a little lower than the value suggested by the single 'best line' shown in Figure 7.37 and the values obtained from Figures 7.35 and 7.36 but reasonably in

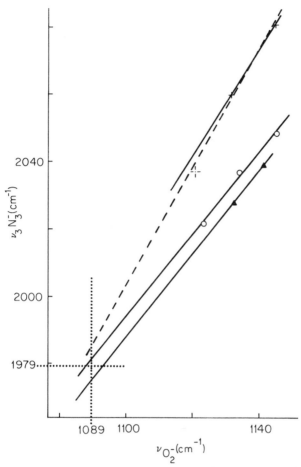

Figure 7.36 A comparison of the $v_3(N_3^-)$ frequencies with the $v(O_2^-)$ frequencies for the ions isolated in the same alkali halides again show a tendency for common cation lines to converge. The free space $v_3(N_3^-)$ frequency is estimated from these plots to be $1979 \pm 5\ cm^{-1}$

keeping with a recent calculation by Cundill[72] which gives $1971\ cm^{-1}$. It may be worth observing however that the region of closest bunching of the lines in Figure 7.37 is for $v_3(N_3^-)$ values of about $2005\ cm^{-1}$, which although much higher than the above values for the $v_3(N_3^-)$ 'free ion' frequency, is close to the earlier estimate of this parameter by Bryant and Turrell.[66] Thus, although the extrapolation of comparative data for two very similar ions (e.g. O_2^- and CN^-) is probably a justifiable method of obtaining information about the 'free ion' frequencies, care must be exercised in this procedure. For example the unitary display presentation favoured by Price *et al.*[63,74]

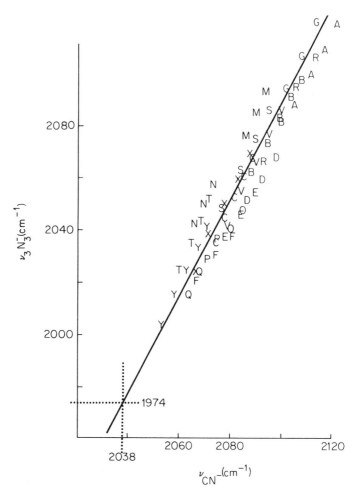

Figure 7.37 A comparison of the $v_3(N_3^-)$ frequences with $v(CN^-)$ frequencies for these ions in different alkali halides. Data is included for twelve different alkali halides at pressures between 0 and 50 kbar. Since the potassium and rubidium halides have a phase change within this pressure range there are eighteen different lines. The single line shown is the best line through the experimental points. Sodium chloride, bromide and iodide data are plotted A, B and C respectively with the equivalent potassium data are plotted D and G, E and M, F and N, the rubidium halide data are plotted O and R, P and S, Q and T. The caesium data is plotted V, X and Y

always gave fairly straight common cation lines which for the sodium chloride structured alkali halides showed a convergence which it was tempting to regard as indicating a free ion frequency, (see Figure 7.38) for the CN^- and $v_3(N_3^-)$ data as presented in reference 74), but the values so obtained

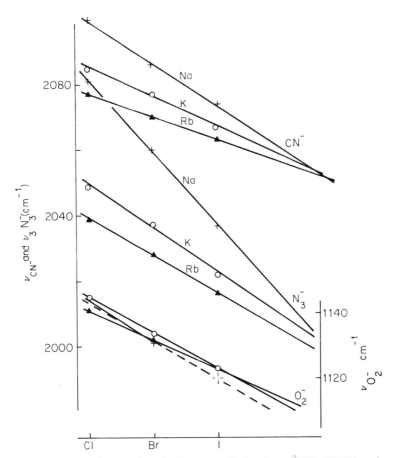

Figure 7.38 Unitary anion displacement display for $v_3(N_3^-)$, $v(CN^-)$ and $v(O_2^-)$ showing the straight common cation lines which for the azide and cyanide ions converge to values which are quite different from the 'free ion' values obtained from the four previous figures. Similar graphs are obtained by plotting the frequencies against the log of the replaced cation radius. The convergence obtained then suggests that the frequency perturbation becomes zero for a 'hole' diameter of about 5 Å. However, the frequency obtained is *not* the 'free space' value

are at variance with the more reliable estimates discussed above. The basic obstacle which undermines any extrapolation method which does not actually calculate all important terms in an equation such as equation (7.8), is that the meaningful contributions from short range and long range potentials are not similarly dealt with over the experimentally determined range and cannot therefore be reduced to zero by the same extrapolation. It is reluctantly concluded therefore that even when a very large amount of comparable data is available for two ions (e.g. CN^- and N_3^- as in Figure 7.37) it is unlikely that this data can be convincingly extrapolated to give 'free ion' frequencies unless the two ions are of very similar size and the vibrations compared are of a similar type. However plots of the type shown in Figures 7.35–7.37 are quite similar to those used in some liquid phase studies of the type initiated by Bellamy, Hallam and Williams,[75] and should give quite reliable predictions about 'free ion' frequencies if the free ion frequency of at least one ion can be regarded as known.

7.3.7 Symmetry considerations

Figure 7.30 and Table 7.5 show the various polyatomic ions which have been studied as isolated impurities within alkali-halides. They are arranged in groups of ions which possess the same symmetry as isolated, or 'free', ions. However, the spectral properties of these ions once they are held within an alkali-halide environment, depend on the resultant symmetry of the ion-lattice system. The high degree of symmetry of both the NaCl type and the CsCl type of lattice, see Figure 7.31, makes it possible to retain all of those symmetry operations which are important to the spectra of the ions, *if they are appropriately orientated within the supporting lattice.* For most ions* so far reported, the observed internal vibrational spectra have been those expected for a rotationally restricted 'free ion', and this fact has often been used to infer information about the orientation of the ion within the lattice. As an example of this, consider Table 7.6 which refers to the azide ion, N_3^-, isolated in alkali-halide lattices. This Table is essentially that shown by Bryant and

Table 7.6 The dependence of the symmetry properties of the vibrations of an azide ion, upon the orientation of its ∞ fold axis within the surrounding alkali halide lattice

Orientation of ∞ fold axis	'Free'	Along a crystal C_4 axis	Along a crystal C_3 axis	Along a crystal C_2 axis	In crystal but off all C axes	Activity
Subgroup / Vibration	$D_{\infty h}$	D_{4h}	D_{3d}	D_{2h} ($\equiv V_h$)	C_i	
ν_1	Σ_g^+	A_{1g}	A_{1g}	A_g	A_g	R
ν_2	π_u	E_u	E_u	B_{2u} B_{3u}	A_u A_u	IR
ν_3	Σ_u^+	A_{2u}	A_{2u}	B_{1u}	A_u	IR
R_x, R_y	π_g	E_g	E_g	B_{2g} B_{3g}	A_g A_g	R
T_z	Σ_u^+	A_{2u}	A_{2u}	B_{1u}	A_u	IR
T_x, T_y	π_u	E_u	E_u	B_{2u} B_{3u}	A_u A_u	IR

* Cases of compound impurities (e.g. $Ca^{2+}SO_4^{-2}$) where the presence of the second impurity has lowered the site symmetry, are notable exceptions to this.

Turrell[66] and the absence of any detectable splitting in v_2 coupled with the calculated unfavourable energy of the C_4 orientation within the NaCl type of lattice led them to propose that the ion lies along the C_3, cube diagonal when isolated in NaCl type lattices. The same type of argument was used by Cundill[72] in selecting the C_4 orientation for this ion in the CsCl type lattices.

Although the above discussion is quite straightforward for use with the internal modes of vibration of a fairly rigidly held impurity ion, there are two other types of phenomena of interest here which need somewhat more careful assessment. These are:

(i) External modes of vibration and their infrared and Raman activity both directly and in combination with other modes.
(ii) The spectra associated with ions which are not rigidly held on, or close to, a high symmetry direction within the crystal, but which for all but the lowest temperatures, are performing large amplitude torsional oscillations or even hindered rotation.

However, phenomena of the above two types are to be discussed in detail later and the relevant points can be made more clearly as part of that discussion.

7.3.8 The cyanate ion, NCO⁻

This ion merits special consideration because it was the first isolated ion to have its internal vibrational spectrum carefully analysed,[62] and it has since then, been the most extensively studied ion. Data have been reported for this ion isolated in twelve different alkali halide lattices,[63] and in some of these lattices 30 different internal vibrational transitions have been identified[76] for the main isotopic species and many of them for the ^{13}C, ^{15}N and ^{18}O species also, with a few reported transitions for ^{17}O, ^{14}C and double-rare-isotope species such as $^{15}N^{13}C^{16}O^-$ or $^{14}N^{13}C^{18}O^-$.

The external modes of vibration of this ion have also been studied via the lattice side band structure on the internal modes.[76] All the above studies have included the temperature dependence of the spectra, and the pressure dependence of both the internal modes and the lattice side-bands has been investigated also.[77,78] The influence of high concentrations of cyanate ions has been studied[76] and the interaction between NCO⁻ and other secondary impurities has been investigated.[71]

From such a large quantity of data only a relatively small selection can be presented and discussed here.

7.3.8.1 Internal modes of vibration of NCO⁻ isolated in alkali halides

In one of the earlier papers on impurity ions in alkali halides, Maki and Decius[62] gave a detailed analysis of the internal vibrational bands of NCO⁻ isolated in KBr and KI. Although later work has revealed new combination

bands,[76] and suggested a change in the interpretation[80] of the Fermi resonance which occurs because v_1 has a value that is very close to twice v_2 the
original paper by Maki and Decius still represents the best overall treatment
of the internal modes of vibration of an isolated impurity ion. A recent
paper by Schettino and Hisatsune[81] has slightly extended the original
analysis by Maki and Decius[62] and reference 81 contains some of the best
data currently available for NCO^-.

The vibrational energy of the isolated NCO^- ion was considered to be
adequately described by the expression used for linear triatomic gaseous
molecules.

$$E - E_0 = \sum_{k=1}^{3} \omega_k^0 v_k + \sum_{k \leqslant k'=1}^{3} X_{kk'} v_k v_{k'} + g_{22} l_2^2$$

Where v_1, v_2, v_3 are the vibrational quantum numbers associated with $v_1, v_2,$
v_3; l_2 is a further quantum number required by the doubly degenerate
bending mode v_2 which indicates the amount of angular momentum to be
associated with the v_2 vibration, and which can take positive values given by
$(v_2 - 2n)$. The ω_k^0, $X_{kk'}$ and g_{22} represent a total of ten parameters which in
principle can be obtained by solving ten simultaneous equations for appro-
priate observed absorption frequencies; but in fact X_{33} had to be chosen
from other considerations because absorption bands such as $2v_3$ which
would have allowed it to be calculated were all too weak to be observed.
Using the above formula it is possible to obtain a fit to the large number of
observed frequencies which is at least as good as that reported for comparable gas molecules, providing that account is taken of the strong Fermi
resonances which occur between ω_1 and $2\omega_2$. The Fermi multiplets are
calculated from the secular determinants

$$\begin{vmatrix} W_1^0 - W & W_{12} & \cdots & W_{1n} \\ W_{21} & W_2^0 - W & \cdots & W_{2n} \\ & & & \\ W_{n1} & W_{n2} & \cdots & W_n^0 - W \end{vmatrix} = 0$$

in which the W_n^0 are the unperturbed energies calculated from the earlier
equation, W is the perturbed energy, and the $W_{nn'}$ are given by:

$$W_{nn'} = g_{122} \int \psi_n Q_1 Q_2^2 \psi_{n'} \, d\tau$$

where g_{122} is one of the cubic terms in the potential energy of the ion when it
is expanded in terms of the normal co-ordinates Q_n, and the ψ's are harmonic
oscillator wave functions. For doublets of the type $\omega_1, 2\omega_2$,

$$W_{(v_1 v_2 v_3);(v_1 - 1, v_2 + 2, v_3)} = \frac{-b v_1^{1/2}}{2^{3/2}} [(v_2 + 2)^2 - l^2]^{1/2}$$

where

$$b = \frac{h^{3/2} g_{122}}{8\pi^{3/2} \omega_1^{1/2} \omega_2}$$

For triplets of the type $2\omega_1$, $\omega_1 + 2\omega_2$, $4\omega_2$, there are two off-diagonal elements identical with that shown above, with the third element $W_{(v_1 v_2 v_3);(v_1 - 2, v_2 + 4, v_3)}$ equal to zero. Thus the Fermi resonances require this eleventh parameter b before these levels can be calculated. Figure 7.39 shows the various energy levels which have been identified for the NCO^- in KBr system,[76] and which can be very closely fitted by the eleven parameter equations described above.

It is instructive to note how good a fit can be obtained in the above way. The model used in constructing the equation completely ignores the environ-

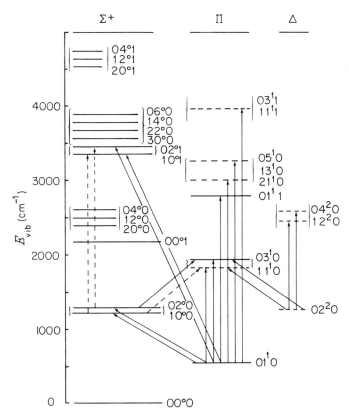

Figure 7.39 The vibrational energy level diagram for NCO^- in KBr (Decius *et al.*) showing the infrared active transitions from thermally excited states. Levels shown with solid lines have been located by the observation of transitions from the ground state

ment and any part played by it in the vibrations. Of course the different host lattices require different sets of parameters to match their individual sets of observed frequencies, but the 'gas-molecule' form of the equations can be taken as a further indication that the most important effect on these internal modes comes from the environmentally-induced distortion of the molecule changing the internal force constants through the anharmonic terms in the potential function. For the frequencies referred to in Figure 7.39, which are between 4 and 30 times the highest natural frequency of the host lattice, the rigid supporting matrix approximation is clearly a very good one indeed.

7.3.8.2 Lattice side bands on vibrational spectra of NCO⁻ isolated in alkali halides

Maki and Decius[62] reported a considerable amount of structure on the sides of the $v_3(NCO^-)$ absorption band for the ion isolated in KBr and KI. They attributed the strongest feature in each spectrum (at about $\pm 94 \, cm^{-1}$ for KBr and about $\pm 80 \, cm^{-1}$ for KI at 300 K) as due to 'sum-and-difference frequencies involving either a translational or rotational motion of the cyanate ion as a rigid unit'. Remember that this was before Cowley et al.[82] published their neutron scattering investigation of the lattice vibrations of the KBr, before Schaefer[3] published his U-centre results, and before the theoretical workers on defect distorted lattice vibrations had any experimental evidence against which to compare their models. However, although the figures showing the lattice sideband spectra for NCO⁻ in KBr and KI were published in 1958, little further interest was shown in this sort of spectrum until 1965 when Decius et al.[76] drew attention to these spectra and compared them with the, by then, available neutron scattering work of Cowley et al.[82,83]

Figure 7.40 shows the lattice sidebands on $v_3(NCO^-)$ when the ion is isolated in KBr. The spectrum is drawn out vertically in this figure so that the phonon dispersion curves[82] with which it is compared can be shown in the usual way. Curves for the $\langle 111 \rangle$ direction are presented since it is the zone boundary values of LA and TO in this direction which define the energy gap between the acoustic and optic modes. The intensity of the strong central absorption can be assessed from that of the 2096 cm^{-1}, $^{15}N^{13}C^{16}O^-$, band which shows about 5 per cent. absorption for this double-rare-isotope species which was present in the sample in the natural relative abundance of about 0·004 per cent. (Since the NCO⁻ total concentration was only about 0·7 per cent. this double-rare-isotope was present in the crystal at a concentration of less than 0·3 ppm.) Ignoring the dotted-under sharp features due to the less abundant isotopic species, the spectrum shown in Figure 7.40 can be seen to be adequately described as due to $v_3(NCO^-)$ plus lattice sideband structure which has the form $v_{(int)} \pm v_{(ext)}$. The difference band structure is progressively more suppressed as it extends away below the v_3

band as the starting levels for the transitions are progressively less thermally populated. Four sections can be clearly seen in the external mode structure: (a) broad bands in the region covered by the host lattice acoustic vibrational energies, (b) a sharp in-gap doublet, (c) broad bands in the region covered by the lattice optic modes, and (d) a fairly sharp doublet just above the top of the lattice optic band.

The broad in-band structure will be discussed later together with similar spectra for other ions but the two doublets in this spectrum are of special interest because they have been carefully examined as a function of isotopic substitution, and this has yielded information about the nature of the vibrations involved in producing these features. Figure 7.41 shows schematically the frequency separations from v_3 at which the in-gap and super-optic doublet features were recorded for the various isotopic species, and Table 7.7

Table 7.7 Isotopic frequency shifts (in cm^{-1}) of the four sharp external mode features found in combination with $v_3(NCO^-)$ isolated in KBr. The observed values are compared with values calculated for pure torsional modes of the NCO^- within an infinitely massive containing well, about two axes perpendicular to the ion axis, one bisecting the N—C bond and the other bisecting the C—O bond. Numbers in brackets represent the above calculated values divided by the scaling factor shown at the foot of the column.

$v_{ext} = (v_{int} + v_{ext}) - v_{int}$ $^{14}N^{12}C^{16}O$ $\Delta v = v_{ext}(14, 12, 16)$ $- v_{ext}$(isotope)	$97.4\ cm^{-1}$	$99.7\ cm^{-1}$	$167.5\ cm^{-1}$	$183.6\ cm^{-1}$	
$\Delta v(14, 12, 16)$–$(15, 12, 16)$	-0.2	0.6	~ 0	3	obs
	$0.2\ (0.03)$	$2.5\ (0.62)$	$0.5\ (0.2)$	$4\ (3.0)$	calc
$\Delta v(14, 12, 16)$–$(14, 13, 16)$	0	0	~ 0	0.4	obs
	$0.2\ (0.03)$	$0.2\ (0.05)$	$0.4\ (0.16)$	$0.4\ (0.3)$	calc
$\Delta v(14, 12, 16)$–$(14, 12, 17)$	0.3	~ 0	~ 1.5	~ 0	obs
	$2.5\ (0.34)$	$0.3\ (0.07)$	$4\ (1.6)$	$0.5\ (0.37)$	calc
$\Delta v(14, 12, 16)$–$(14, 12, 18)$	0.7	~ 0	3	~ 0	obs
	$5\ (0.71)$	$0.6\ (0.15)$	$8\ (3.2)$	$1\ (0.7)$	calc
Estimated reliability of observed frequency shifts	± 0.15	± 0.15	± 0.3	± 0.3	
Scaling (delocalization) factor	~ 7	~ 4	$\sim 2\frac{1}{2}$	$\sim 1\frac{1}{3}$	

shows observed and calculated shifts from the values recorded for the most abundant isotopic species. The isotopically produced frequency shifts shown in Figure 7.41 are apparently only consistent with an interpretation which associates motion of an essentially torsional nature with each of the four absorption peaks. The calculated shifts quoted in Table 7.7 refer to the simplest possible model, that of a rigid NCO^- ion held in a rigid lattice.

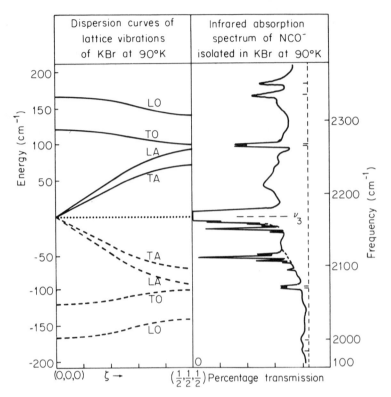

Figure 7.40 The IR absorption spectrum for KBr/NCO⁻ in the v_3 vibration frequency region (Cundill and Sherman). External mode vibrational structure in sum and difference combination with the v_3 vibration is compared with the dispersion curves for pure KBr. Note the relatively sharp 'in-gap' and 'super-optic' structure and the broader 'in-band' features

Such a simple model would not show so many normal vibrations of this sort, but it does allow an estimate to be made of the frequency shift of such a vibration which would result from an isotopic substitution. The 97.4 cm^{-1} band and the 167.5 cm^{-1} band are associated with a torsional motion about an axis which perpendicularly bisects the C—N bond; whereas the 99.7 cm^{-1} band and the 183.6 cm^{-1} band represent torsional oscillations about an axis which bisects the C—O bond. This calculation produces shifts which have the form shown by Figure 7.41 because a given oscillation is nine times more sensitive to a mass change in the nucleus farthest from its torsional axis than it is to an equivalent change in either of the other two masses. Inevitably the calculated shifts are larger than the observed values, but if a scaling factor is associated with each absorption band then agreement is easily obtained between the calculated and observed values. The scaling factors are included at the foot of Table 7.7 and indicate the extent of the

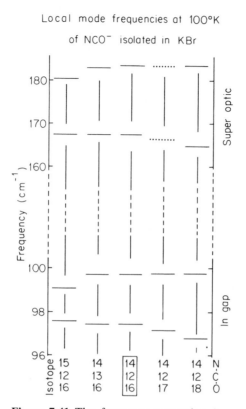

Figure 7.41 The frequency separation between v_3 and the 'in-gap' and 'super-optic' doublet features which occur in the lattice sideband structure for various species of NCO^- in KBr. Note (a) ^{13}C substitution has no appreciable effect, (b) only the upper component of each doublet is affected by the ^{15}N substitution, and (c) the lower components shift on ^{17}O and ^{18}O substitution. Only torsional oscillations can produce these effects (Cundill and Sherman)

environmental participation in the motion. For example the scaling factor 7 associated with the 97.4 cm^{-1} band shows that six-sevenths of the vibrational energy of the oscillation responsible for this band is carried by the environment and only one-seventh resides on the NCO^- ion itself.

The size of the scaling factors show that at least for the gap, modes reasonably large amplitudes of vibration can be associated with the environment. Figure 7.42 shows a $\langle 111 \rangle$ orientated NCO^- ion in an undistorted KBr lattice and calculations of the force constants operating between the NCO^-

NCO⁻ in KBr

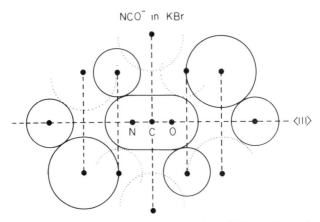

Figure 7.42 An NCO⁻ ion isolated in a KBr lattice and oriented in the ⟨111⟩ direction. The full circles represent ions in the plane of the drawing and the dotted circles the positions that out of plane ions would occupy if they rotated around the ⟨111⟩ axis into the plane. Large circles represent Br⁻ ions and small circles K⁺ ions. Note the ring of three nearest neighbours which hold each end of the NCO⁻ ion (Cundill and Sherman)

and the individual ions of its supporting lattice show that a very high percentage of the torsional restoring force comes from the six nearest neighbours. These six nearest neighbours can be considered as two groups of three between which the NCO⁻ is held. Since both the 183·6 cm⁻¹ band and the 99·7 cm⁻¹ band were described as due to torsional oscillations about an axis perpendicularly bisecting the C—O band, the difference between them, which is almost a factor of four in the restoring force constant, must come from the movement of the environment, which in turn means essentially the movement of the nearest neighbours. If only the force constant from the three nearest neighbours to the nitrogen is considered for the large amplitude nitrogen oscillations (99·7 cm⁻¹, 183·6 cm⁻¹) then the amplitude of these three neighbours moving together, relative to that of the nitrogen nucleus, can be called α for the 183·6 cm⁻¹ band and β for the 99·7 cm⁻¹ band, and the relationship $(183\cdot6/99\cdot7)^2 = (1 - \alpha)/(1 - \beta)$ is obtained. Since both α and β must be small numbers ($-0\cdot20 < \alpha < 0\cdot20$, $-0\cdot42 < \beta < 0\cdot42$) to stay within the observed degree of localization, then α can be seen to be a small *negative* number and β can be seen to be a small *positive* number. Thus the three nearest neighbours must move with the nitrogen nucleus during the 99·7 cm⁻¹ vibration and against it during the 183·6 cm⁻¹ vibration. Similar results are obtained for the three nearest neighbours to the oxygen nucleus for the other two bands at 97·4 cm⁻¹ and 167·5 cm⁻¹ (cf. local mode and gap mode in Figure 7.7).

The nature of the vibrations associated with the four sharp bands in the lattice sideband structure can now be more fully described. The $97 \cdot 4 \, \mathrm{cm}^{-1}$ band is due to a torsional oscillation of the NCO^- ion about an axis perpendicularly dividing the $N—C$ bond, accompanied by an in-phase movement of the three potassium nearest-neighbour ions which lie on a ring round the oxygen nucleus. Similarly the $99 \cdot 7 \, \mathrm{cm}^{-1}$ band corresponds to a large nitrogen amplitude torsional oscillation of the NCO^- ion accompanied by an in-phase movement of the nitrogen's three nearest potassium ion neighbours. The $167 \cdot 5 \, \mathrm{cm}^{-1}$ band is due to a large oxygen amplitude torsional oscillation opposed by an anti-phase motion of the oxygen's three nearest neighbours, whereas the $183 \cdot 6 \, \mathrm{cm}^{-1}$ band is due to the equivalent large nitrogen amplitude oscillation. Figure 7.43 shows schematically four

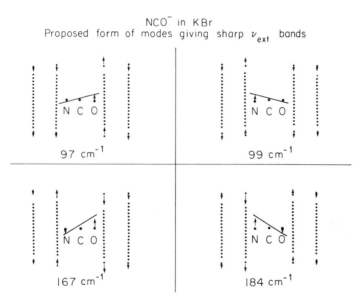

Figure 7.43 The predicted form of the torsional vibrations which give rise to sharp bands in the lattice side bands in combination with the v_3 frequency for NCO^- isolated in KBr. The dotted lines represent the planes of nearest and next nearest neighbours illustrated in Figure 7.42
(Cundill and Sherman)

vibrations which fit the above descriptions and which have been shown[78] to be capable of giving the observed frequencies and delocalization factors, to have amplitude envelope functions which decay exponentially with distance from the large amplitude NCO^- nucleus, whilst satisfying the requirements of no net translation and zero net angular momentum.

Two points of particular interest are illustrated by the above isotopic substitution investigation of the KBr/NCO$^-$ lattice side band structure:

(i) The four sharp bands in the spectrum are all due to vibrations which involve torsional motion of the NCO$^-$ ion.

(ii) The degree of delocalization of the gap modes is considerably greater than that of the superoptic localized modes.

The first point underlines the importance of torsional oscillations of the impurity in the study of spectra of external modes of polyatomic impurity ions. It is also sufficient to explain the single in-gap and single super-optic features in the equivalent spectra[77] of KBr/N$_3^-$ and KBr/BO$_2^-$. For these two latter systems, Table 7.6 can be extended to show that it is the torsional modes (R_x, R_y) which would be expected to be active in combination with v_3 although they would be inactive directly in the far infrared.

The delocalization factors found for these bands are also of a more general interest. If the vibrations illustrated in Figure 7.43 are considered with the object of making them all as highly localized as possible it will be evident that the stationary centre of mass requirement could be met by the three nearest neighbours to the large amplitude impurity nucleus for the super-optic modes; but for the in-gap modes where these three move in the same direction as the impurity nucleus, appreciable movement of more distant lattice ions is required. This partial 'acoustic' nature of gap modes is a general effect (see for example Figure 7.7) and such modes are always going to be spatially less localized than analogous super-optic modes, even though, as in this case, they may show much sharper spectral features. The delocalization factor of about 7 for the lower component of the gap mode doublet in the KBr/NCO$^-$ system can be compared with the same value, 7, which would be obtained by treating the gap mode in the KI/Cl$^-$ system[84] in this way and the value $\geqslant 6$ for the KI/Br$^-$ gap mode.[84] The super-optic modes in the KBr/NCO$^-$ system show a very high degree of localization with even the 167.5 cm^{-1} vibration which is almost coincident with the top of the optic band, being about twice as highly localized as either of the gap modes, and the 183.6 cm^{-1} band, only 16 cm^{-1} clear of the optic lattice modes, is almost completely localized. This last point indicates how good an approximation the infinitely massive rigid containing lattice model of Figure 7.1 can be expected to be for any super-optic localized modes which occur at more than a few wave numbers above the top of the host lattice optic band.

7.3.8.3 Pressure effects on the spectra of NCO$^-$ isolated in alkali halides

(i) *Internal modes.* The v_3 absorption band of NCO$^-$ isolated in NaCl was one of the very earliest infrared spectra to be investigated under high pressure.[85] This band is sharp, even at room temperature the half bandwidth is only just over 1 cm^{-1}, and it shifts steadily upward in frequency as pressure is applied to the sample. Drickamer *et al.*[85] measured the position of this band at various pressures up to about 54 kbar. (1 kbar $\equiv 10^9$ dyn cm^{-2} $\equiv$

10^8 Nm^{-2}.) Initially the frequency increases at the rate of about 1 cm^{-1}/kbar and for a band of half width about 1 cm^{-1} this makes quite a good pressure calibrant allowing pressures to be quoted to an accuracy of about 0·1 kbar. This band has therefore been used for pressure calibration purposes by other workers.[67,72,77,78,86]

The above steady increase in the frequency of v_3(NCO$^-$) with increasing pressure found by Drickamer et al.[85] is quite typical of the equivalent absorption band for NCO$^-$ isolated in other alkali halides. Price, Sherman and Wilkinson[87] investigated this v_3 absorption and many of the strong combination bands in the near infrared region for this ion isolated in each of the chlorides, bromides and iodides, of sodium, potassium, rubidium and caesium. For ions in the potassium and rubidium halides the frequency shift with pressure relationships show a discontinuity at the pressure at which the alkali halide changes from a low pressure, sodium chloride type, structure to a high pressure, caesium chloride type, structure. Typical of such discontinuities is that for v_3(NCO$^-$) isolated in KCl which was first reported by Drickamer and Slykhouse[88] and is shown in Figure 7.44.

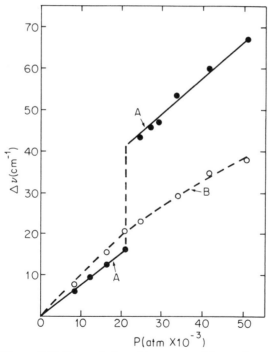

Figure 7.44 The variation with pressure of the v_3 vibrational frequency of NCO$^-$ isolated in KCl (Line A). Note the discontinuity in frequency when KCl undergoes a NaCl → CsCl type phase transition (Drickamer and Slykhouse)

A detailed interpretation of these shifts in the internal vibrational frequencies of the NCO^- ion isolated in various alkali halides, as a function of pressure, was hampered by the lack of a proper understanding of the interaction between the ion and its supporting lattice. It was in order to clarify some of these points that the more straightforward problems of first the CN^- ion[67] and then the N_3^- ion[72,87] were investigated. The absence of a centre of symmetry in the NCO^- ion increases the number of cubic terms in the potential function compared with the N_3^- case and it is still not clear how much useful information can be extracted from an analysis of the NCO^- data. It is fairly clear, however, from the other two cases, CN^- and N_3^-, that the dominant effect is the distortion of the molecular ion causing a change in frequency through the anharmonic terms in the internal potential function of the ion.

Although the analysis of the NCO^- data is still not completed it should be said that this data has already been of considerable value as pressure calibrants. The CN^- and N_3^- studies used $v_3(NCO^-)$ in the appropriate alkali halide as the pressure calibrant as did the lattice side band work reported later and several other studies not discussed here.

(ii) *Lattice side band structure.* The lattice side band structure on $v_3(NCO^-)$ isolated in many of the alkali halides has been studied as a function of pressure.[78] For example Figure 7.45 shows the effect of pressure on the lattice side band spectrum of $v_3(NCO^-)$ in KBr shown in Figure 7.40. Although the in-gap doublet of the low pressure phase is shown as a single line on the Figure 7.45 graph it is in fact a quite well-resolved doublet right up to the phase change, and only the lack of space on this small scale graph has caused it to be plotted in this way. There is no sign of any gap in the high pressure phase spectra and certainly no sharp in-gap structure, and these facts were taken[78] as clear indications that no gap exists in the KBr high pressure (CsCl type) density of vibrational states. Since the gap mode part of the spectrum shifted with pressure but showed no appreciable change in character, for pressure up to the phase change, it seems very probable that the gap-defining frequencies of the host lattice also shift at a rate very close to that shown by the gap modes. The only in-band feature that it was possible to measure reliably as a function of pressure was the lowest energy broad feature within the acoustic band. This can be seen in Figure 7.45 to *decrease* in separation from v_3 as the pressure is increased from zero up to the phase change, and is the only sign of lattice instability which is shown by this spectrum as the phase change is approached. An analogous feature in the high pressure, CsCl type, structure can be seen to occur at almost twice the frequency, and to increase slowly with pressure. The super-optic doublet separation from v_3 can be seen to increase rapidly with pressure in the low pressure phase then to suffer a discontinuous decrease in frequency accompanied by a large increase in doublet spacing at the phase change, and

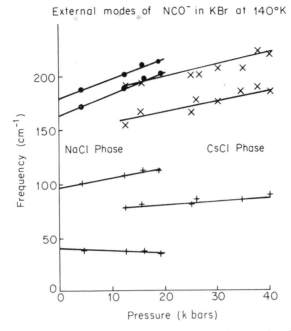

Figure 7.45 The pressure dependence of the frequencies of the prominent features in the lattice sideband structure on v_3 of NCO^- isolated in KBr (cf. Figure 7.40). The two components of the in-gap doublet could still be distinguished at pressures just below the phase change, but only the mean frequency could be plotted on this scale. Note the negative slope of the lowest energy 'in-band' feature in the NaCl phase and the absence of 'in-gap' features in the CsCl phase (Cundill and Sherman)

thereafter to increase at a more modest rate with pressure. If the special position of the lower component of the super-optic doublet, which coincides with the top of the KBr optic band at zero pressure, is retained as a function of pressure then this feature monitors this host lattice parameter. Cundill and Sherman[78] investigated the implications of this assumption and used it to obtain a value of $(\partial S/\partial P)_0 = -0.9 \times 10^{-3}\,kbar^{-1}$ for the initial rate of change of the effective charge for KBr with pressure.

Figure 7.46 shows the $v_3(NCO^-) \pm v_{ext}$ spectrum for the $RbBr/NCO^-$ system at two different pressures in each phase, and Figure 7.47 shows the graphs of v_{ext} against pressure that were obtained from such spectra.[78] Again the super-optic doublet can be seen, and its behaviour as a function of pressure is very similar to that exhibited by these features in the KBr/NCO^- system. The low energy feature within the acoustic band of the lattice can be seen again to decrease as the pressure is applied up to the phase change,

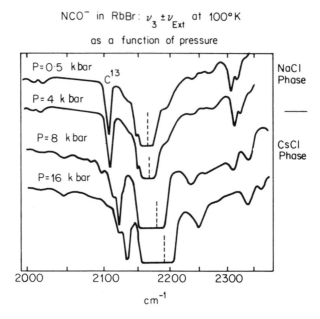

NCO⁻ in RbBr: $\nu_3 \pm \nu_{Ext}$ at 100°K

as a function of pressure

Figure 7.46 The effect of pressure on the $\nu_3 \pm \nu_{ext}$ spectrum of NCO⁻ isolated in RbBr. The phase transition NaCl → CsCl structure occurs at about 5 kbar (Cundill and Sherman)

whereafter it seems to be replaced by a band at about twice the frequency separation which increases its separation from $\nu_3(NCO^-)$ only very slowly with increasing pressure within the CsCl structured, high pressure phase of RbBr.

Other examples of lattice side band spectra which have been observed on NCO⁻ internal vibration bands will be discussed in the next section together with external modes found in combination with internal modes of other impurity ions.

7.3.9 Lattice side bands on vibrational spectra of impurity ions isolated in alkali halides

As implied above, many different ions when isolated in alkali halides show spectra which have lattice side band structure. A full discussion of all this data is not to be attempted, but in the following two sub-sections, first one lattice, KI, is taken and the spectra of many ions isolated in this one lattice are compared, then the spectra of one ion, NCO⁻, in many different lattices are compared.

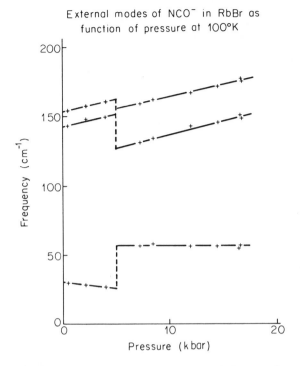

Figure 7.47 The pressure dependence of the prominent features in the lattice sideband structure on $\nu_3(NCO^-)$ isolated in RbBr. Note the negative slope of the lowest energy feature in the low pressure phase and the behaviour of the super optic doublet at the phase charge (Cundill and Sherman)

7.3.9.1 Comparison of far-infrared spectra and lattice side band spectra of ions in KI

The ions for which lattice side band data has been published for KI are: NCO^-, CN^-, N_3^-, BO_2^-, NCS^-, NO_3^-, OH^-, H^-. There is also some far infrared absorption data for some of these ions, NO_3^-, OH^-, H^-, CN^-, and some far-infrared data on other ions, F^-, Cl^-, Br^-, NO_2^-, Na^+, Ag^+, Tl^+, Cs^+ isolated in KI, which can usefully be compared. Figure 7.48 shows lattice side band spectra for impurity ions in KI, and Figure 7.49 shows far-infrared absorption spectra for ions in this same lattice.

Consider first what general observations can be made about the lattice side band spectra. Super-optic features are shown by the NCO^-, N_3^-, BO_2^-, NCS^- and NO_3^- spectra, and in-gap features by the NCO^-, N_3^-, BO_2^-, NCS^-, NO_3^-, CN^- and H^- spectra. The KI lattice sets the limits within

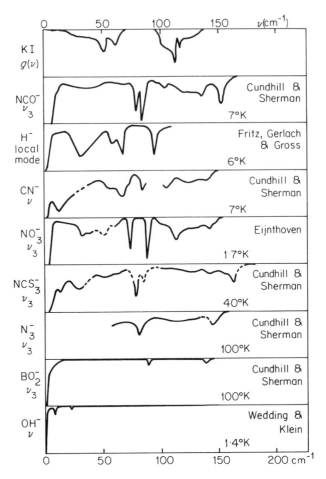

Figure 7.48 Lattice sidebands for several different impurity ions, each isolated in KI

which these various types of absorption can be located but beyond this, each individual impurity ion must be separately considered in explaining the number of and energy at which these bands appear. Cundill and Sherman[78] used the similarity in size, shape and mass of the three ions NCO^-, BO_2^- and N_3^-, to draw some comparisons between these spectra, particularly as regards the doublets in the NCO^- spectra being replaced by singlets in the spectra of the other two ions, but beyond this it is difficult to see any valuable comparisons. Similarly the broad in-band features shown by these spectra are limited by the host lattice to those energy regions covered by its vibrational bands, but the form they show is clearly determined separately in each case.

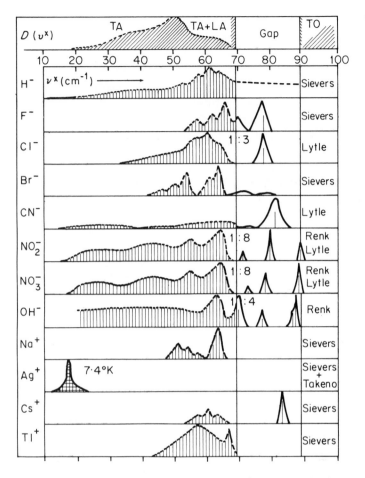

Figure 7.49 The far-infrared absorption due to different impurity ions isolated in KI (Genzel)

The same comments are relevant to the far-infrared spectra, where again it is the lattice which dictates what sort of feature, if any, appears at a particular energy, but each 'lattice impurity ion system' must be considered individually to decide whether or not any strong absorption is to be observed at a particular frequency.

In the earlier sub-section on symmetry considerations, the relevance of these to external modes was deliberately passed over until after the spectra shown in Figures 7.48 and 7.49 had been presented. This was so that the experimental results should be available to guide the discussion which is complicated by the fact that the ideal picture of a very large crystal built up from identical unit cells is lost once a random doping of impurity ions is implanted. However, even if the idealized model used for the lattice dynamical calculation of the host lattice vibrational properties is clearly invalidated by the inclusion of even a very dilute doping of impurity ions, the overall vibrational properties of the host crystal cannot be expected to be greatly changed by such a dilute doping. It is not intended to imply by this that the observed marked changes in spectra, low temperature specific heat, low temperature thermal conductivity, etc., are in any way surprising, but only that these effects are emphasizing such changes as have taken place, whilst the vast majority of the vibrational properties have gone almost unchanged. In fact the linear increase of these various effects with doping concentration shows quite clearly the very large 'reservoir' of essentially undistorted lattice vibrations which is retained within the crystal.

Super-optic localized modes, and in-gap modes can be adequately pictured in terms of the motion of the impurity itself and a few of its nearest neighbours, and no conceptual difficulties arise in understanding the symmetry assigned to such vibrations. Take as an example the gap-mode in the side band structure on the U-centre KI/H$^-$ local mode. It is assigned as the A_{1g} 'breathing' vibration of the six nearest neighbours to the H$^-$ ion with the H$^-$ itself stationary.[34] This is easy to visualize and ties in with experimental observation of this band in an allowed combination with the F_{1u} local mode, and its absence from the fundamental far-infrared absorption of the system. The A_{1g} mode would be expected to be Raman active. In a less easily visualized way in-band resonance modes which involve relatively large changes in the amplitudes of vibration only in the vicinity of an impurity, can also effectively be assigned a symmetry. This leaves the vast majority of the vibrations of the impurity distorted lattice which have been only very slightly changed from those of a pure crystal. Strictly speaking the random doping of impurities, in removing the periodic nature of the lattice, has reduced the size of the zone to its absolute minimum and all vibrations have become zone centre vibrations. This can be seen as the limiting case of Figure 7.50 in which the effects on the dispersion curves of progressively doubling the size of the unit cell, and halving the size of the first Brillouin zone, is illustrated. The limit is clearly reached when the unit cell is increased to the size of the crystal (i.e. all periodicity is lost) and the crystal is simply a large molecule with no symmetry (other than the inevitable identity E) and all normal modes of the molecule are both infrared and Raman active. Even in this limit when all modes are active they are *not* all *equally* active, and no spectroscopist would look at a spectrum (whether fundamental infrared, combinations with a dominant fundamental, Raman, vibronic, or any other type), and conclude that its intensity as a function of frequency must be showing a one-to-one correspondence with the density of vibrational states. However this sort of claim has been made on several occasions for spectra of the lattice side-band type, and although in some specific instances

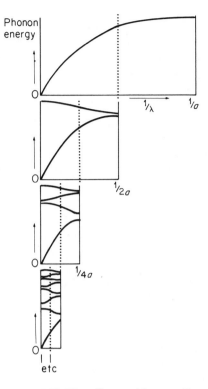

Figure 7.50 The effect on phonon dispersion curves v against k of small distortions of mass, force constant, or geometry, progressively doubling the size of the unit cell and halving the size of the first Brillouin zone. Although the first few stages in this procedure add considerably to the structure shown by the equivalent density of states functions, if this process is continued towards the limit when the unit cell is as large as the crystal and all vibrations are zone centre vibrations then the structure becomes almost continuous and a density of vibrational states very similar to the original is re-established

this type of spectrum may show a marked similarity to calculated density of states curves this should be regarded as fortuitous unless a detailed justification can be given.

Thus we can say in general that the bulk of the crystal vibrations although theoretically robbed of their propagating properties by the presence of impurities are likely to show a very similar density of vibrational states as a function of energy. The propagation of vibrational energy (i.e. thermal conductivity) at other than the lowest temperatures would not be expected to be greatly affected by their change in description because of the very strong coupling between vibrational levels in regions of high vibrational density of states. All of these vibrations would be expected to show a certain amount of vibration-dependent infrared, Raman and lattice side-band activity. Modern Green's-function methods which derive the impurity distorted vibrational properties of crystals from those of the pure crystal, using the model of a single impurity, have been very successful in accounting for the spectra of the U-centre and other point defects (see §7.2) and it seems reasonable to suppose that they could be extended to cover the external mode spectra of these polyatomic ions. Without going this far it is worth noting that the disruptive effect of the randomly distributed impurities is apparently causing only minor effects on the vibrations of the host crystal and the observed spectra are adequately described in terms of a small impurity distorted volume within a large, otherwise undistorted, crystal.

7.3.9.2 Comparison of the lattice side band spectra on $v_3(NCO^-)$ isolated in twelve different lattices

Figure 7.51 shows the $v_3(NCO^-) + v_{ext}$ spectra for this ion isolated in twelve different lattices, as published by Cundill and Sherman.[78] The four most extensive spectra (those for NaCl, NaBr, KCl and RbCl) show an additional band at about 2400 cm^{-1} which is the lowest energy component of a Fermi triplet, $2v_1$, $v_1 + 2v_2$, $4v_2$. This concentration-broadened line in NaBr may be obscuring some relevant v_{ext} structure. These Figure 7.51 spectra were all obtained from samples held at about 100 K, although most of these spectra have been investigated at temperatures down to 7 K. Lowering the temperature caused a general slight sharpening of most of the v_{ext} features except for the lowest energy broad band which usually became less prominent as the sample temperature was reduced below 100 K. Figure 7.52 shows the 7 K spectrum for NCO$^-$ in KI which is to be compared with the equivalent 100 K spectrum in Figure 7.51 to see the sharpening of the in-gap and super-optic features and the apparent reduction of the acoustic in-band modes. The inset in Figure 7.52 shows the super-optic region at elevated pressure (15 kbar) to illustrate the clearer resolution of the super-optic doublet under these conditions. A better resolution of the lower component of the super-optic doublet from the optic band structure can also be effected for the CsI spectrum by lowering the temperature or increasing the pressure.

Several general trends can be identified in this series of spectra:

(i) No really sharp features are shown by the v_{ext} structure on the four chloride spectra.

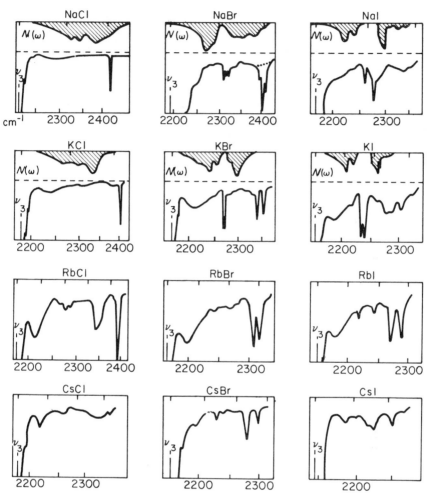

Figure 7.51 Lattice sideband structure on v_3 of NCO$^-$ isolated in twelve different alkali halides. The spectra were recorded for samples held at 100 K. The density of vibrational states for the pure host lattices is shown for the cases where the data is available (Cundill and Sherman)

(ii) The four lattices which possess a gap between their acoustic and optic modes show sharp doublet structure in this gap, although in NaBr another peak is also to be seen.

(iii) No super-optic features are to be seen in the sodium halide spectra.

(iv) The bromides and iodides of potassium, rubidium and caesium all show super-optic doublets although at zero pressure the lower component is imperfectly separated from the optic band for both potassium and caesium iodide.

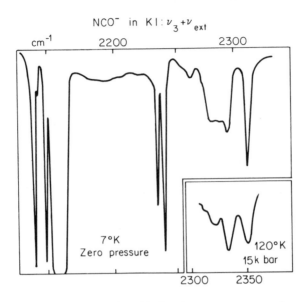

Figure 7.52 Lattice side band structure on ν_3 of NCO$^-$ isolated in KI. The main figure shows the spectrum recorded at 7 K and zero pressure. Note by comparison with Figure 7.51 how the reduction in temperature has sharpened both the 'in-gap' and 'super-optic' doublet structures, but has reduced the prominence of the acoustic band structure. The insert shows the 'super optic' doublet region of the spectrum recorded at 120 K and 15 kbar pressure when the doublet nature is more clearly revealed (Cundill and Sherman)

Thus the essentially torsional assignment given earlier for the NCO$^-$ in KBr external mode spectrum can probably be extended to include most of the localized modes and gap modes of these other spectra, since these generally seem to occur in the expected doublet form. The absence of clearly defined gap modes in some spectra suggests that for these lattices no gap exists, but against this it must be noted that regions of very low side band absorption (e.g. near $+85$ cm^{-1} for CsCl) could equally well be indicating the presence of an energy gap. For a super-optic mode to exist, the motion must be highly localized as indicated in the earlier detailed discussion of the KBr/NCO$^-$ system. Therefore the absence of such modes for NCO$^-$ in the sodium halides can be understood in terms of the lightness of the sodium ions, which results in relatively large amplitudes of vibration of the impurity's nearest neighbours, involves further ions, delocalizes the motion, and thus holds down the frequency to within the optic band of the lattice.

As well as the twelve spectra shown in Figure 7.51 there could be included the high pressure spectra such as that shown in Figure 7.46 which refer to the potassium and rubidium halides in their high pressure, CsCl type, structure. These are available for the bromides and iodides of both potassium and rubidium and all four show what appear to be super-optic doublets, and none of them show any obvious in-gap structure.

7.3.9.3 Other $v_{int} \pm v_{ext}$ spectra for impurity ions in alkali halides

The emphasis on torsional motion for the sharp features in the lattice side band spectra for NCO^- led Cundill and Sherman[78] to investigate more closely the lattice side band spectra for the CN^- ion which was known (see next section) to be held by a very small barrier to rotation. Figure 7.53 shows the 7 K lattice side band spectra on the CN^- fundamental vibration band. The species NCO^- was also always present in these samples and it is unfortunate that its v_3 absorption always falls in the gap region of the CN^- external mode spectra. With the exception of the $154 \, cm^{-1}$ band in NaBr, the three sodium halide spectra have not changed appreciably as a result of lowering the temperature from 90 K, at which temperature the stronger features in the external mode spectrum could also be seen in the difference band. This has allowed the gap modes in NaI and NaBr to be drawn-in with

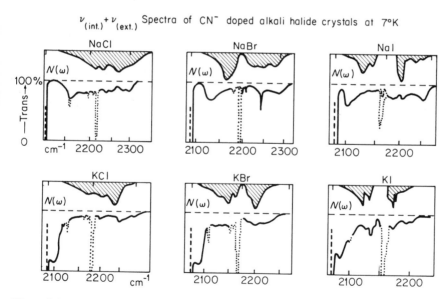

Figure 7.53 Lattice side band structure on the summation side of the stretching vibration of CN^- isolated in various alkali halides. The absorption spectra were recorded at 7 K. The density of vibrational states are shown shaded for comparison with the infrared absorption. Features due to NCO^- are shown by dotted lines (Cundill and Sherman)

some confidence although the exact 7 K frequency of the obscured modes is undetermined. The external modes in the potassium halide spectra only began to assume the shape shown in Figure 7.44 below about 20 K, so no assistance from the difference band can be expected for most of the features shown there, but this same temperature dependence is in itself an indication of their authenticity. Figure 7.54 shows the lattice side band on the cyanide ion stretching fundamental for the KBr/CN^- system compared with the equivalent structure on the $v_3(NCO^-)$ for the KBr/NCO^- system and the integrated density of states for KBr.[82] This figure shows more clearly the broad low separation absorption on the high energy side of CN^- band by showing how sharply the low energy side of this band falls away. The peak of this low energy broad feature in v_{ext} is confused by the presence at about $v_{(CN^-)} + 12\,cm^{-1}$ of the residual rotational-structure peak, which

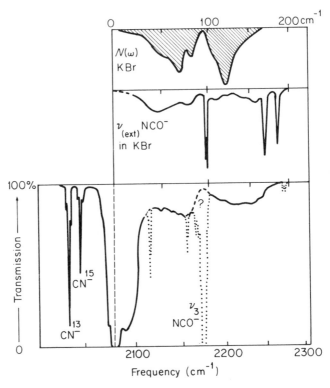

Figure 7.54 Lattice side band structure on $v(CN^-)$ and $v_3(NCO^-)$ in KBr compared with the density of vibrational states for KBr. Features due to NCO^- in the CN^- doped sample are shown by dotted lines, and unfortunately obscure the interesting gap region of the spectrum. Note that there are few similarities between the absorption spectra and the density of vibrational states (Cundill and Sherman)

sharpens noticeably on further cooling of the sample.[90] It is quite reasonable to explain a very low temperature sharp band at this energy as a summation band involving the torsional oscillation of the CN^- ion (or equivalently as the $F_{1u}(J = 1)(v = 0) \rightarrow F_{1g}(J = 2)(v = 1)$ transition of a vibrating hindered rotator obeying a Devonshire model[91] but this must not be allowed to hide the presence of the underlying broad band which can be seen to extend up to about $40 \, cm^{-1}$ above $v(CN^-)$. The hindered rotation of the CN^- ion will be further considered in the next section, but the assignment of the sharp feature at a separation of about $12 \, cm^{-1}$ in the very low temperature lattice side band spectrum of the CN^- ion, to a motion which is analogous to that proposed for the $167 \, cm^{-1}$, $183 \, cm^{-1}$ doublet in the NCO^- spectrum should be noted. Not only are the sharp peak positions very different in the two spectra compared in Figure 7.54 but also the broad features are almost as different as possible within the limitation that they must both be within the bands of the pure KBr lattice.

Comparing the other CN^- and NCO^- lattice side band spectra in Figures 7.51 and 7.53 it will be noticed that points of contrast are more obvious than points of similarity. Perhaps the most marked point of contrast is to be found at about $154 \, cm^{-1}$ in the NaBr spectra, where the NCO^- spectrum shows a quite limited region of very low absorption, and the CN^- spectrum shows its sharpest, strongest absorption. This $154 \, cm^{-1}$ band in the CN^- spectrum is relatively broad and ill-defined at 90 K and only below 25 K does it start to sharpen and by 7 K it has come to dominate the lattice side band spectrum. The behaviour of this band, and the very low absorption at the equivalent frequency in the NCO^- lattice side band suggest that contrary to calculation there is in fact a very low density of vibrational states for NaBr at this frequency. Against this must be stated that the two other lattice side band spectra available for NaBr (for $NaBr/H^-$,[92] and for $NaBr/NO_3^-$[93,94]) do not show any very special features at this frequency. However, all the H^- lattice side bands so far reported show very low optic band intensities and the NO_3^- spectrum is not available at very low temperature.

7.3.10 Torsional and hindered rotational motion of impurity ions in alkali halides

As was indicated above for the case of the CN^- ion, some of the impurity ions isolated in alkali halides show spectral features which can apparently only be explained in terms of relatively free rotation of the ions within their crystalline environment. Since the CN^- ion illustrates most of the relevant points it will be considered first in some detail and then other ions will be dealt with later. It is also interesting to compare this rotational behaviour with that observed in low-temperature molecular lattices (Chapter 3) and in clathrates (Chapter 8).

7.3.10.1 *Rotational motion of the* CN⁻ *ion isolated in alkali halides*

In 1958 Maki and Decius[62] reported a 'strong broad band with a sharp maximum' at 2070 cm⁻¹ in the IR absorption spectrum of the KBr/CN⁻ system, and Price *et al.*[63,74] 1958–60, investigated this fundamental CN⁻ vibrational frequency for the ion isolated in twelve different lattices. By 1963[95] the broad structure of this absorption was being discussed in terms of the rotational motion of the ion and by the end of 1964[96] the first published vibration–rotation analysis of this band had appeared. Later papers[97,67,78]

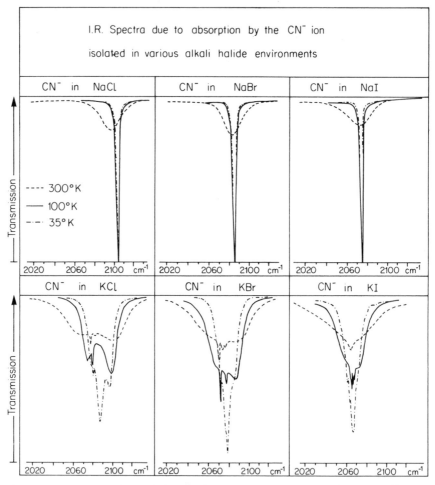

Figure 7.55 The infrared absorption in the fundamental C-N stretching region of twelve different alkali halides doped with CN⁻. The spectra which were recorded at 300 K, 100 K and 35 K show the appreciable changes of band contours with temperature that occur with the cyanide ion (Field and Sherman)

gave details of more careful investigations of these fundamental CN^- vibrational spectra. However, the suggestion that molecular groups within solids could be expected, under some circumstances, to exhibit rotational motion predated these papers by about 30 years. Pauling[98] in 1930 made this suggestion and Devonshire,[91] 1936, published calculations of the expected changes in the rotational energy levels of a linear molecule held centrally in an octahedral potential. Devonshire's calculations, which were presented as a function of an increasing potential barrier hindering the rotational motion, have been the basis of most of the interpretations so far presented for observed spectral features associated with the rotation of impurity ions.

Figure 7.55 shows the spectra published by Field and Sherman[67] of the fundamental CN^- absorption recorded at three different temperatures,

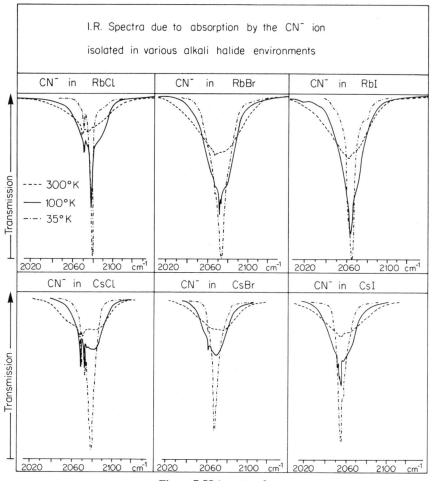

Figure 7.55 (*continued*)

300 K, 100 K and 35 K for the CN^- ion isolated in each of the chlorides, bromides and iodides of sodium, potassium, rubidium and caesium. The KCl/CN^- spectrum can be seen to be the one which most clearly suggests a poorly resolved vibration–rotation band, but the sharpness of the extra band (at about 2081 cm^{-1} and due to the cyanide associating with another impurity, possibly lead) shows that the spectra were recorded under sufficient resolution to show clearly any structure occurring at the spacing (about 4 cm^{-1}) expected for a completely free CN^- rotator. At lower temperatures the Q branch is seen in Figure 7.56 to become completely dominant,[97] and

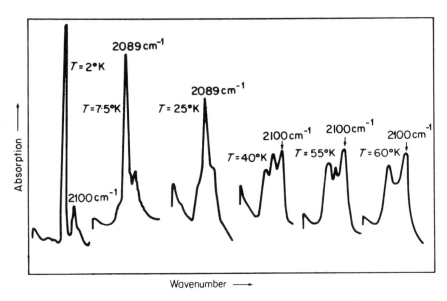

Figure 7.56 Absorption spectrum of KCl/CN^- at several low and very low temperatures (Narayanamurti[96])

at the lowest temperature, 2 K, only one small sharp feature at $Q + 12$ cm^{-1} is left from all the rotational structure which dominated the spectrum at higher temperatures. This temperature dependence of the band contour, coupled with the fact that the overtone band shows a band contour which is almost identical to that of the fundamental at all temperatures,[97] strongly suggests an interpretation in terms of rotational motion of the CN^- ion.

Devonshire's calculations[91] are illustrated by Figure 7.57 which shows how the free rotator levels are split by the octahedral field, and then begin to regroup at high fields into the torsional levels. Positive k values are relevant to the CN^- case where the low temperature, uniaxial pressure, preferred orientation, experiments of Seward and Narayanamurti[97] show there to be six equivalent ⟨100⟩ type preferred orientations of this ion in NaCl type

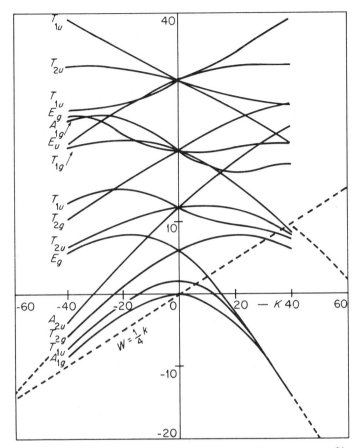

Figure 7.57 This shows graphically the result of Devonshire's[91] calculation of the perturbing effects of an octahedral field on the rotational energy levels of a linear molecule (reproduced by permission of the Royal Society)

lattices. Note the three levels A_{1g}, T_{1u} and E_g converging to give the sixfold degenerate ground state expected for this example (cf. the eightfold degeneracy from A_{1g}, T_{1u}, T_{2g}, A_{2u} levels for the negative k, $\langle III \rangle$ minima case). Both the energy of the distorted levels and k the barrier constant (minimum classical energy of $1 \cdot 25 \, k$ required to move from one position of minimum energy to another) are in units of B, the rotational constant of the ion.

If any very-far-infrared (1–20 cm^{-1}) absorption lines for the KCl/CN^{-} system had been recorded then Figure 7.57 could have been used directly, and a k value selected which best fitted the frequencies and intensity temperature dependence of these lines. Unfortunately no such lines have yet been recorded and if this model is to be used to describe the vibration–

rotation band shown in Figures 7.55 and 7.56 then the complication of requiring two such pictures, which might well be appreciably different from one another, one for the ground state and one for the vibrational state, needs to be faced. Even if the similarity of the band shape of the first overtone to that of the fundamental is taken as sufficient evidence to justify considering the rotational barrier to be almost independent of vibrational quantum number, there still remains the problem of the correct value of B to be used in applying Figure 7.57. This is illustrated by the discussion of Figure 7.58

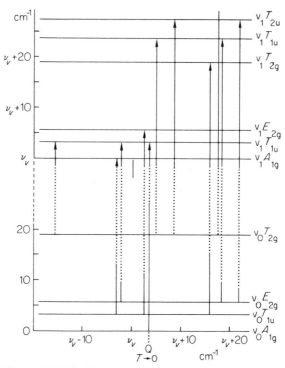

Figure 7.58 The hindered rotational levels of the ground and first excited vibrational states of a CN^- ion calculated using the Devonshire model. Allowed transitions from the ground and first three excited hindered rotational levels to similar levels in the vibrationally excited state are shown. A B value of $1\cdot94\,cm^{-1}$ and a Devonshire barrier constant $k = 13\cdot5$ has been used in both vibrational levels to obtain a $+12\,cm^{-1}$ side band in the infrared absorption spectrum at low temperatures. At the lower ends of the dotted lines showing the allowed transitions a solid part has been drawn of length equal to the calculated intensity of the transition at $5\,K$. Note in particular the transition $v_0 T_{1u} \rightarrow v_1 A_{1g}$ which leads to a side band with the same temperature dependence as the $+12\,cm^{-1}$ band

which shows similar hindered rotational levels on the ground state and first vibrationally excited state of a Devonshire hindered rotator. The lengths of the solid parts of the lines showing the transitions from the three lowest levels indicate their calculated intensities for a temperature of about 5 K. A value of $B = 1.94 \text{ cm}^{-1}$ has been used together with Figure 7.57 so as to get a high energy side band at about 12 cm^{-1} above the very low temperature absorption frequency, indicated by $Q_{T \to 0}$. Using the same hindered rotational levels makes the selection of the appropriate k value quite straightforward if only one frequency spacing ($+ 12 \text{ cm}^{-1}$ in this case) is to be fitted, since the spacing can be seen from Figure 7.58 to be given by $(T_{2g}) - 2(T_{1u})$. However Figure 7.58 has been constructed using free rotation selection rules, which should still be at least qualitatively fairly good at these small barrier heights, and the required $+ 12 \text{ cm}^{-1}$ line can be seen to be one of three transitions originating at the $v_0 T_{1u}$ level all of which are of comparable intensity and all of which should show the same temperature dependence. Although the $v_0 T_{1u} \to v_1 E_{2g}$ transition at $(Q_{T \to 0}) - 2T_{1u} + E_{2g}$ (i.e. about $(Q_{T \to 0}) - 1 \text{ cm}^{-1}$ in Figure 7.58) might be expected to be hidden by the main band, the $v_0 T_{1u} \to v_1 A_{1g}$ transition at $Q_{T \to 0} - 2T_{1u}$, (or about $(Q_{T \to 0}) - 7 \text{ cm}^{-1}$ in Figure 7.58) should be clearly visible. Using a B value of 1.94 cm^{-1}, discussed below, there would appear to be no way of constructing a diagram like Figure 7.58, even using different values for the barriers in the two vibrational states, which would allow a low temperature side band at $+ 12 \text{ cm}^{-1}$ but which would not show an equivalent side band at least 3 cm^{-1} below the main band.

Seward and Narayanamurti[97] used a much smaller B value, 1.25 cm^{-1}, which allowed them to bring the low temperature, difference, side band sufficiently close to the main band, $- 2.8 \text{ cm}^{-1}$, for them to propose that it had become merged with it. However, their value for B is very difficult to believe, in that it requires a C—N distance of 1.4 Å, whereas Elliott and Hastings[97] report a value of $1.16 \pm 0.02 \text{ Å}$ for the C—N bond in pure KCN from neutron-diffraction data, and Bijvoet and Lely[100] quote 1.06 Å as the best value for this distance from their low temperature X-ray diffraction studies of NaCN and KCN. Also, the method used to obtain the value 1.25 cm^{-1}, is in itself highly suspect since it depends on drawing the best straight line through the origin on a graph of the separation of the P and R branch maxima versus $(T)^{1/2}$. Field and Sherman[67] find that their experimental points on such a graph do not extrapolate back through the origin but in fact intersect with a much steeper line from the origin, corresponding to $B = 1.94 \text{ cm}^{-1}$, at about 35 K, which is the lowest temperature at which the dominating Q branch allows the P to R spacing to be measured, and this is shown in Figure 7.59.

There is no reason to suspect that the CN^- ion isolated in alkali halides has a bond length very different from that which it possesses in NaCN or

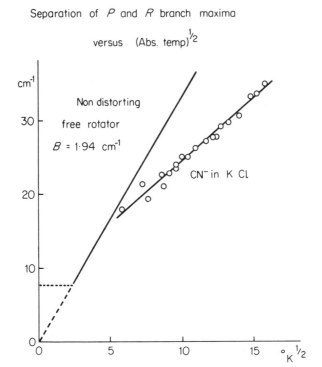

Separation of P and R branch maxima

versus   (Abs. temp)$^{1/2}$

Figure 7.59 The separation of the maxima in the P and R branches of the vibration hindered rotation band of CN^- isolated in KCl plotted versus $T^{1/2}$. The theoretical relationship for this P_{max} to R_{max} spacing for a rigid free rotator of B value $1.94\,cm^{-1}$ is also shown (Field and Sherman)

KCN. On the contrary the calculations[67] described earlier show that even very small changes in this length cause marked changes in the C—N bond stretching frequency, and since this is the dominant mechanism for changing the frequency, the frequency is a reliable guide as to the bond length. (The frequency increases about $60\,cm^{-1}$ for a compression of the C—N bond by $0.01\,\text{Å}$ therefore a B value of 1.25 requiring an r value of about $1.44\,\text{Å}$ would be expected to be associated with a very low C—N frequency.)   Thus the best B value to use for the CN^- ion would seem to be that calculated using $r = 1.16\,\text{Å}$; that is $B = 1.94\,cm^{-1}$.

This leaves the anomalous situation that the very low temperature spectra are better explained by the simple two-well cosine potential used by Seward and Narayanamurti[97] in the preliminary discussion of their spectra than by the three-dimensional potential of octahedral symmetry. The high quality experimental work in the above paper seems most unlikely to be seriously in error and with the exception of the difficult-to-measure P_{max} to R_{max} spacing, it has only received further confirmation from later papers. There is also the success of the Devonshire model in accounting for the very low temperature $(T < 0.3 \text{ K})$ thermal conductivity and specific heat of the cyanide doped alkali halides.[97] The reason for this anomaly may well be found in the lattice distortion which Field and Sherman[67] found to be necessary in their frequency shift calculations discussed earlier. If in fact the lattice does distort in the way suggested then the site on which the CN^- ion sits, is no longer one of octahedral symmetry, and the impurity requires a coincident, (E_g) motion of the surrounding ions if it is to find one of the four adjacent $\langle 100 \rangle$ type orientations equally energetically favourable. Should the ion rotate unaccompanied by such a movement of the surrounding ions, then it would in fact only find one other potential minimum (at 180° to its initial position) in which to rest, which begins to suggest that the two-well cosine potential may be quite a reasonable model for very low temperature, low rotational energy states. High rotational states cannot expect sympathetic lattice motion to follow them in general,* and they could be expected to re-establish the octahedral symmetry of their lattice site. This mixed picture, although more complicated than the composite parts because of the transition region between them, does have several points to commend it.

Firstly, the problem of the low temperature side band on the negative side is avoided by the use of a non-Devonshire model for the rotational ground state which can easily be made to reduce the tunnelling splitting down to the level where the difference band is merged into the broadened Q branch.

Secondly the dependence of the barrier to rotation, and hence the tunnelling splitting of the level, on the presence of lattice vibrations, would be expected to: (a) broaden transitions which include tunnelling (i.e. give breadth to $Q_{T \to 0}$) and (b) make the tunnelling transition itself a good scatterer of phonons, as found by Seward and Narayanamurti.[97]

Thirdly, the torsional oscillation, in which the CN^- ion remains within the same potential minimum would not be expected to be a good scatterer of phonons, whereas the somewhat higher levels near the top of the barrier and at the energy associated with the change in model description would be good scatterers.[97]

Fourthly, the higher hindered rotational levels, above the barrier height, although split by the Devonshire type field as shown in Figure 7.57, blurring the rotational structure, and reducing the P to R branch maxima spacing, would not in general be expected to be very efficient phonon scatterers.

Thus the Devonshire model of a molecule rotating centrally within a fixed octahedral field, although of great value in providing an understanding of the

---

* There is some evidence, see for example Field and Sherman,[67] that particular rotational states do find enhancement due to sympathetic lattice motion, i.e. resonate with a high density lattice mode.

perturbing effects on the rotational levels of such a field, must not be expected to provide an exact description for molecules rotating in real solid environments. A real molecule in a real solid *must* distort the environment in which it is held, and a real rotation *must* be accompanied by a changing distortion of this lattice. The part played by the lattice in transitions involving the changing of impurity ion orientation at energies covered by the lattice vibrational bands must *not* be expected to be adequately described by a static potential, and such transitions must be expected to be good scatterers of phonons, and to be broadened accordingly.

7.3.10.2 Rotational motion of the NO_2^- ion

The rotational motion of the diatomic CN^- ion described above is much easier to visualize than that of the NO_2^- ion which has three principal axes about which rotation can be assumed to take place. However, Narayanamurti *et al.*[101] give a very convincing analysis of the low temperature vibration-hindered rotation spectra of this ion isolated in several alkali halides. Very low temperature thermal conductivity data for these systems is also included in the above paper and is used to supplement the spectroscopic data. The orientation of the ion was found from the degree of alignment of the ions at very low temperature under uniaxial stress. A $\langle 110 \rangle$ orientation was found for the NO_2^- dipole, which coincides with the C_2 axis of the ion, which suggests that a twelve minimum potential should be considered for this ion, but virtually all the wealth of low temperature structure, see Figure 7.60, was found to be explained by simpler models of the rotations of the ions about its three principal axes treated independently.

With the two fold axis of the ion called B, the other principal axis in the plane of the ion called A, and the axis perpendicular to the ion called C, values for the barriers to rotation about these axes, V_B, V_A, V_C, respectively, were estimated for the NO_2^- ion in the chloride, bromide and iodide of potassium. For KCl/NO_2^- the values quoted were: $V_A = 11.8$ cm^{-1}, $V_B = 11$ cm^{-1}, $V_C = 40$ cm^{-1} or 50 cm^{-1}. For KI/NO_2^- the values quoted were: $V_A = 290$ cm^{-1}, $V_B = 0$ cm^{-1}, $V_C = 100$ cm^{-1}, with the centre of mass of the NO_2^- displaced from the centre of the cavity along the C_2 axis to allow a closer approach of the nitrogen to one particular I^- ion. From the less fully explained KBr/NO_2^- spectra were quoted values of $V_A = 21$ cm^{-1}, $V_B = 1$ cm^{-1} or 2 cm^{-1}, V_C probably in the 50–100 cm^{-1} range, with a slightly off-centre positioning of the ion favoured.

It will be appreciated that the large values of V_C in each case, and more particularly the large value of V_A for the KI/NO_2^- system, are associated with side bands which are quite widely spaced from the fundamental vibrational transition frequencies. In fact the bands used to define V_A and V_C in the KI/NO_2^- system are the gap-mode lattice side bands on $\nu_3(NO_2^-)$. Gap-modes for this system have been observed directly in the far-infrared[102,103] (see

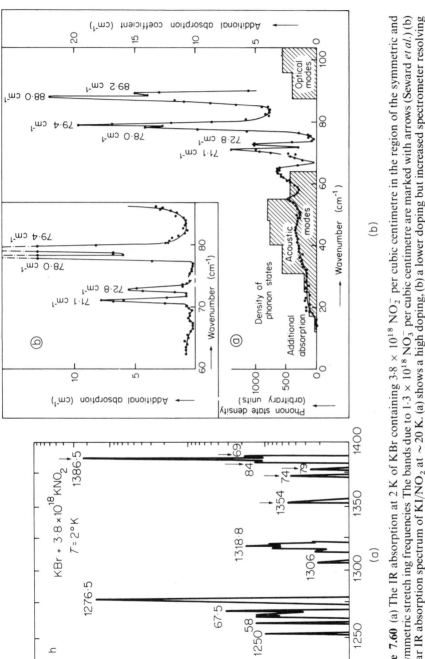

Figure 7.60 (a) The IR absorption at 2 K of KBr containing 3.8×10^{18} NO_2^- per cubic centimetre in the region of the symmetric and antisymmetric stretching frequencies The bands due to 1.3×10^{18} NO_3^- per cubic centimetre are marked with arrows (Seward *et al.*) (b) The far IR absorption spectrum of KI/NO_2 at ~ 20 K. (a) shows a high doping, (b) a lower doping but increased spectrometer resolving power. The density of vibrational states for pure KI are shown shaded in histogram form (Renk)

Figure 7.60b) and also as vibrational structure on the lowest energy electronic absorption.[104] Various models have been used to explain these bands, cf. refs. 101, 102, 103, 104, 94 and 19, and the true physical significance of them must still be regarded as uncertain. However, the above interpretation giving a torsional motion as the part played by the NO_2^- in the vibration is of particular interest as the first published suggestion that this type of motion should be important for these bands.

Consider now this interpretation in the context of the KBr/NCO^- work[70] discussed in an earlier sub-section. The KBr/NCO^- gap modes were found from isotopic substitution measurements to involve torsional motion of the NCO^- ion, but up to 85 per cent. or more of the vibrational energy was seen to reside in the vibrating environment, with only a relatively small fraction being carried by the impurity ion itself. This can be seen to support the idea that the rotational degrees of freedom of impurity ions can be incorporated into the lattice vibrational structure of the host in the form of torsional oscillations of relatively high frequency but shows quite clearly the dangers of trying to describe the part played by the lattice in these vibrations as that of a static containing potential.

Thus, as claimed by Narayanamurti et al.,[101] the alkali halide/NO_2^- systems do provide examples of impurity ion motion of a most general nature, with internal modes of vibration, and external modes which include not only those of a translatory type, but also those of a rotatory type which experience barriers which can be almost zero about one axis whilst being very large about another. However, the apparent ability of the Pauling type model to adequately describe the rotational motion of molecules in solids must be questioned for those energies which lie in the range covered by the vibrational bands of the solid where it is unreasonable to expect the solid to remain rigid and inactive during the motion. The Green's function methods could surely be extended to include at least the torsional modes of the rotational degrees of freedom. This would mean that the response function (or mechanical susceptibility) of the supporting lattice would be required not only for the 21 degrees of vibrational freedom of the point-like impurity +6 nearest neighbours system as is currently commonly evaluated, but for the 24 degrees of freedom of the three-dimensional impurity +6 nearest neighbours system.

7.3.10.3 Rotational degrees of freedom of other impurities

Most recent investigations of polyatomic impurity ions have given some consideration to the torsional or rotational motion of the ion. However only the OH^- ion will be mentioned here since it shows some points of special interest, and this type of motion for the NH_4^+ ion will be considered later.

Supposedly pure alkali halides often contain several parts per million of the OH^- ion, see for example Klein, Kennedy, Gie and Wedding,[105] and the references therein. Although the OH band just below the fundamental ultra-violet absorption edge of the alkali halides has been known for some time,[106] the infrared evidence for OH⁻ in these crystals was much more difficult to locate because of the breadth of the bands at other than very low temperatures,[107] and the low oscillator strengths.[107,108]

Lattice side band structure on the OH⁻ stretching vibration band contains two unusual types of feature as well as the more common broad, in-band, structure. These are: (a) sharp very low temperature in-band peaks and (b) a band which although separated from the main band by about twice the highest lattice frequency, is still quite broad (for $KCl/OH^- \Delta v_{1/2} \approx 25 \, cm^{-1}$) even at very low (5 K) temperature.[107] This lattice side band structure, and such equivalent direct far-infrared absorption as the crystals allow to be investigated has been discussed in some detail by Wedding and Klein,[109] (this paper contains 72 references relevant to the infrared spectra of OH⁻) and Klein et al.[110] These papers review the various models used to explain low frequency modes and broadly assign frequencies as due to Devonshire or non-Devonshire type transitions, without reaching any very positive conclusions about the form of the non-Devonshire levels. This need to use more than one of the currently available models in order to explain all of the observed features is almost certainly due to the inability of the much favoured Devonshire model to allow for the participation of the lattice in so-called torsional and hindered rotational motion of the impurity. As emphasized in the two previous sub-sections this sort of composite description can only be expected to be resolved when the whole lattice impurity system is treated fully in each case. The question, 'what is the motion of the impurity during this vibration?' is as incomplete as the question, 'what is the motion of the lattice during that vibration?' Only when the question in each case is, 'what is the motion of the lattice impurity system?' can a comprehensive model be expected to develop.

Before leaving this section on the OH⁻ torsional modes the Raman work by Fenner and Klein[111] on alkali halides containing this ion must be mentioned. The librator (Devonshire torsional mode) band at $313 \, cm^{-1}$ in KBr and $305 \, cm^{-1}$ in KCl were observed for liquid helium temperature samples as well as the main OH stretching mode which, in marked contrast to the infrared band, remained sharp right up to room temperature. NaCl/OH⁻ and NaCl/OD⁻ samples were also investigated and the strength of the OD⁻ stretching mode was found to be the same in the Raman as for the OH⁻ mode, whereas in the infrared the OD⁻ shows an apparently zero oscillator strength which is quoted as less than 1 per cent. of that for the OH⁻.

Callender and Pershan[112] have also recorded Raman spectra for the OH⁻ ion, which are in agreement with those described above, and for the CN⁻ and NO₂⁻ ions. A more comprehensive account of this work has recently been published by these same two authors.[113] Low energy bands were recorded for KCl/CN⁻, KBr/CN⁻ and NaCl/CN⁻, which show certain similarities

with the IR lattice sidebands published by Cundill and Sherman[78] and shown
in Figure 7.53. A suitable choice of polarization allowed rotational structure
to be seen on the side of the C—N stretching mode, for example, Figure 7.61

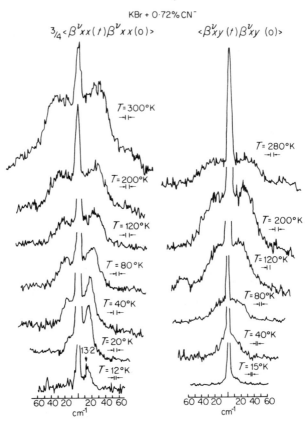

Figure 7.61 Polarized Raman spectra KBr/CN⁻ in the
vicinity of the CN⁻ vibrational frequency at different
temperatures (Callender and Pershan)

shows the Raman spectrum[113] of the KBr/CN⁻ system which can usefully
be compared with Figures 7.55 and 7.56. In that paper[113] a very low B value
was again used in fitting the very low temperature spectra to the Devonshire[91] model, but this is unjustified as discussed in detail earlier.

7.3.11 The ammonium ion NH_4^+

This ion is considered separately because as well as being the only polyatomic cation to have been studied in alkali halides,* it also possesses some

* Some data are now also available for PH_4^+ (Sherman and Smulovitch unpublished work).

unique properties when isolated in CsCl type lattices due to its ability to orientate itself so as to retain its full T_d symmetry.

The vibrational spectra of the ammonium ion were first studied as the dominant features in the spectra of the ammonium halides[114,115] Later this ion was also studied as an isolated impurity within the alkali halides.[63,74,116]

Vedder and Hornig[116] concentrated on the NaCl structured alkali halides, and investigated the hindered rotational structure shown by the vibrational bands of the isolated NH_4^+ at temperature down to 5 K. This was the first isolated ion to have its hindered rotational structure studied, and although some of the features shown by their spectra could be re-interpreted in the light of the more recent studies on CN^{-}[96,97] and NO_2^-,[101] nevertheless most of the points of interest were brought out in this early paper.

Price, Sherman and Wilkinson[63,74] studied the NH_4^+ ion, and many of its isotopically substituted equivalents, in the CsCl structured alkali halides where the vibrational transitions showed as sharp single bands. The reason for the qualitative difference between the spectra of this ion isolated in the two different types of lattice lies in the differences between the barriers to rotation in the two cases. Various types of potential term make contributions to these barriers, but since both the electrostatic attraction of the protons to the negative near neighbours, and the high polarizability of these near neighbours, favour an alignment of the N—H bonds with nearest neighbour anions, it is to be expected that the ion should seek to orientate itself so as to meet this condition. Thus in the CsCl type structure where all four N—H bonds can be simultaneously satisfied, a rotationally stable situation is found with relatively high barriers to rotation about all axes. However, in the NaCl type structures, in which if one N—H bond is satisfied there are three which are not, a complex rotatory motion can be expected, at all but the very lowest temperature, in the presence of relatively low barriers to rotation at least for a C_3 ion axis and probably for a C_2 ion axis also, with the barrier to the change of N—H bond which lies along a crystal $\langle 100 \rangle$ type direction also probably very small.

Only CsCl structured alkali halides will be considered here in which the NH_4^+ ion orientates itself so as to align all four N—H bonds with nearest neighbour anions, in which orientation it is held by fairly high barriers to rotation and retains its full T_d symmetry. v_1 the symmetric N—H stretch has A_1 symmetry and is Raman active and infrared inactive, as is the E type v_2 symmetric bending mode. The antisymmetric N—H stretching mode v_3 and antisymmetric bend v_4 are both of F_2 type, which is IR and Raman active, as are the free-ion non-genuine translations. However the rotations are of F_1 type and are IR and Raman inactive. Although overtones of the A_1 and E modes are IR inactive as are combinations between them and between A_1 and F_1, this still leaves many other combinations and overtone modes which are IR active through their F_2 sub-levels. Since only F_2 levels are IR active for this system, the entire IR spectrum is one vast Fermi polyad, although in general the levels are sufficiently far displaced from one another to retain a fairly easily identifiable character.

In the region 2800 cm^{-1} to 3500 cm^{-1} Sherman and Smulovitch[117] have identified at least 10 transitions in the IR absorption of the CsBr/NH$_4^+$ system, and in some similar systems even more features can be easily observed in this region[118] particularly at lower temperatures.

Figure 7.62 shows the spectrum of the CsBr/NH$_4^+$ system in the above region for two different NH$_4^+$ concentrations, each for sample temperatures

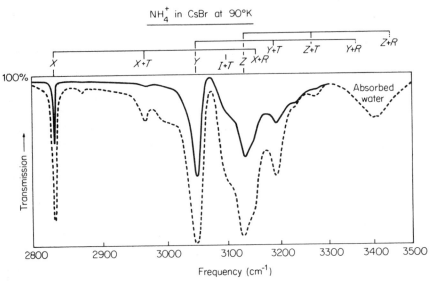

Figure 7.62 The N–H stretching region of the IR absorption spectrum of NH$_4^+$ isolated in CsBr is shown for two samples of different concentration. Fermi resonance between the different levels involved makes a precise assignment of the bands very difficult, but the bands are all due to transitions starting from the ground state. To a first approximation band X is principally due to a transition to the $2v_4$ level and bands Y and Z are due to transitions to the two levels which results from a close coupling of v_3 and $(v_2 + v_4)$, I, T and R refer to bands discussed in the text. Evidence for $Y + R$ and $Z + R$ comes from the spectra of very heavily doped samples (Sherman and Smulovitch)

of 90 K. The spectrum can be seen to be dominated by three strong bands X, Y, Z in Figure 7.62 which have been widely reported for NH$_4^+$ containing systems, and are usually interpreted as the $2v_4$, $(v_2 + v_4)$, v_3 Fermi triplet. Six smaller features were assigned[117] as the combination of two external modes $v_T \approx 142$ cm^{-1} and $v_R \approx 315$ cm^{-1} with each of the strong features X, Y and Z, and the tenth feature was assigned as $v_1 + v_T$. With no precise data available on the lattice dynamics of CsBr it is not possible to be sure about the highest optic mode frequency but from $\omega_{LO} \approx 105$ cm^{-1}[119] and the lattice side band 'super-optic' local modes on v_3NCO in CsBr[78] at 119 cm^{-1} and 138 cm^{-1} it seems certain that v_T, above, is well clear of the optic band, and v_R is almost three times the highest lattice frequency, almost as high as the U-centre local mode (363 cm^{-1}). Thus both these modes, but more particularly v_R, are going to be highly localized modes which to a good approximation can be treated as motions of the ion with respect to a rigid infinitely massive containing potential. The frequency of v_R or its equivalent for NH_4^+ in other environments, has been fairly widely reported but in particular the paper by Price et al.[63] gave data for this mode for the ND_4^+ and $^{15}NH_4^+$ species. These allow frequency ratios $v_R(NH_4^+)/v_R(ND_4^+) = 1.34$ and $v_R(^{14}NH_4^+)/v_R(^{15}NH_4^+) = 1.00$ to be calculated which indicate the rotational nature of the motions of the NH_4^+ ion during this mode. Note that if the $2(V_0 B)^{1/2} = v_R$ equation is used to evaluate the approximate height of the barrier to rotation then the barrier V_0 is calculated to be about 4000 cm^{-1}. Thus even though the F_1 fundamental of this mode is inactive there would seem to be reason to hope that higher overtones might be found through their F_2 sublevels. Although the v_T combinations are showing more clearly in the v_3 region of Figure 7.62 than the v_R combinations, this situation is strongly reversed in the combinations with v_4 and v_2 which were analysed by Price et al.[63] to get their values for v_R and so no isotope data are yet available to aid in the interpretation of v_T.

The ten (or more) component polyad shown in Figure 7.62 is of interest for several reasons,[117,118] but we will consider some of the implications of only one of these. Summation lattice side band levels, noticeably the super-optic features v_T and v_R, in summation with X or Y, gain in intensity if they are close to one of the other strongly active levels Y or Z because of the resonance between all observable F_2 transitions from the ground state. Lattice side bands are in general fairly sensitive to applied pressures, and so by using pressures of up to 50 kbar, many of the lattice side bands can be scanned over regions in which their intensities change markedly due to changing resonance with the stronger features.

Sherman and Smulovitch[117] recorded the spectrum (Figure 7.62) as a function of pressure up to 53 kbar at 100 K using a modified Drickamer[85] high pressure optical cell.[86] There were several marked changes in the spectrum over this pressure range, such as a threefold increase in intensity of each of the three strong bands X, Y, Z, which all moved to lower frequency with pressure, a general increase in frequency of the smaller intensity components which was accompanied by an increase in intensity if the small feature was approaching a larger one, but a decrease in intensity if it was moving away from the nearest strong feature. However in this first paper[117] the emphasis was on only two bands, $(v_1 + v_T)$ and Z of Figure 7.62, which were pressure scanned right across the region of maximum resonance. Figure 7.63 shows the observed frequencies, labelled v_+ and v_-, and also the calculated hypo-

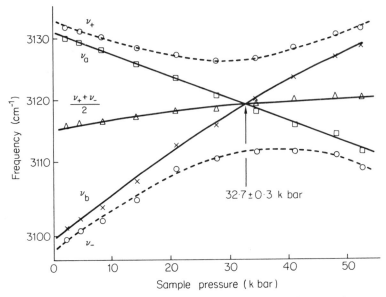

Figure 7.63 This diagram illustrates the Fermi resonance between the two levels responsible for the peaks marked ($I + T$) and Z in the IR spectrum in Figure 7.62 when the $CsBr/NH_4^+$ sample is subjected to pressures of up to 50 kbar and 100 K. The experimentally determined frequencies are those marked v_+ and v_- and the pressure dependence of the two calculated non resonant levels are shown by the lines v_a and v_b. Note also the change of relative intensities with pressure of v_+ and v_- in the next diagram (Sherman and Smulovitch)

thetically-non-resonant frequencies v_a and v_b, plotted as a function of sample pressure. This figure shows that v_b moving up in frequency as a function of pressure, crosses v_a, moving downwards, at about 32·7 kbar, and Figure 7.64 shows the change-over of intensity that takes place (superimposed on the general increase with pressure) as the pressure is raised; the components having equal intensity at about 29·1 kbar. The fact that the intensities are not equal at the pressure at which $v_a = v_b$ means that the weaker component v_b (i.e. $v_1 + v_T$) does have a non-zero intensity of its own[117,120] and a value of $\mu = 0·12$ was obtained. Where μ is the ratio of the transition moments between the ground state and the hypothetically non-resonant levels a and b.

The results shown in Figures 7.63 and 7.64 are of general interest, because of the example they show of Fermi-resonance with direct experimental data as a function of changing coupling which is essentially continuously variable right through the region of maximum coupling. They are of particular interest to the study of impurity ion motion in alkali halides because of the facility they offer to amplify features in the lattice side band structure. At maximum resonance in the above example, the absorption coefficient of the lattice

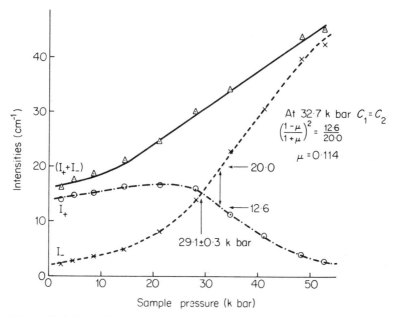

Figure 7.64 I_+ and I_- are the intensities of the computed resolved components v_+ and v_- of the CsBr/NH$_4^+$ IR absorption spectrum shown in Figure 7.62. The intensities are plotted as a function of sample pressure and are seen to be equal at 29.1 ± 0.3 kbar, which is unmistakably different from the value of 32.7 ± 0.3 kbar at which the maximum frequency resonance occurs. This allows a value to be calculated for the ratio of the transition moments from the ground state to the two levels in the absence of this resonance (Sherman and Smulovtich)

side band feature was increased by a factor of nearly 70. This gain is with respect to a level which is already amplified by the other resonances within the polyad, and an overall gain of about 1000 would seem to be a more reasonable figure.

Spectra analogous to that shown in Figure 7.62 have been investigated as a function of pressure[118] for the NH$_4^+$ ion isolated in the chlorides, bromides and iodides of caesium ($P = 0 \rightarrow 50$ kbar) rubidium ($P \approx 5 \rightarrow 50$ kbar) and potassium ($P \approx 20 \rightarrow 50$ kbar). As well as examples of resonances like that described above for sharp localized mode features passing peaks Y or Z, which are apparently adequately described by the crossing of two line-like levels a and b, more complex behaviour is also observed which can be at least qualitatively described in terms of a 'band with high frequency cut off and low frequency tail' scanning past a line level. This latter effect could arise if a summation band including a feature due to a Van Hove singularity is scanned past peak Y or Z. A rigorous treatment of the Fermi-resonance of such an example is being considered and the experimental work is being extended to lower temperatures (5 K) and higher pressures (300 kbar at 100 K). Apart from the continuing analysis of these spectra as indicated above, the two questions for which answers are currently being sought are:

(i) Why is the v_3 region of the NH_4^+ spectrum, shown in Figure 7.62 not more complicated still? (Except for A_2 levels, all allowed levels under T_d symmetry combine with F_2 levels in such a way as to give rise to F_2 sub-levels).

(ii) Can the anomalous higher temperature band shapes of the NH_4^+ vibrational transitions within CsCl type structures be understood in terms of these Fermi resonance effects?

7.3.12 Multiple impurities

This sub-section has been included in the polyatomic section although most of the experiments which could be quoted under this heading have been concerned with modifications to spectra associated with monatomic impurities. It is correctly placed in this section however, because even if the bonding between the multiple impurities is not sufficiently great to produce 'internal modes' which are above the host lattice modes, at least the symmetry of the system is reduced in the way that is characteristic of the more normally considered polyatomic impurities discussed in the earlier sub-sections.

One such experiment was briefly described earlier, when the effect of Rb^+ impurities in KCl/H^- was discussed, see Figure 7.20. The lowering of the symmetry is seen in the single F_{1u} band for the O_h KCl/H^- system being split into two components v_α, v_β when one nearest neighbour K^+ is replaced by Rb^+. This reduces the local symmetry to C_{4v} and the local mode splits into an A_1 mode (v_β) and an E mode (v_α) which have the expected relative intensity ratio of about 1 : 2. Two Rb^+ neighbours for the H^- ion could (with a probability of about one fifth if it is not energy dictated) cause the symmetry to reduce to C_{4v} again with the two Rb^+ being co-linear with the H^-, and it is the A_1 mode of this configuration which is too high in frequency to be shown in Figure 7.20. The E mode is almost certainly part of v_ϵ. If the two Rb^+ and the H^- form a triangle (with probability four fifths) then the resultant C_{2v} symmetry would split the local mode into three components, A_1 along the C_2 axis and B_1 parallel to the line of the two Rb^+ ions probably have very similar frequencies (v_δ) and B_2 perpendicular to the $Rb^+H^-Rb^+$ plane (part of v_ϵ).

It can be seen in the above example that the treatment of this case as an atomic impurity subjected to minor perturbation is more realistic than its treatment as a polyatomic impurity, when for the single Rb^+ near neighbour the model would be a $Rb-H$ stretching mode for the A_1 and a $Rb-H$ torsional oscillation for the E mode. However, it is worth looking at it in this light, if only to draw the analogy with the OH^- ion, which in several alkali halides shows an E type vibration at about 300 cm^{-1}, directly in the infrared, as a side band to the A_1 O—H stretch in the infrared, and directly in the Raman, which is generally described as the Devonshire type librator mode. (See §7.3.10.3). Together with the generally accepted off-centre position of the OH^- centre of mass, with the oxygen displaced outwards along a $\langle 100 \rangle$ direction, the analogy with the Rb^+H^- E type vibration suggests that the oxygen in the OH^- vibration moves in-phase rather than anti-phase with its proton. That is the vibration can be regarded either as a composite rotation plus translation of the OH^- ion or, equivalently as a rotation of the ion about a point along the line of the OH^- axis but *beyond* the oxygen nucleus. This is essentially one of the models investigated by Klein *et al.*[110] and found to be far superior to the Devonshire model for this system.

Another interesting experiment on multiple impurities is shown by the work of Clayman and Sievers[121] on the Li^+ resonance mode in mixed alkali halides. Figure 7.65 shows the effect on the resonant mode frequency of the KBr/Li^+ system in crystals which also contain minor concentrations of KCl or KI. This resonance mode shows a sensitivity to the other impurity which is quite different to that shown by the U-centre local mode discussed above, and can only easily be discussed in terms of a perturbed monatomic effect.

The CaF_2/H^- system is another one which has been studied using multiple impurity methods. For example Jones *et al.*[122] present evidence of the

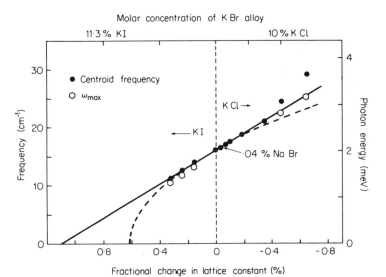

Figure 7.65 The dependence of the absorption frequency of the KBr/Li^+ resonance mode upon the mean lattice spacing of the KBr which also contains minor concentrations of KI or KCl. As well as shifting in frequency, the absorption peak also broadens and becomes asymmetrical and the peak absorbance frequency has therefore been plotted when it was found to be appreciably different from the centroid frequency (Clayman and Sievers)

vibrations of rare-earth/H^+ ion pairs in CaF_2 whereas Newman and Chambers[123] discuss also the more complicated $O^{2-}-Yb^{3+}-H^-$ triply substituted impurity in CaF_2.

Charge compensation is an important consideration in the above two references, as it must always be in examples where introduced impurities carry a charge different from that carried by the ion they replace. This is clearly as true for polyatomic impurities as it is for monatomic impurities, and therefore any examples such as CO_3^{2-} or SO_4^{2-} are automatically multiple impurity problems if they are introduced into halide lattices. Decius[124] identified all the eight fundamental modes of the SO_4^{2-} which are allowed under C_{2v} selection rules for KBr doped with $CaSO_4$ and $BaSO_4$. This indicates the formation of ion pairs in which the M^{2+} and SO_4^{2-} ions occupy adjacent substitutional positions in the lattice and distort the symmetry of the system accordingly.

Figure 7.66 shows the IR absorption of a lithium fluoride crystal which contained OH^- impurities associated with other crystal defects.[38] However,

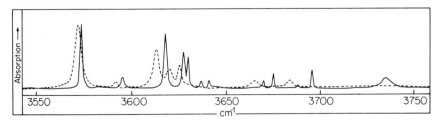

Figure 7.66 Infrared absorption of OH^- in LiF at 300 K is shown by the broken line and at 100 K by the full line, displaying a large number of peaks due to different types of impurity centres

as might be expected for a spectrum of this complexity, it is very difficult to associate the observed peaks with specific multiple impurity configurations.

Another form of multiple impurity which inevitably occurs is that due to aggregation of a desired impurity when for some reason high concentrations are required. This effect has had to be considered for many impurity ions because often they can be introduced into alkali halides up to about the 1 per cent. level, by which time about 10 per cent. of the impurity ions will have other impurity ions on at least one of the 12 nearest 'like-ion' sites. For example Decius, et al.[76] made a careful study of the satellite lines which grow beside the main vibrational lines of the cyanate ion as the concentration is increased.

Conant and Decius[79] studied complexes formed between alkaline earth ions and cyanate ions isolated in alkali halides, and Morgan[125] has made some very interesting studies on lead cyanide pairs in alkali halides where the migration rate of the lead through the alkali halide has been studied at different temperatures by measurements on the rate of growth of the appropriate Pb—CN vibrational absorption band in quenched samples initially showing only isolated CN ions.

7.4 Conclusion

The last decade has seen considerable progress in our understanding of the vibrational spectra of impurities in solids. Already the main features in the infrared spectra of doped crystals can be understood in principle. Effects due to localized modes, resonant band modes and gap modes have been clearly identified in many different studies on ionic and covalent crystals. Considerable work, however, remains to be done before detailed theories are available for many important experimental systems. The solution of these problems demands more realistic theoretical models and very detailed infrared and Raman spectroscopic studies in which temperature, pressure, uniaxial stress and isotopic substitution are all varied, so that the validity of different models may be carefully tested.

From a chemical point of view infrared spectroscopy has proved to be one of the best methods available for the detection of impurities in crystals. Analytical work of this type will become much more important with the introduction of new Michelson interferometers of high light grasp which can cover the whole of the infrared region and are capable of giving very high signal to noise ratios especially when used with cooled detectors. Consequently very weak impurity absorption may be successfully measured even in the presence of quite intense host absorption.

In this article only the simplest of ionic and covalent host lattices have been discussed; however, it is important to emphasize that the techniques reviewed here are applicable for the study of impurities in a very wide range of chemical compounds.

7.5 References

1. I. M. Lifshitz, *J. Phys. USSR*, **7**, 215, (1943).
2. E. Mollwo, *Z. Physik*, **85**, 56, (1933).
3. G. Schaefer, *Phys. Chem. Solids*, **12**, 233, (1960).
4. R. L. Mossbauer, *Z. Physik*, **151**, 124, (1958).
5. A. A. Maradudin, *Solid State Phys.*, **18**, 271, (1966); **19**, 1, (1966).
6. *Proc. International Conference on Lattice Dynamics held at Copenhagen 1963*, ed. R. F. Wallis, Pergamon Press, 1965.
7. 'Phonons in perfect lattices and in lattices with point imperfections', *Scottish Universities 6th Summer School, 1965*, ed. R. W. H. Stevenson, Oliver and Boyd, 1966.
8. 'Localized Excitations in Solids', *Proc. of 1st International Conference on localized excitations in solids, University of California, 1967*, ed. R. F. Wallis, Plenum Press, 1968.

9. 'Localized Modes and Resonance States in Alkali Halides', by M. V. Klein in *Physics of Colour Centres*, ed. W. Beall Fowler, Academic Press, 1968.
10. P. Dean, p. 109–116 and 123–131 in ref. 8 and the references contained within these papers.
11. E. W. Montroll and R. B. Potts, *Phys. Rev.*, **100**, 525, (1955).
12. P. Mayer, E. W. Montroll, R. B. Potts and J. Walsh, *Acad. Sci.*, **46**, 2, (1956).
13. R. L. Bjork, *Phys. Rev.*, **105**, 456 (1957).
14. R. Brout and W. M. Visscher, *Phys. Rev. Letters*, **9**, 54, (1962).
15. Yu. M. Kagen and Ya-A. Iosilershii, *Sov. Phys. J.E.T.P.*, **15**, 842 (1962).
16. S. Takeno, *Progr. Theor. Phys. (Kgoto)*, **29**, 191, (1963).
17. P. G. Dawber and R. J. Elliot, *Proc. Phys. Soc.*, **81**, 453, (1963); *Proc. Roy. Soc.*, **A273**, 222, (1963).
18. R. Loudon, *Proc. Phys. Soc.*, **84**, 379, (1964).
19. A. J. Sievers, p. 29, ref. 8.
20. K. F. Renk, *Z. Physik*, **201**, 445, (1967).
21. R. Weber, Ph.D. Thesis, Physikalisches Institut der Universitat, Freiberg im Breisgau, 1967.
22. R. W. Poll, *Z. Physik*, **64**, 606, (1930).
23. C. Smart, Ph.D. Thesis University of London, 1962, (unpublished but partially included in ref. 24).
24. W. C. Price, G. R. Wilkinson and C. Smart, Final Technical Report No. 2 (1960) on U.S. Army Contract DA-91-591-EUC-130801-4201-60.
25. M. Born and J. E. Mayer, *Z. Physik*, **75**, 1, (1932).
26. B. G. Dick and T. P. Das, *Phys. Rev.*, **127**, 1053, (1962).
27. B. Fritz, p. 485 of ref. 6.
28. A. J. Sievers, p. 1170, *Proc. 9th International Conference on Low Temperature Physics, Ohio, 1964*, Vol. L.T. 9 (Part B), ed. J. G. Gaunt *et al.*, Plenum Press Inc., 1965.
29. T. Timusk, E. J. Woll and T. Gethins, p. 538, ref. 8.
30. G. Benedek and B. F. Nardelli, *Phys. Rev.*, **155**, 1004, (1967).
31. K. Patnaik and J. Mahanty, *Phys. Rev.*, **155**, 987 (1967).
32. G. Dolling, R. A. Cowley, C. Schittenhelm and I. M. Thorson, *Phys. Rev.*, **147**, 577, (1966).
33. G. Gilat and L. J. Raubenheimer, *Phys. Rev.*, **144**, 390 (1966).
34. B. Fritz, J. Gerlach and U. Gross, p. 506, ref. 8.
35. T. Timusk and M. V. Klein, *Phys. Rev.*, **141**, 664, (1966).
36. B. Fritz, U. Gross and D. Brauerle, *Phys. Stat. Sol.*, **11**, 231, (1965).
37. J. A. Harrington, R. T. Harley and C. T. Walker, *Solid State Communications*, **8**, 407 (1970).
38. G. Hirst, Ph.D. Thesis University of London, 1970.
39. D. N. Mirlin and I. I. Reshina, *Sov. Phys. Sol. State*, **C**, 2454, (1965).
40. R. J. Elliot, W. Hayes, G. D. Jones, H. F. MacDonald and C. T. Sennett, *Proc. Roy. Soc.*, **A289**, 1, (1965).
41. J. L. Jacobson, Ph.D. Thesis, University of London, 1965.
42. T. Gethins, T. Timusk and E. J. Woll, *Phys. Rev.*, **157**, 744, (1967).
43. W. Hayes and H. F. Macdonald, *Proc. Roy. Soc.*, **A297**, 503, (1967).
44. S. Takeno, p. 91, ref. 8.
45. A. J. Sievers and S. Takeno, *Phys. Rev.*, **140**, 3A, 1030, (1965).
46. R. D. Kirby, I. G. Nolt, R. W. Alexander and A. J. Sievers, *Phys. Rev.*, **168**, 1057, (1968).
47. B. P. Clayman and A. J. Sievers, p. 54, ref. 8.
48. I. G. Nolt and A. J. Sievers, *Phys. Rev.*, **174**, 1004, (1968).

49. H. Bogardus and H. S. Sack, *Bull. Amer. Phys. Soc.*, **11**, 229, (1966).
50. G. Lombardo and R. O. Pohl, *Bull. Amer. Phys. Soc.*, **11**, 212, (1966).
51. J. P. Harrison, P. P. Peressini and R. O. Pohl, *Phys. Rev.*, **171**, 1037, (1968).
52. F. C. Baumann, J. P. Harrison, R. O. Pohl and W. D. Seward, *Phys. Rev.*, **159**, 691, (1967).
53. A. J. Sievers, A. A. Maradudin and S. S. Jaswal, *Phys. Rev.*, **138**, A272, (1965).
54. R. Weber and F. Siebert, *Z. Physik*, **213**, 273, (1968); and R. Weber, P. Nette, *Phys. Let.*, **20**, 493, (1966).
55. M. V. Klein, p. 504, ref. 9.
56. R. Robertson, J. J. Fox and A. E. Martin, *Phil. Trans. Roy. Soc.*, **A232**, 463, (1934).
57. M. Balkanski and W. Nazarewicz, *J. Phys. Chem. Solids*, **27**, 671, (1966).
58. A. R. Goodwin and S. D. Smith, *Phys. Lett.*, **17**, 203, (1965).
59. W. Hayes, p. 142, ref. 8; and *Phys. Rev. Lett.*, **13**, 275 (1964).
60. I. Maslakowez, *Z. Physik*, **51**, 696, (1928).
61. J. A. A. Ketelaar, C. Hass and J. Van der Elsken, *J. Chem. Phys.*, **24**, 624, (1956).
62. J. C. Decius and A. Maki, *J. Chem. Phys.*, **28**, 1003, (1958).
63. W. C. Price, W. F. Sherman and G. R. Wilkinson, *Disc. Roy. Soc.*, **Dec. 1958**; *Proc. Roy. Soc.*, **A255**, 5, (1960).
64. A. M. Benson and H. G. Drickamer, *J. Chem. Phys.*, **27**, 1164, (1957).
65. A. D. E. Pullin, *Spectrochim. Acta*, **13**, 125, (1958).
66. J. I. Bryant and G. C. Turrell, *J. Chem. Phys.*, **37**, 1069, (1962).
67. G. R. Field and W. F. Sherman, *J. Chem. Phys.*, **47**, 2378, (1967).
68. A. D. Buckingham, *Proc. Roy. Soc.*, **A248**, 169, (1958).
69. E. Bauer and M. Magat, *J. Phys. Radium*, **9**, 319, (1938).
70. E. M. Layton, R. D. Kross and V. A. Fassel, *J. Chem. Phys.*, **25**, 135, (1957).
71. M. Margoshes, F. Fillwalk, V. A. Fassel and R. E. Rundle, *J. Chem. Phys.*, **22**, 381, (1954).
72. M. A. Cundill, Ph.D. Thesis University of London, 1968.
73. W. Holzer, W. F. Murphy, H. J. Bernstein and J. Rolfe, *J. Mol. Spectroscopy*, **26**, 543, (1968).
74. W. C. Price, W. F. Sherman and G. R. Wilkinson, *Spectrochim. Acta*, **16**, 663, (1960).
75. L. J. Bellamy, H. E. Hallam and R. L. Williams, *Trans. Faraday Soc.*, **54**, 1120, (1958).
76. J. C. Decius, J. L. Jacobson, W. F. Sherman and G. R. Wilkinson, *J. Chem. Phys.*, **43**, 2180, (1965).
77. M. A. Cundill and W. F. Sherman, *Phys. Rev. Letters*, **16**, 570, (1966).
78. M. A. Cundill and W. F. Sherman, *Phys. Rev.*, **168**, 1007, (1968).
79. D. R. Conant and J. C. Decius, *Spectrochim. Acta*, **23A**, 2931 (1967).
80. J. C. Decius and D. J. Gordon, *J. Chem. Phys.*, **47**, 1286, (1967).
81. V. Schettino and I. C. Hisatsune, *J. Chem. Phys.*, **52**, 9, (1970).
82. R. A. Cowley, W. Cochran, B. N. Brockhouse and A. D. Woods, *Phys. Rev.*, **131**, 1030, (1963).
83. A. D. B. Woods, B. N. Brockhouse and R. A. Cowley, *Phys. Rev.*, **131**, 1025, (1963).
84. A. J. Sievers, p. 33, ref. 8.
85. H. G. Drickamer, R. A. Fitch and T. E. Slykhouse, *J. Opt. Soc. Amer.*, **47**, 1015, (1957).
86. W. F. Sherman, *J. Sci. Inst.*, **43**, 462, (1966).
87. W. C. Price, W. F. Sherman and G. R. Wilkinson, Final Technical Report No. 4 (1964) on U.S. Army Contract DA-91-591-EUC-2127 01-26489-B.
88. H. G. Drickamer and T. E. Slykhouse, *J. Chem. Phys.*, **27**, 1226, (1957).

89. M. A. Cundill and W. F. Sherman, to be published.
90. W. D. Seward and V. Narayanamurti, *Phys. Rev.*, **148**, 463, (1966).
91. A. F. Devonshire, *Proc. Roy. Soc.*, **A153**, 601, (1936).
92. B. Fritz, U. Gross and D. Brauerle, *Phys. Stat. Sol.*, **11**, 231, (1965).
93. R. Metselaar and J. van der Elsken, *Phys. Rev. Letters*, **16**, 349, (1966).
94. R. Metselaar and J. van der Elsken, *Phys. Rev.*, **165**, 359, (1968).
95. W. F. Sherman, G. R. Wilkinson and J. L. Jacobson. Abstracts from 7th European Molecular Spectroscopy Conference, Budapest, 1963.
96. V. Narayanamurti, *Phys. Rev. Letters*, **13**, 693, (1964).
97. W. P. Seward and V. Narayanamurti, *Phys. Rev.*, **148**, 463 (1966).
98. L. Pauling, *Phys. Rev.*, **36**, 430, (1930).
99. N. Elliot and J. Hastings, *Acta. Cryst.*, **14**, 1018, (1961).
100. J. M. Bijvoet and J. A. Lely, *Rec. Trav. Chem.*, **59**, 908, (1940).
101. V. Narayanamurti, W. D. Seward and R. O. Pohl, *Phys. Rev.*, **148**, 481, (1966).
102. A. J. Sievers and C. D. Lytle, *Phys. Letters*, **14**, 271, (1965).
103. K. F. Renk, *Phys. Letters*, **14**, 281, (1965).
104. T. Timusk and W. Staude, *Phys. Rev. Letters*, **13**, 373, (1964).
105. M. V. Klein, S. O. Kennedy, T. I. Gie and B. Wedding, *Mat. Res. Bull.*, **3**, 677, (1968).
106. S. Akpinar, *Ann. Physik*, **37**, 429, (1940).
107. C. K. Chan, M. V. Klein and B. Wedding, *Phys. Rev. Letters*, **17**, 521, (1966).
108. B. Fritz, F. Lutz and J. Anger, *Z. Physik*, **174**, 240, (1963).
109. B. Wedding and M. V. Klein, *Phys. Rev.*, **177**, 1274, (1969).
110. M. V. Klein, B. Wedding and M. A. Levine, *Phys. Rev.*, **180**, 902, (1969).
111. W. R. Fenner and M. V. Klein, p. 497, 'Light Scattering Spectra of Solids', *Proc. Int. Conf. New York, 1968*, ed. G. B. Wright, Springer Verlag, New York, 1969.
112. R. H. Callender and P. S. Pershan, 'Light Scattering Spectra of Solids', *Proc. Int. Conf. New York, 1968*, ed. G. B. Wright, Springer Verlag, New York, 1969.
113. R. Callender and P. S. Pershan, *Phys. Rev.*, **A2**, 672, (1970).
114. E. L. Wagner and D. F. Hornig, *J. Chem. Phys.*, **18**, 296, (1950).
115. L. F. H. Bovey, *J. Opt. Soc. Amer.*, **41**, 836, (1951).
116. W. Vedder and D. F. Hornig, *J. Chem. Phys.*, **35**, 1560, (1961).
117. W. F. Sherman and P. P. Smulovitch, *J. Chem. Phys.*, **52**, 5187, (1970).
118. W. F. Sherman and P. P. Smulovitch, unpublished data.
119. R. P. Lowndes, *Phys. Letters*, **21**, 26, (1966).
120. G. Amat, p. 383, *Proc. 9th European Mol. Spec. Conf. Madrid 1967*, Butterworths (1969).
121. B. P. Clayman and A. J. Sievers, p. 54, ref. 8.
122. G. D. Jones, S. Yatsir, S. Peled and Z. Rosenwachs, p. 512, ref. 8.
123. R. C. Newman and D. N. Chambers, p. 520, ref. 8.
124. J. C. Decius, *Spectrochim. Acta*, **21**, 15, (1965).
125. H. Morgan, private communication.

8 Infrared and Raman spectra of clathrates

D. C. McKEAN

Contents

8.1 Introduction

The term 'clathrate' is used to describe a compound of variable stoichiometry formed when a small molecular species, the 'guest,' is completely enclosed within a cage formed by the crystal lattice of the 'host' species.[1] Some of the best known examples of clathrates are afforded by the β-quinol compounds with methane, carbon dioxide, sulphur dioxide etc., Dianin's compound with various organic adducts, nickel-II-cyanide ammonia, $Ni(CN)_2 \cdot NH_3$, with aromatic compounds like benzene and pyrrole, and the gas hydrates. Figure 8.1 shows the cage formed by 6 quinol molecules in β-quinol.

The features which perhaps most distinguish these compounds from systems involving species dissolved in inert gas matrices or in ionic crystals, are firstly the stability of the guest crystal involved and secondly, the small degree of interaction between the host and guest molecules. The latter accompanies the feature of clathrates, that only molecules small enough to

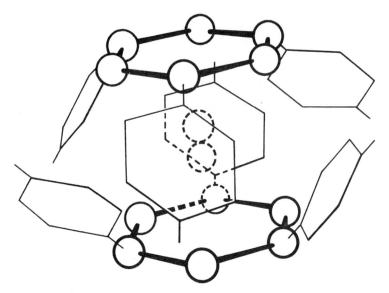

Figure 8.1 Structure of a β-quinol clathrate cage, containing a trapped diatomic molecule. (Reproduced from reference 20 with the permission of the Journal of Chemical Physics and H. Meyer and J. C. Burgiel).

fit easily into the cavities in the lattice are in fact incorporated in the usual method of preparation by recrystallization. The crystal stability makes it readily accessible for examination by X-ray diffraction; there is usually then no doubt as to the size and shape of the cavity occupied by the absorbing molecule. It also means that the system can be studied at ordinary temperatures, when the absorbing molecule will possess considerable thermal energy.

A closer consideration of the term clathrate reveals no basic physical distinction between a system such as that described above and the one in which there may be some freedom of motion for the guest molecule, as in so-called *channel-adducts* such as those formed between urea and hydrocarbons, or in *interlayer* compounds such as those formed by graphite or clay minerals. In fact, by adding zeolites to this series, one can establish a gradation of properties all the way from clathrate compounds to simple physical adsorption. The scope of the present chapter can conveniently be limited by the existing monographs on the latter subject.[2,3]

8.2 Historical survey

8.2.1 Near-infrared region ($> 200 \, \text{cm}^{-1}$)

The first infrared study of clathrates was reported by Hexter and Goldfarb in 1957, on the systems β-quinol with HCl, H_2S, SO_2 and CO_2 respectively.[4]

In the first two of these, no absorption by the guest molecules could be detected due to the intense absorption by the host. In 1962 Ball and McKean[5] carried out a study of the CO-quinol clathrate, while in 1965 Davies and Child[6] made a much more extensive survey of eight quinol clathrates (CH_3OH, CH_3CN, $HCOOH$, CH_3Cl, CH_3F, SO_2, HCl and DCl) and another eight formed by Dianin's compound (C_2H_5OH, $HCOOH$, CH_3OH, CH_3COOH, $CH_3(CH_2)_4COOH$, CH_3CN, CH_3I and NH_3). The quinol-formic acid clathrate has also been studied by Gosavi and Rao,[7] together with a urea-n-heptylazide system.

Another extensive investigation is the recent one by Casellato and Casu,[8] who investigated the spectra of a large number of mono- and disubstituted benzenes clathrated in Ni^{II} or Co^{II} thiocyanate-γ-picoline$_4$ cages. This followed earlier work on similar systems by Aynsley *et al.*,[9] Drago *et al.*,[10] Hart and Smith[11] and Leysen and Van Rysselberge.[12] The work of Bhatnagar[13] also falls in this group. Some isolated studies are those of SF_6 in Dianin's compound,[14] of the SO_2 hydrate,[15] and of the thiourea-camphor channel adduct.[16] The latter is related to a number of investigations of the adducts of urea with n-paraffins (reference 17 and references therein).

Finally, reference must be made to the related studies[18,19] of the interlayer compounds formed by the clay mineral montmorillonite with aromatic compounds.

8.2.2 Far-infrared region ($< 200 \text{ cm}^{-1}$)

An extensive survey of β-quinol clathrates including the guest species Ar, Kr, Xe, CO, N_2, NO, O_2, CH_4, CH_3F and NF_3 was carried out at low temperatures by Burgiel, Meyer and Richards,[20] while a detailed study of the HCl-quinol compound was made by Allen.*[21] More recently, the study of the HCl compound has been repeated by Davies[22] and by Barthel, Gerbaux and Hadni.[23] The former also investigated the N_2, CO, HBr, H_2S, CO_2, SO_2, HCN and CH_3CN quinol clathrates; the latter H_2S, CH_3CN and CH_3OH.

8.3 Techniques

Since most of the compounds involved are stable in the atmosphere at ordinary temperatures, the usual mull or pressed disc techniques for preparing specimens for infrared study may be employed.

However, the SO_2 hydrate was prepared by condensing the guest and water vapours together on to a cold window in the manner usual for the investigation of polycrystalline films of volatile materials.[15]

Apart from their advantage of stability, clathrates have one serious disadvantage as compared for instance to noble gas matrices: the absorption spectrum of the host is usually extensive and may altogether obscure that of

---

* Allen's paper also contains spectra of the SO_2, N_2 and HBr clathrates, of indifferent quality.

the guest species. Moreover, if the mull or potassium bromide disc technique is used, mismatching of the refractive indices of the clathrate crystals and the paraffin or potassium bromide medium may lead to variations in the scattering of the infrared beam with frequency. Thus Christiansen effects are marked in the β-quinol series when the disc material is potassium bromide.[5] Use of the appropriate solid solution of potassium bromide and sodium bromide could have eliminated them altogether.

In the far-infrared study of Allen,[21] who employed a modulated lamellar grating interferometer, an elaborate approach was adopted to eliminate scattering and reflection effects.

First, the transmission of the sample powder dispersed in paraffin was compared with that of a similar thickness of pure paraffin, to eliminate reflection loss and a weak absorption due to paraffin itself. Then the contribution of scattering by the crystals to the attenuation was estimated theoretically for a random distribution of particle sizes up to the maximum permitted by the sieve employed.

Finally the remaining attenuation coefficient was interpreted in terms of the real and imaginary parts of the dielectric constant of the clathrate, by means of a Hilbert transform, due allowance being made for the modification by the paraffin of the field inside the clathrate crystals.

By contrast with the above, in the repetition of this work both by Davies[22] and Barthel et al.,[23] a Michelson interferometer was employed. In both of these latter investigations, the spectra appear to be free of the noise-like features seen in Allan's spectra and therefore of better quality.

Davies also introduced an advance in sampling technique by his success in forming pressed discs of *pure clathrate*, as opposed to clathrate dispersed in potassium bromide or similar medium, thereby achieving a higher proportion of absorbed to scattered light.*

In all four far-infrared investigations, and in one near-infrared one,[5] spectra were obtained at low temperatures.

8.4 Molecular motion in clathrates

The stability of clathrates has enabled a wide range of thermodynamic and other physical studies to be carried out, which bear on the nature of the rattling, rotational or even diffusional motions of the guest molecules. Earlier reviews on this field are those of Staveley[24] and of Child.[25] Thus heat capacity data have shown for instance that barriers to rotation of small molecules such as CH_4, O_2 and N_2 in the quinol cage lie in the range 0–400 cm^{-1} per molecule. More recently, NMR studies[26] have provided useful information.

In considering the value of infrared studies, by far the most important to the subject of molecular motion have been the far-infrared investigations of the quinol clathrates.

* Unfortunately, Barthel et al. do not state what sampling arrangement they employed.

8.4.1 Far-infrared studies

Table 8.1 lists nearly all of the far-infrared β-quinol clathrate frequencies so far studied.

Table 8.1 Some far-infrared absorptions in β-quinol clathrates

Guest species	Freq. (cm^{-1})	Temp. (K)	Assignment	a^e	d_0^e (Å)	$2V_0^f$ (cm^{-1})	Ref.
Ar	35·5	1·2	rat.	23·2	2·4		a
Kr	36·0	1·2	rat.	35·5	2·0		a
Xe	43·5	1·2	rat.	high	~2		a
N$_2$	53·5	1·2	rat.	35·5	2·8		a
	52·5	90	rat.	12	1·7 (ass.)		c
O$_2$	40·0	1·2	rat.	16·9	2·1		a
NO	46·5	1·2	rat.	30·5	2·7		a
	33·0	1·2	libr.			160	a
CO	55·2	1·2	rat.	109·5	4·9		a
	51·5	90	rat.	16	2·0 (ass.)		c
	81·5	1·2	libr.			860	a
CO$_2$	74	90	rat.	18	1·4 (ass.)		c
SO$_2$	78	4·2	rat.	12	0·9 (ass.)		c
	30	4·2	libr.		?		c
HCN	20–90		free rotn?				c
HBr	43	90	rat.	18	1·2 (ass.)		c
HCl	52	1·2	rat.				b
HCl	52	18	rat.				d
HCl	52·5	4	rat.				c
HCl	39	90–300	$J = 1 \to 2$				c
HCl	'continuum'	80	$J = 1 \to 2$ and higher transitions				d
HCl	29	4·2 (66%)^g	$J = 0 \to 1$ (or libr.)				c
HCl	25	18 (73·5%)^g	$J = 0 \to 1$				d
HCl	18–27	1·2 (7–77%)^g	$J = 0 \to 1$				b
H$_2$S	71·5	90	?				c
H$_2$S	75	18–80	rat.				d
H$_2$S	55	90	?				c
H$_2$S	57	18–80	rat.				d
CH$_4$	82	1·2	?				a
	31	1·2	?				a
CH$_3$OH	73	80	rat.				d

a. Ref. 20.
b. Ref. 21.
c. Ref. 22.
d. Ref. 23.
e. Parameters of the modified Pöschl–Teller equation, $E_n = (h^2/8md_0^2)/(a + n)^2$.
f. Barrier height, from $2V_0 = v_{rot}^2/4B_e$, $V = V_0 (1 - \cos 2\theta)$.
g. Extent of cage occupancy.
rat. = rattling, libr. = librational, ass. = assumed.
N.B. Data for the CH$_3$CN clathrate (see text and refs. 22, 23) are omitted, as it is likely that its frequencies arise from lattice modes.

In the work of Burgiel et al.,[20] absorptions attributed to *translational* or rattling motions of the guest species, Ar, Kr, Xe, CO, N_2, NO and O_2 were found in the spectral range 35–55 cm^{-1} (see Figure 8.2 and Table 8.1); while for CO and NO, a second peak, at 81·5 cm^{-1} and 33·0 cm^{-1} respec-

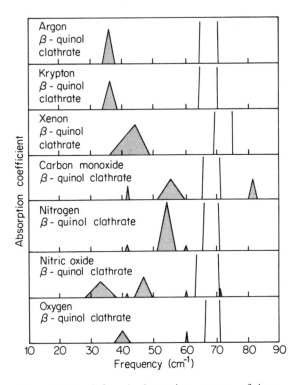

Figure 8.2 Far-infrared absorption spectra of inert gases and diatomic molecules in β-quinol clathrates between 15 cm^{-1} and 90 cm^{-1}, at 1·5 K. The shaded lines represent absorptions attributed to the trapped molecules, the unshaded ones being those of the host lattice. Specimens—crystals in rubber cement. (Reproduced from reference 20 with the permission of the Journal of Chemical Physics and H. Meyer and J. C. Burgiel)

tively at 1·2 K, was found which was assigned as the restricted rotational or *librational* mode of the trapped molecule. All spectra contained in addition the strong β-quinol lattice band near 67 cm^{-1}.

Several arguments were used to support the above assignments. Thus the rattling frequencies for CO, N_2, NO and O_2 reflected approximately the molecular weights of these molecules.

The librational frequency of 33 cm^{-1} for NO agrees excellently with the value of 31 cm^{-1} calculated from magnetic susceptibility measurements,[27] while the relative intensities of the 81·5 cm^{-1} and 33 cm^{-1} bands in CO and NO agreed with the ratio calculated from their molecular dipole moments.*

However in a given situation, these two types of motion may be coupled together, as Burgiel *et al.* suggest may occur in the case of the N$_2$ clathrate, where the librational frequency deduced from nuclear quadrupole resonance measurements is 52 cm^{-1}, in almost exact coincidence with the infrared frequency at 54 cm^{-1}, assigned above to the rattling mode.†

The identification of a rattling mode naturally prompts an attempt to describe the potential function governing this motion. Burgiel *et al.* employ a one-dimensional modified Pöschl-Teller potential function in the form

$$V(x) = h^2 a(a - 1)/8md_0^2 \sin^2 \pi[(x/d_0) - \tfrac{1}{2}]$$

where m is the mass of the particle, d_0 the diameter of the box (or the distance through which the molecule can move) and a is a parameter governing the stiffness of the walls whose minimum value is 1. The energy levels for this potential are given by

$$E_n = (h^2/8md_0^2)(a + n)^2 = B(a + n)^2$$

The two parameters of this function, B and a, were obtained by assigning the absorption maximum at low temperatures to the $n = 0 \rightarrow 1$ transition, and then fitting the broadening and shift to higher frequency of the rattling mode which they observed when the temperature rose. As seen in Table 8.1, the a values so obtained lay between 16 and 110, indicating a nearly harmonic oscillator potential. The d_0 values which their results for B imply, tend to be rather high, the value for CO of 4·9 Å, being unrealistically so.

With this model for the rattling motion and a conventional one for the libration, heat capacities were calculated for these clathrates and found to be in fair agreement with the observed ones, deviations being ascribed to the failure either to consider correlations between the rattling motions of neighbouring molecules or to take into account rattling-librational coupling.

The Pöschl–Teller model is also used by Davies in his treatment of the rattling frequencies in the N$_2$, CO, CO$_2$, HCl, HBr, H$_2$S and SO$_2$ quinol clathrates. However he adopts the more cautious approach of assuming values of d_0 based on van der Waals radii and calculating the second parameter a from the appropriate peak frequency (Table 8.1). For the N$_2$, CO

---

* It may be noted that the barrier height of 860 cm^{-1} in the CO clathrate which is implied by $v_{rot} = 81·5$ cm^{-1} agrees better with the high temperature calorimetric value obtained by Stepakoff and Coulter[28] of about 1000 cm^{-1}, than with their low temperature value of 260 cm^{-1}, or that of Grey and Staveley[29] of 420 cm^{-1}.

† However, common sense would argue that if these two assignments for the N$_2$ clathrate are simultaneously correct, then the motions must be remarkably *un*coupled.

and CO_2 clathrates these a values are 12, 16 and 18 respectively. Of the two peaks found in the SO_2 clathrate at 4 K, at $30\,cm^{-1}$ and $79\,cm^{-1}$ respectively, the first yields an a value of 2, the second, one of 12, for $d_0 = 0.9$ Å. Thus the second, higher, frequency is likely to be the rattling one and the lower, a librational one. That the latter is a restricted rather than a free rotation follows from the relatively slight effect of changing the temperature from 300 K to 90 K. In an asymmetric molecule like SO_2, the type of librational mode involved in the band at $30\,cm^{-1}$ is not immediately obvious. Taking an approximate B value of $2\,cm^{-1}$, and N_1 the number of potential minima per revolution of the molecule as 2 or 3, the barrier height would be roughly $120\,cm^{-1}$ or $50\,cm^{-1}$. Dielectric relaxation studies[30] merely indicate a value less than $350\,cm^{-1}$. The two absorptions described in H_2S at $71.5\,cm^{-1}$ and $55\,cm^{-1}$ yield a values of 7 and 5 respectively, which does not permit an easy decision as to which is the rattling motion. Possibly both are librational modes and the rattling frequency is $> 100\,cm^{-1}$ and unobservable.*

Undoubtedly the clathrate of greatest interest is the HCl-β-quinol one, studied in the separate investigations of Allen,[21] Davies[22] and Barthel et al.[23]

In addition to the peak near $55\,cm^{-1}$, which is due to a rattling mode, since it is almost unaffected both by change in temperature and by deuteration, all three investigations show a band near $20\,cm^{-1}$, assigned to the $J = 0 \rightarrow 1$ transition of freely rotating HCl molecules. Davies also observes a second peak near $39\,cm^{-1}$ (at 90 K and 300 K) which he attributes to the $J = 1 \rightarrow 2$ transition, since it disappears on cooling to 4 K. Barthel et al. state that at 18 K the transition $J = 1 \rightarrow 2$ broadens the high frequency wing of the band, and that at higher temperatures only a broad continuous absorption is found resulting from rotational transitions from the J levels 0, 1, 2 and 3. There is thus a significant disagreement between their experimental results and those of Davies, which may perhaps be attributable to a difference in sampling technique. Spectra from the first two sources are compared in Figures 8.3 and 8.4.

In Allen's work, the $0 \rightarrow 1$ band is altered in both peak frequency and half width when either the temperature or the extent of occupation of the cages was altered. Thus at about 2 K the frequency increases from $18\,cm^{-1}$ at 7 per cent. occupancy (half width $10\,cm^{-1}$) to about $28\,cm^{-1}$ at 77 per cent. occupancy (half width $25\,cm^{-1}$). At the same occupancy of 77 per cent. the frequency fell from $28\,cm^{-1}$ at 2 K to $24\,cm^{-1}$ at 17 K (half width $38\,cm^{-1}$). This fall in frequency with rise in temperature is confirmed by Davies, but not the change in width, although this may have been due to the cut-off just below $20\,cm^{-1}$ in the spectral range of the latter's interferometer. Davies was also unable to detect any rotational absorption below about 40 per cent.

---

* Barthel et al.[23] merely express the opinion that these two bands are probably due to translational modes. An additional feature near $20\,cm^{-1}$ shown in their Figure 3, is not commented on in their text.

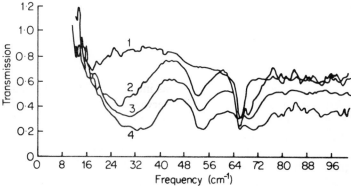

Figure 8.3 Far-infrared transmission in the HCl-β-quinol clathrate, in paraffin mull at 1·2 K, for varying HCl concentration (cage occupancy): (1) 7 per cent., (2) 40 per cent., (3) 62 per cent., (4) 77 per cent. (Reproduced from reference 21 with the permission of The Journal of Chemical Physics and S. J. Allen)

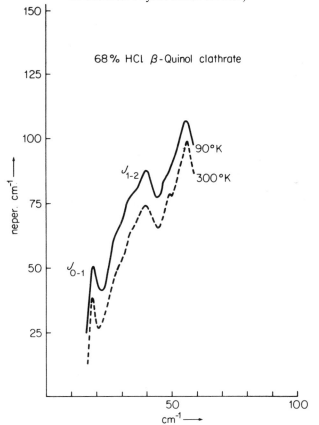

Figure 8.4 Far-infrared absorption spectra of HCl-β-quinol clathrate, in pure pressed disc (cage occupancy 68 per cent.). (Reproduced from reference 22 with the permission of The Faraday Society)

occupancy, in marked contrast with Allen. Allen's HCl absorption intensities are in general very much higher, which raises doubts about their authenticity.

Allen has interpreted his results in terms of two types of interaction.

The observation of a $0 \to 1$ transition of 18 cm^{-1} at 1.2 K in a clathrate of 7 per cent. occupancy, which is about 3 cm^{-1} lower than the gas value, he attributes to translation–rotation coupling, and applies the theory of Friedmann and Kimel (§4.2.3). The changes in frequency with change in concentration or temperature he attributes to dipole–dipole interactions between HCl molecules in adjacent cages, which cause a particular molecule in an excited J level to lose its rotational quantum to a neighbouring ground state molecule. The number of the latter kind increases both with fall in temperature and with rise in concentration so that these two effects modify the peak frequency in the same way.

The alternative suggestion has been made by M. Davies[38] that the band near 26 cm^{-1} seen below 20 K is not in fact to be described as due to the $J = 0 \to 1$ transition of a free rotor, but to the librational motion of a restricted rotor, the change from restricted to free rotation with change in temperature arising from a change in population of the rattling mode levels. The present author considers that this might come about if the rattling motion were governed by a double-minimum potential energy function having a low barrier greater than or equal to 50 cm^{-1}.

For the HBr quinol clathrate, Davies observes only one absorption near 43 cm^{-1}, which he ascribes to the rattling motion (see Table 8.1). A similar assignment is given by Barthel *et al.* to the single band at 73 cm^{-1} in the CH_3OH clathrate, at 80 K. For the HCN clathrate, Davies sees a broad absorption extending from 20 cm^{-1} to 90 cm^{-1}, with a number of minor peaks, which may represent the contour of unresolved rotation lines.

Several quinol clathrates containing larger molecules than the above were also investigated by Burgiel *et al.*, namely CH_3F, CHF_3, NF_3 and CH_4. It would appear that part at least of the absorptions found are due to the distortion of the lattice resulting from too large a guest molecule; indeed the absorption in several cases becomes continuous throughout the range 20–100 cm^{-1}. The CH_3CN molecule is even bulkier than those mentioned above, and Davies finds in the CH_3CN quinol clathrate a lattice spectrum quite different from that of the normal β-quinol one, which may indicate the presence of an alternative γ phase.* In concluding this section it seems clear that the considerable experimental difficulties encountered in making these measurements are still a handicap to detailed interpretations of the spectra, although the basic features of the spectra are not in doubt.

---

* For this clathrate, Barthel *et al.* list five new bands, 40 cm^{-1}, 46 cm^{-1}, 57 cm^{-1}, 77 cm^{-1} and 85 cm^{-1} at 80 K, in addition to the usual one near 65 cm^{-1}.[23]

8.4.2 Near-infrared studies

By contrast with the above, the information so far derived from near-infrared vibrational spectra is much more limited. This is due in part at least to the difficulty of finding solute molecules having vibration bands of suitable frequency and intensity to be clearly distinguishable from the absorption of the cage itself. This consideration for instance has made it so far impossible to study HX molecules in β-quinol. For heavier molecules, the much smaller B values mean that the resolution of separate rotational transitions, under conditions where these are likely to have considerable width, is very unlikely.

The most that can be expected in a spectrum is an unresolved envelope of transitions, and how to determine whether such absorption arises from rotational, librational or translational motion is not obvious.

These considerations are illustrated by the study of the fundamental vibration band of CO in β-quinol.[5]

Despite the high value of $d\mu/dr$ for this molecule (about $2\cdot4\,\text{D}/\text{Å}$) the infrared spectrum in Figure 8.5 of a sample of about 60 per cent. occupancy shows only a weak band attributable to the carbon monoxide molecules. At room temperature this band has a strong central maximum or Q branch at $2133\,\text{cm}^{-1}$, with, in addition (Figure 8.6), weak shoulders on each side distant about $40\,\text{cm}^{-1}$ from the centre, an appearance strongly reminiscent of the spectra of CO and other diatomic molecules in liquid solutions.[24]

As the sample was cooled progressively to $-130\,°\text{C}$, the Q branch became narrower and the shoulders less pronounced, a narrower high frequency band at $2180\,\text{cm}^{-1}$ being still present at the lowest temperature studied. The latter was assigned as a combination band involving the rattling motion of the CO molecule in its cage. Bearing in mind the difference in temperature,

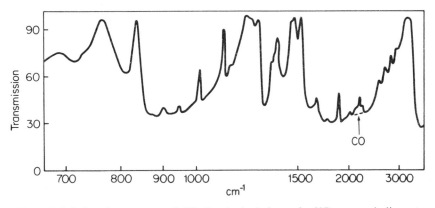

Figure 8.5 Infrared spectrum of CO-β-quinol clathrate in KBr pressed disc, at 20 °C. (Reproduced from reference 5 with the permission of Spectrochimica Acta)

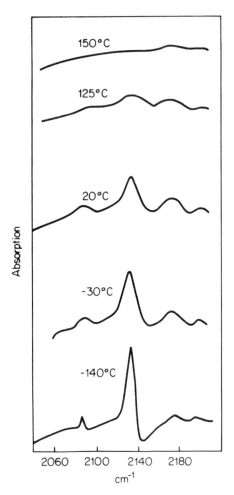

Figure 8.6 Infrared spectra of CO-β-quinol clathrate in KBr pressed disc, at various temperatures. (Reproduced from reference 5 with the permission of Spectrochimica Acta)

the value of 47 cm^{-1} for the rattling frequency is in excellent agreement with the far infrared values in Table 8.1.

The wings prominent at room temperature may also be combination and difference bands involving the rattling frequency, in which case the latter has fallen to about 42 cm^{-1} at room temperature. However they *may* also arise from vibration–rotation transitions, as has been postulated generally for wings seen in liquid solutions. The $P - R$ separation of about 85 cm^{-1}, if such it is, would then be rather greater than the value of 55 cm^{-1} found in the gas phase at room temperature, a feature also found in the liquid solutions.[31]

Ball and McKean suggested that this might be due to a barrier to rotation of the order of magnitude of the rotational energy $BJ_m(J_m + 1)$ for the J_m value corresponding to the maximum of the P or R branch.* The value of the barrier height, $2V_0$, would then be about 240 cm^{-1}. This is much lower than the value of 860 cm^{-1} indicated by the far-infrared torsional frequency of 82 cm^{-1}, and the assignment of these side bands to transitions involving the rattling frequency is perhaps more plausible.

However in a discussion of the internal rotation of XH_3 groups in the solid state, Leech et al.[32] suggest that a more appropriate expression for the barrier height is $2BJ_m(J_m + 1)$, citing the discussion of this problem by Pitzer.[33] It seems to the present author that in Pitzer's discussion the fact that the quantum numbers of states above the barrier must reflect the number of states below it, right down to the ground torsional state, is ignored. It is however hard to say what the effect of this would be without making detailed calculations for a specific model, since the lowest torsional states are here spaced far more widely than the free rotational ones of low J would have been. A further difficulty lies in assessing contributions to the wing intensity from combinations and difference bands due to the torsional frequency. It seems unlikely that the transition moment falls discontinuously to zero as the barrier is approached from above.

The only other clathrate systems in which comparable side bands have been found are those of the HCOOH-quinol compound[6] where wings lie at about 25 cm^{-1} from the central peak of the carbonyl stretching band (31 cm^{-1} according to Gosavi and Rao[7]); and the SO_2 hydrate[15] where shoulders 10 cm^{-1} from the centre peak due to v_3 are seen. The first of these yields a barrier of 420 cm^{-1} [6] or 1230 cm^{-1},[7] depending on (a) whose data are assumed, (b) whether $BJ_m(J_m + 1)$ or $2BJ_m(J_m + 1)$ be taken as the measure of the barrier height. The latter estimate is quite close to the value of 1430 cm^{-1} from dielectric relaxation.[6] The SO_2 hydrate wing frequencies indicate barriers of about 70 cm^{-1} or 140 cm^{-1}, which lie within the range expected from dielectric studies.[6]

In several other quinol clathrates, where free rotation of the guest molecules might have been expected, it is apparently absent. Thus in the CH_3F

---

* It is worth here recording the observation of Davies and Child,[6] that the approximate barrier height

$$2V_0 = BJ_m(J_m + 1)$$
$$= 2\pi^2 Ic(v_R - v_Q)^2/h$$

is the same expression as that used to relate V_0 to the librational frequency, if $v_R - v_Q$ is replaced by v_{rot}. Thus, the barrier height could be determined from the sidebands, whether their origin was free-rotational or torsional (not, of course, if they are due to rattling).

one,[6] the CF stretching mode has a half width of only $10 \, cm^{-1}$ at room temperature, compared with the value of $25 \, cm^{-1}$ in liquid solution which has been interpreted[39] as signifying free rotation about axes perpendicular to the three-fold one. This is a little surprising in view of the fact that the van der Waals length of the CH_3F molecule is similar to that of CO; its 'thickness' is of course greater.

Similarly in CH_3CN, the narrowness of the perpendicular band at 1036 cm^{-1} [6] implies loss of freedom to rotate about the three-fold axis which otherwise is present in some liquid solutions.

In the spectrum of CO_2 in quinol[4] the intense band due to v_3 has a contour described as a broadened Q branch, again implying restriction of rotation.

8.5 Vibration frequencies in clathrates

Apart from the feat of actually distinguishing an infrared vibration band in a clathrate due to a guest molecule, some interest attaches to the precise frequency found. This will bear of course on the state of the molecule.

Thus the OH and CO stretching frequencies in CH_3OH and HCOOH quinol clathrates lie in the ranges expected for monomeric species, indicating that the guest molecules are not hydrogen bonded to the quinol walls.[6]

For CH_3COOH in Dianin's compound, two $v_{C=O}$ bands appear to be present, which may indicate the presence of both monomeric and dimeric forms.[6]

In general, however, the frequencies found are similar to those in liquid solutions; bond stretching modes are nearly always lower in clathrates than in the gas, as might be expected from the relative freedom they enjoy.*

The knowledge of the shape and size of the cage prompts a test of theories such as the KBM one which presupposes an oscillating dipole moment situated at the centre of a cavity in a dielectric medium. If an absolute calculation from the KBM equation is made for the CO-frequency in quinol, a downwards shift of about $4 \, cm^{-1}$ is predicted, compared with the observed one of $10 \, cm^{-1}$ ($2143-2133 \, cm^{-1}$).[5] The relatively much greater shift of $32 \, cm^{-1}$ in the CF stretch of CH_3F ($1048-1016 \, cm^{-1}$) is still in line with the shifts found in liquid solutions if, in the 'relative' use of the KBM equation in which $\Delta v/v$ is plotted as a function of $(\varepsilon - 1)/(2\varepsilon + 1)$, a dielectric constant of 3·2 is used for the quinol.[6]

The much smaller shift of $3 \, cm^{-1}$ found for the CCl stretch in CH_3Cl is likely to be due to the greater size of the latter molecule, which will increase the importance of the repulsion forces.[6] A similarly small shift of $9 \, cm^{-1}$ is found for v_3 of CO_2.[4]

---

* Exceptions to this generalization are v_{C-O} in HCOOH and v_{C-C} in CH_3CN, both in quinol, both of which are slightly higher than in the gas.[6]

A comparison of the frequency shifts exhibited by CO in quinol and in noble gas matrices may bear on the assignment of the bands in the latter. The size of the quinol cage is about 4 Å, very close to the diameter of a xenon atom. Thus similar shifts might be expected for the clathrate and for the CO on a substitutional site in xenon. This would entail assignment of the broader, lower frequency band at $2133{\cdot}4$ cm^{-1} in xenon to the CO on a substitutional site, in agreement with judgment of Charles and Lee.[34] However, other evidence[35] now suggests a different assignment for this band (i.e. to a CO aggregate), so that the comparison may not be valid.

Amongst the studies of large molecules in clathrates, the work of Casellato and Casu[8] is essentially devoted to the frequency shifts found in the out-of-plane CH bending modes of mono and di-substituted benzenes in passing from CS$_2$ solution to the Ni- or CO(NCS)$_2$-γ-picoline$_4$ clathrates. These γ_{CH} modes rise in frequency by amounts varying from 1 cm^{-1} to 11 cm^{-1}. The sum of the shifts $\Delta\nu$ for a particular mono-substituted benzene can be correlated with the sum of the Hammett constants $\sigma_m + \sigma_p$ for the substituent. This is taken to be an indication that charge-transfer forces exist between the host γ-picoline and the guest benzene compound, and that the former is the electron-releasing partner of the complex. However, the possibility of 'amphoteric' behaviour by each partner is also suggested.

Similar rising of out-of-plane CH bending mode frequencies have been found in other systems.[10,11,12] Splittings of these bands have recently been examined by Borkowska et al.[40]

8.6 Infrared intensities in clathrates

Some qualitative observations of intensity changes between the liquid and clathrate phases are of interest. In the urea-u-paraffin adducts, the intensity of the CH$_2$ rocking mode near 730 cm^{-1} relative to the CH stretching band is not enhanced on passing from liquid solution to the adduct, as it is on going to the pure crystal state. Aynsley, Campbell and Dodd,[9] in the spectrum of benzene in nickel-cyanide-ammonia, comment on the reduction of CH stretching intensity passing from liquid to clathrate. They also ascribe strong bands at 1573 cm^{-1} and 1166 cm^{-1} to the E_{2g} modes which are infrared inactive in both gaseous and crystalline phases. Some other instances of the appearance of otherwise infrared-inactive bands are mentioned by Bhatnagar.[13]

Clearly there is a limited potential use of clathrate spectra in assisting the task of vibrational assignments. It seems relevant to mention at this point the studies of Farmer et al. on the interlayer compounds formed between aromatic molecules such as nitrobenzene and benzoic acid and the clay mineral montmorillonite.[17,18] In the specimens of the latter prepared for infrared examination, the silicate planes are nearly parallel to the supporting window.

By studying the changes in intensity of the guest absorption bands when the angular position of the window relative to the light beam is changed, information can be obtained on the orientation of the planes of the molecules relative to the silicate layers. Alternatively, when this orientation is known, the direction of the dipole moment change within the absorbing molecule may be found for particular bands and important evidence thereby gained as to their assignment.

The feasability of such a study requires a clathrate with a highly asymmetric cavity, together with an ability to orient the crystal or crystals, but the more general application of this technique should not be overlooked.

8.7 Raman spectra of clathrates

The study of the Raman spectra of crystals has been made very much easier since the advent of the laser source, and one recent paper by J. E. D. Davies,[36] on N_2, O_2, HCl, HBr, CO_2, SO_2, H_2S, C_2H_2, HCOOH, CH_3OH, CH_3CN and CD_3CN in β-quinol demonstrates the advantages of this technique in studying clathrates.

These include not only the normal ability to study infrared-inactive modes, but also the very different intensities of the host bands. Thus in β-quinol, the hydrogen-bonded OH stretching bands which are so strong in the infrared spectrum that they mask absorption by guest HCl or H_2S molecules are weak in the Raman effect, thereby allowing the stretching modes of these species to be seen. Elsewhere in the Raman spectrum of β-quinol, the host bands appear to be much sharper than in the infrared, which provides greater opportunities for observing guest molecule bands.*

In the Raman spectrum of the HCl clathrate, a considerable decrease is seen in the half width of the HCl band at 2805 cm^{-1}, from 16 cm^{-1} at room temperature to 7 cm^{-1} at 78 K, and this confirms the diagnosis of free rotation made from the far-infrared spectrum. The HBr clathrate behaves in a similar manner.

A further advantage of the Raman technique is that the same spectrometer can be used to study also the 0–100 cm^{-1} region where $\Delta J = +2$ transitions may be found for molecules like HCl. Such studies are currently being carried out.[37]

The author is greatly indebted to Professor Mansel Davies and Dr. J. E. D. Davies for communicating details of their far infrared and Raman studies in advance of publication.

---

* This may be associated with the well-known phenomenon that combination, difference and overtone bands are usually much weaker in the Raman effect than in the infrared.

8.8 References

1. *Non-Stoichiometric Compounds*, ed. L. Mandelcorn, Academic Press, New York, 1964.
2. L. H. Little, *Infrared Spectra of Adsorbed Species*, Academic Press, London, 1966.
3. M. L. Hair, *Infrared Spectroscopy in Surface Chemistry*, M. Dekker, New York, 1967.
4. R. M. Hexter and T. D. Goldfarb, *J. In. Nucl. Chem.*, **4**, 171, (1957).
5. D. F. Ball and D. C. McKean, *Spectrochim. Acta*, **18**, 933, (1962).
6. M. Davies and W. C. Child, *Spectrochim. Acta*, **21**, 1195, (1965).
7. R. K. Gosavi and C. N. R. Rao, *Indian J. Chem.*, **5**, 162, (1967).
8. F. Casellato and B. Casu, *Spectrochim. Acta*, **25A**, 1407, (1969). See also ref. 40.
9. E. E. Aynsley, W. A. Campbell and R. E. Dodd, *Proc. Chem. Soc.*, 210, (1957).
10. R. S. Drago, J. T. Kwon and R. D. Archer, *J. Am. Chem. Soc.*, **80**, 2667, (1958).
11. M. I. Hart and N. O. Smith, *J. Am. Chem. Soc.*, **84**, 1816, (1962).
12. R. Leysen and J. Van Rysselberge, *Spectrochim. Acta*, **19**, 237, (1963).
13. V. M. Bhatnagar, *J. prakt. Chem.*, **311**, 302, (1969).
14. L. Mandelcorn, N. N. Goldberg and R. E. Hoff, *J. Am. Chem., Soc.*, **82**, 3297, (1960).
15. K. B. Harvey, F. R. McCourt and H. F. Shurvell, *Can. J. Chem.*, **42**, 960, (1964).
16. V. M. Bhatnagar, *Chem. Ind.* (London), 38, (1965).
17. R. A. Durie and R. J. Harrisson, *Spectrochim. Acta*, **18**, 1505, (1962).
18. S. Yariv, J. D. Russell and V. C. Farmer, *Israel J. Chem.*, **4**, 201, (1966).
19. V. C. Farmer, *Spectrochim. Acta*, **23A**, 728, (1967).
20. J. C. Burgiel, H. Meyer and P. L. Richards, *J. Chem. Phys.*, **43**, 4291, (1965).
21. S. J. Allen, *J. Chem. Phys.*, **44**, 394, (1966).
22. P. R. Davies, Ph.D. Thesis, University of Wales, (1969); *Disc. Faraday Soc.*, **48**, 181, (1969).
23. C. Barthel, X. Gerbaux and A. Hadni, *Spectrochim. Acta.*, **26A**, 1183, (1970).
24. L. A. K. Staveley, ref. 1, Chapter 10.
25. W. C. Child, *Quart. Rev.*, **18**, 321, (1964).
26. E.g. J. P. McTague, *J. Chem. Phys.*, **50**, 47, (1969); C. A. McDowell and P. Raghunathan, *Mol. Phys.*, **15**, 259, (1968).
27. J. H. Van Vleck, *J. Phys. Chem. Solids*, **20**, 241, (1961).
28. G. L. Stepakoff and L. V. Coulter, *J. Phys. Chem. Solids*, **24**, 1435, (1963).
29. N. R. Grey and L. A. K. Staveley, *Mol. Phys.*, **7**, 83, (1963).
30. M. Davies and K. Williams, *Trans. Faraday Soc.*, **64**, 529, (1968).
31. C.f. M. O. Bulanin and N. D. Orlova, *Optics and Spectroscopy*, **15**, 112, (1963), and earlier papers.
32. R. C. Leech, D. B. Powell and N. Sheppard, *Spectrochim. Acta*, **21**, 559, (1965).
33. K. S. Pitzer, *Quantum Chemistry*, p. 243, Prentice-Hall, 1953.
34. S. W. Charles and K. O. Lee, *Trans. Faraday Soc.*, **61**, 614, (1965).
35. J. B. Davies and H. E. Hallam, *J. Chem. Soc., Faraday Trans. II*, **68**, 509, (1972).
36. J. E. D. Davies, *J. Chem. Soc. Dalton*, 1182, (1972).
37. J. E. D. Davies, private communication.
38. M. Davies (private communication) from unpublished work of M. Davies, P. R. Davies and H. A. Poulis.
39. H. E. Hallam and T. C. Ray, *Trans. Faraday Soc.*, **59**, 1983, (1963).
40. Z. Borkowska, J. Lipkowski, W. Wolfram and B. Moszynska, *J. Mol. Structure*, **12**, 265, (1972).

9 Matrix isolation laser Raman spectroscopy

G. A. OZIN

Contents

9.1 Introduction

The number and types of problems involved in Raman spectroscopy of low temperature matrices are considerable, and in this chapter we will attempt to outline some of the problems encountered in the development of the technique. The contents are primarily based on our own experiences, as, at the time of writing, there are only six brief mentions in the literature of the matrix Raman technique,[1-6] dealing mainly with stable species and generally with little or no experimental detail to which we can refer. We will also be concerned here with the production and stabilization by trapping of highly reactive inorganic species in rigid matrices (generally noble gas matrices) at low temperatures and their detection by laser Raman spectroscopy.

The technique of matrix infrared spectroscopy has been known and utilized with notable success since 1954 and the need for reliable Raman data on comparable systems is obvious, for example, in making reliable structural and vibrational assignments for either new chemical systems or for systems whose available data has either been incomplete or ambiguous.

The principal factors responsible for the relatively slow advancement in the field of matrix Raman spectroscopy as compared to the numerous outstanding breakthroughs in matrix infrared spectroscopy, are the intricacies of the necessary experimental procedures in the Raman which are generally of a different nature from those in the infrared. However, we have now had sufficient successes to feel confident that matrix Raman is a viable technique and provided the experiment is performed under closely controlled conditions, data concerning many chemically important systems can be obtained.

In the hope of presenting the following material clearly we divide the chapter into five main parts: (a) a brief history of the matrix method and the techniques which are widely used, (b) the advantages and limitations of the matrix Raman technique as compared to other spectroscopic methods, (c) the types of study which can be undertaken and the information which may be obtained, (d) the development of matrix Raman apparatus and experimental techniques, and (e) recent matrix Raman results and discussion.

9.2 Brief history of matrix isolation techniques

As early as 1885 Widemann and Dewar[7] noted that complex molecules emit phosphorescence when irradiated at low temperatures. Vergard[8] in 1928 studied the luminescence from solid N_2 at 21 K after electron bombardment. Luminescence changes were observed with temperature rise and were referred to as thermoluminescence, that is, the radiation emitted from the reactions of chemically active species upon controlled diffusion. Lewis and Lipkin[9] in 1942, photolysing organic molecules suspended in rigid glassy media at 77 K and 20 K, succeeded in producing stable radicals in the hydrocarbon glass.

The modern era of study began in 1951 when Rice and Freamo[10] reported a blue paramagnetic substance on condensing at 77 K the products of HN_3 which had passed through a glow discharge. Seven laboratories then independently reported their attempts in 1954 to detect free radicals suspended in some sort of rigid matrix.[11] The detection methods used were those of ESR, IR, UV-visible spectroscopy, magnetic susceptibility and calorimetry.

Norman and Porter,[12] and independently Whittle et al.,[13] proposed what is now called the matrix isolation method for studying free radicals or highly reactive molecules. Pimentel *et al.*[14] were the first to report the detection, by infrared spectroscopy, of a free radical (HCO) suspended in a CO matrix support. Both these groups anticipated the wide applicability of the method and defined the technique 'as a means of accumulating a reactive substance under environmental conditions which prevent reaction'. Pimentel's group was first to place the matrix technique on a semi-quantitative basis.[15] They used various matrix supports for a selection of strongly hydrogen bonded solutes (for example, HCl, H_2O, NH_3, etc.) and determined the matrix to solute ratios ($M : A$) and temperatures for which efficient isolation could be achieved.

Linevsky[16] first applied the matrix isolation method to Knudsen cell vapours and showed that many of the difficulties of high temperature infrared work, such as possible reactions with furnace walls, complexity of spectra caused by hot bands and difficulty of achieving a sufficiently high concentration for a long period of time, could be reduced or eliminated by matrix trapping. Weltner's[17] elegant work on matrix isolated C_2, C_3, C_4, etc., in which carbon was vaporized at approximately 3000 K and co-condensed with a matrix gas at 4 K, was an impressive example of the versatility of the technique, which has been discussed in Chapter 6.

The matrix processes of diffusion; the cage and blanket effects; interstitial, substitutional and multiple trapping sites; the theoretical treatment of matrix shifts and secondary reactions have been well documented and are the centre of much interest.[18] Matrix isolation studies involving diffusion controlled reactions have proven to be most useful in yielding mechanistic and structural information. It is found that chemical changes of reactive species suspended in an inert matrix can be traced spectroscopically as the temperature is raised, the onset of reaction indicating that diffusion is no longer prevented by the rigidity of the matrix. Experiment has shown that diffusion becomes rapid at temperatures near $T_M/2$ where T_M = m.p. of matrix material, and that small molecules (as expected) diffuse more rapidly than larger ones. The choice of matrix material is determined by its melting point, reactivity, interfering spectral lines and the size of its substitutional and interstitial sites (see Chapter 2).

Thus highly reactive free radicals or molecules formed in a matrix environment are probably lost by diffusion and reaction if the temperature of the matrix ever reaches or exceeds about $T_M/2$. The need for liquid helium or hydrogen temperatures, especially in laser excited matrix Raman spectra (see later) is of prime importance. The noble gases, Ne, Ar, Kr, Xe, or N_2 generally have been used as matrix supports and reactive matrices such as CO, SF_6, CO_2, C_6H_{12}, $Si(CH_3)_4$, etc., are also becoming widely used. In these cases reaction of the solute species with the *matrix material* may be followed spectroscopically at the onset of a diffusion controlled reaction.

9.3 Advantages and limitations of matrix isolation Raman spectroscopy

The most convenient physical techniques for the detection and study of highly reactive species have been IR, UV-visible, and ESR spectroscopy. Although IR spectroscopy generally lacks the sensitivity of UV spectroscopy it has the advantage of yielding directly the vibrational spectrum of the reactive species. However, for a full vibrational and structural assignment, data from both the IR and Raman spectra are invariably required (notably in the cases of centrosymmetric systems; see, for example, $(SeO_2)_2$, $XeCl_2$ and Br_3 described later) and up to very recently Raman data has not been available.

The primary reasons for the slow development of matrix Raman spectroscopy are the following:

(i) In the pre-laser period of Raman spectroscopy, there were limitations imposed by mercury arc excitation. Generally speaking, one was restricted to colourless or pale yellow matrices, the most suited of which were crystalline rather than opaque. Also, sample handling under cryogenic conditions was experimentally difficult.

(ii) The application of lasers to Raman spectroscopy had certain special advantages in low temperature studies, which related mainly to the physical nature of the sample excitation, and sample handling no longer presented a formidable experimental problem. Even so, the inherent weakness of the Raman effect (approximately 10^{-6} of the intensity of the Rayleigh scattering) and the need to work at concentrations less than 1 per cent. for efficient isolation (often with high background scattering—see later) has undoubtedly slowed down rapid advancement in the field of matrix Raman spectroscopy.

The usefulness of matrix Raman spectroscopy will have far-reaching effects on many types of study, for example:

(i) The study of stable molecules in the absence of perturbing crystal field effects.

(ii) To clarify the ambiguities which often arise in matrix IR studies and the determination of vibrational spectra and molecular structure of new species.

(iii) The study of thermally unstable compounds.

(iv) The study of molecular association and diffusion controlled polymerization processes.

(v) The study of unstable free radicals and their reactions both with other matrix isolated species and with the matrix itself.

Some advantages of the matrix Raman technique over other methods of optical spectroscopy are listed below:

(i) The complete vibrational spectroscopic range ($c.$ 30–3500 cm^{-1}) can be covered in one scan without the need for changing window materials, matrix supports or spectrometer optics. Also, there is no interference from carbon dioxide or water bands, as often found in infrared recordings.

(ii) Vibrational and structural information may often be directly obtained from the frequencies of totally symmetrical Raman active stretching modes without the requirement of detailed analysis. Totally symmetrical modes are easily identified, being usually the strongest and sharpest bands in the Raman spectrum.

(iii) One can monitor different areas of the matrix by scanning the laser beam over the surface of the matrix. In this way one can check the uniformity of the matrix and investigate areas for differing degrees of isolation.

(iv) Polarization measurements are easily made in Raman spectroscopy and, in principle, depolarization ratios may be obtained for species randomly oriented in the host matrix (see for example paragraph 9.6 and the depolarization measurements made on Cl_2^- in alkali halide–alkali borate glasses[19]). Also the possibility of oriented matrix isolated species may be realized by making polarization measurements on the matrix.

(v) Gas phase Raman data which are not accessible because of the high temperatures (> 1000 K) required to attain the necessary vapour pressures (usually > 100 mmHg) may be obtained by high temperature molecular beam techniques (where low vapour pressures of material are required) and subsequent quenching at low temperatures in a matrix environment.

(vi) The determination of Raman data in a matrix environment which is not normally accessible in the gas phase because of resonance fluorescence phenomena which often dominate and obliterate vapour phase Raman spectra. The determination of essentially gas phase Raman data in a matrix environment but in the absence of crystal field perturbations and hot bands.

There are five main factors which *may* limit the success and applicability of matrix isolation laser Raman spectroscopy:

9.3.1 Matrix: solute concentration limit

We have been able to detect 100 ppm of solute in a matrix under favourable conditions (see below) although matrix to solute ratios of less than 1000 : 1 will not generally be accessible with conventional detection techniques. However, ratios of about 500–100 : 1 are often sufficient to obtain successful isolation which can be tested by: (a) absence of solute lattice modes, (b) removal of crystal field effects, (c) observation of monomer which polymerizes on the onset of diffusion.

Before we commenced our programme of matrix isolation Raman spectroscopy it was crucial to ascertain the approximate concentration limits of the technique. This was achieved by studying CS_2, SF_6, $CHCl_3$, SO_2, S_2Cl_2 and Cl_2 in various matrix supports at low solute concentrations. The detection capabilities of laser Raman spectroscopy were found to be very encouraging. From our experiences, two types of test were found to be most useful. In the case of CS_2, we were able to record a matrix Raman spectrum of reasonable quality (Figure 9.1) at a $M : A = 10\,000 : 1$. The results indicated that as little as 1 ppm of CS_2 in CO_2 could probably be detected. A second rather stringent test was to detect the extremely weak lines of v_1 of CS_2 corresponding to the ^{32}S and ^{34}S isotopic species (these had not previously been observed and are shown in Figure 9.2). The third test involved chlorine, in which we were able to observe and easily resolve the first overtone of Cl_2 at $M : A = 100 : 1$. Shirk and Claassen[6] at the Argonne National Laboratory were also studying the problems of extreme dilution and were able to report that with careful technique it is possible to obtain good Raman spectra of samples in noble gas matrices deposited on a platinum mirror cooled to either 4·2 K or 20 K. They were able to obtain a Raman spectrum of SF_6 in argon at $M : A = 500$ (Figure 9.3). All three Raman active fundamentals were observed close to known gas phase frequencies. They also observed Raman spectra of $CHCl_3$ isolated in argon at dilutions of $M : A = 100–200$ and concluded that it was clearly possible to observe Raman spectra of molecules in a matrix at frequencies comparable to those observed in the gas phase.

Even though it was felt that these particular choices of matrix isolated molecules might represent favourable cases, in the sense that they were inherently strong Raman scatterers, nevertheless the results were significant as they were *within* the concentration ranges required for efficient isolation

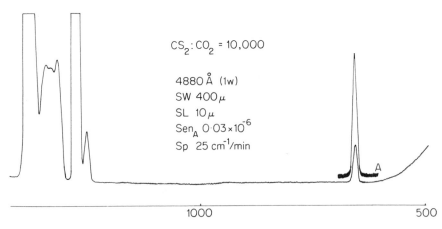

Figure 9.1 The matrix Raman spectrum of CS_2 in CO_2 at $M : A = 10\,000 : 1$ and a temperature of 4·2 K

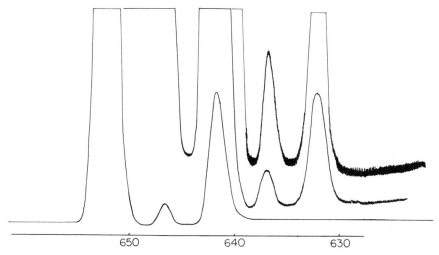

Figure 9.2 High resolution and high sensitivity Raman spectrum of v_1 of solid CS_2 (77 K) showing some of the low abundance ^{13}C, ^{32}S and ^{34}S isotope components

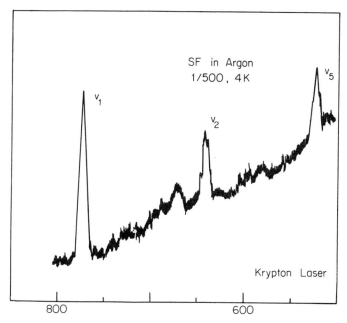

Figure 9.3 Raman spectrum of SF_6 in an Ar matrix at $M : A = 500 : 1$ and 4·2 K (reproduced by permission from J. Shirk and H. H. Claassen, ref. 6)

(approximately 0·1–1 per cent.) and most of the species in which we are interested (i.e. oxides, sulphides, halides, etc.) are generally excellent Raman scatterers.

9.3.2 Colour of the solute or matrix

Absorption of the laser light rather than scattering will cause local heating effects in the matrix material. The subsequent reduction in the rigidity of the matrix may be sufficient to help destroy the delicate species under investigation. This softening effect is easily recognized by sudden increases in pen noise, reduction in signal intensity and baseline variations. Often the result of absorption of the laser beam by the sample can be seen visually as a hole in the matrix. Local heating effects (and reduced Tyndall scattering— see later) may be avoided in favourable cases by working with a defocused laser beam but sometimes with considerable losses in signal intensity. Alternatively, the local heating effect may be reduced to some extent by careful selection of the power output and wavelength of the incident laser light (remembering that signal intensity is proportional to $(v_0 + v_i)^4$).

For matrix Raman studies it is highly recommended that the operator has at his disposal laser outputs of various wavelengths. We have found that the krypton (red and yellow) and argon (blue and green) laser lines are adequate for most experiments.

9.3.3 Power output of the laser

Laser Raman spectra are generally recorded using a diffraction limited point focused laser beam. In matrix Raman spectroscopy a profitable modification of the conventional point focused laser beam system may be made by bringing the laser beam to a line focus on the sample matrix by means of a cylindrical lens (see for example, the optical arrangement of Figures 9.4a and 9.4b). The scattered light from the illuminated line is imaged on the entrance slit of the monochromator. Use of the cylindrical lens dilutes the light at the sample so as to reduce the amount of heating. Also illumination of a greater volume element of the matrix increases the number of scattering centres in the beam with a corresponding increase in the Raman signal intensity. Even so, it will always be of primary importance in all matrix isolation Raman experiments (particularly of reactive species) initially to scan the spectrum at the lowest possible laser energy to obtain a useful signal. In this way, a minimum of energy is dissipated in the matrix and as a result any temperature rise in the matrix is minimal. Once the desired Raman spectrum has been obtained, it is recommended that spectral scans are repeated at gradually increasing laser powers, using either current regulation of the laser or the correct combinations of neutral density filters (0·1 → 99·9 per cent. transmittance). Any changes which are observed in the Raman spectra on repeat scans can then be attributed to diffusion and reaction in

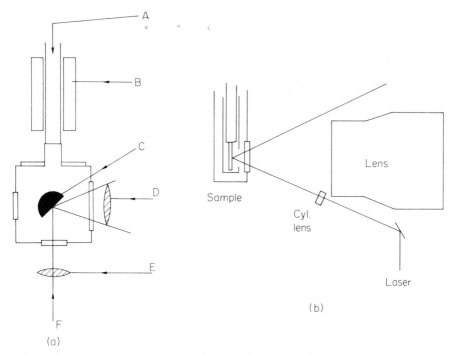

Figure 9.4 Arrangements of sample, lens and laser beam for matrix Raman spectroscopy[6,20]: A = sample inlet; B = microwave cavity; C = cold tip; D = collecting lens;
E = laser focusing lens; F = laser

the matrix. This laser method of initiating self diffusion in the matrix may
prove to be a bonus of the laser Raman technique. However an essential
stage of most matrix experiments is to achieve controlled warm up of the
matrix and to observe diffusion controlled reactions spectroscopically.

It is our experience in matrix Raman spectroscopy, that the best experimental device for these diffusion controlled reactions is undoubtedly the
commercially available liquid helium transfer Cryo-Tip systems. These
often compact and easily manageable Cryo-Tips (described later) which
allow for a fine temperature control may be easily adapted for matrix
Raman experiments and give temperature stability of ± 0.1 K anywhere in
the range of 2–300 K.

9.3.4 Resolution

High resolution will invariably be difficult to achieve for matrix isolated
species; the low concentration required for efficient isolation will usually
result in a weak Raman spectrum and to achieve an acceptable signal-to-
noise ratio, large spectral slit widths are often necessary. The criterion for
high resolution is intimately related to the light scattering properties of the

matrix (discussed below) and the degree of sophistication of the Raman detection system.

9.3.5 Light scattering properties of the matrix

Almost every matrix has different scattering effects to laser light. It appears that the matrix preparation for Raman studies could well be a more critical stage of the experiment than in the comparable infrared experiment. This is primarily because the Raman technique relies mainly on specular reflection from the matrix rather than by transmittance through it. We find that the background scattering is extremely sensitive to the physical nature of the matrix. Among the factors to be considered is the physical appearance of the matrix; that is, whether the matrix is opaque or transparent; amorphous, microcrystalline, highly crystalline or glassy, etc. The amount of background scattering may be controlled and the quality of the spectra considerably enhanced by experimenting with different types of matrix (produced by varying the rates of deposition, temperature of deposition, thickness of the matrix, matrix support material, etc.) and by recording spectra with different angles of incidence of the laser beam on the matrix. Quite often, laser excitation, near grazing incidence, rather than 45–60° scattering can reduce the background scattering (which may also allow closer approach to the exciting line) and improve the signal to noise. In addition, the uniformity of the matrix over tip surface and thickness of the matrix can limit the technique. If the matrix is too thick, it is possible that the outer surface of the matrix is not at the same temperature as the tip itself permitting undesired cracking and diffusion effects in the matrix. In one extreme case where the matrix was too thick, the matrix fell off the tip during controlled warm-up experiments even though the temperature was well below $T_M/2$ of the material. We have also observed formation of matrix blisters during deposition. It is recommended that during the critical deposition stage, close observation of the matrix formation be maintained.

Our experiences indicate that the most favourable matrix for Raman spectroscopy is one that is transparent and deposited on a highly polished mirror surface (platinum, silver, aluminium and copper have all been used). This arrangement generally produces lowest background scattering with lowest noise level and is particularly useful for low concentrations. A possible advantage of the transparent matrix-polished mirror combination is a multiple internal reflection effect of the laser beam in the matrix with a resultant gain in Raman signal intensity.

9.4 Apparatus and experimental

We began our matrix Raman experiments with a conventional double walled helium Dewar, designed specifically for the sample optics of the Spex

1401 spectrophotometer. The Dewar was of an all glass construction, the main difference between it and ones used for IR studies, is the construction of the cold tip and the sample illumination and Raman light collection arrangement.

For matrix Raman studies, we found that the most convenient method for matrix deposition and spectral recording was to have a fixed cold tip (without a radiation shield) and an adjustable glass sample shroud. Temperature measurement of the matrix was initially achieved with a copper-constantan thermocouple, attached to the surface of the cold tip by means of a thermally conducting epoxy resin. The reference junction of the thermocouple was maintained at 4·2 K inside the Dewar. We found that temperature control of the matrix was most simply obtained by employing a fine manganin heating element wound externally around the cold tip. With this arrangement we could measure and control the temperature of the tip to an estimated ± 1 K and were able to investigate the matrix-tip temperatures with (a) increasing laser energy, (b) increasing rates of matrix deposition and (c) increasing microwave powers in our discharge experiments where the sample was discharged before deposition. Laser energy was found to be the most serious cause of undesired matrix temperature rise. It was immediately concluded from various experiments, that focused laser beams with power levels in excess of 200 mW would generally cause local heating effects, sufficient to cause diffusion and reaction of the species in the matrix.

Although the 'home-made' glass helium Dewar provides an economical route to matrix Raman experiments, it is fragile, difficult to handle and uneconomical. We have no doubt that the recently developed and commercially available liquid helium transfer Cryo-Tip systems, which incorporate a flexible transfer line, are a most convenient arrangement for matrix Raman studies. The flexible line allows complete manoeuverability of the cold end. The construction of the cryostat permits independent rotation of the Cryo-Tip with respect to the vacuum shroud during liquid helium transference. The consumption of liquid helium is very low, varying from 0 to 1 liquid litres per hour, and makes the experiment more economical. Temperature control of the cold tip is easily achieved by a combination of needle valves (which vary the helium liquid *and* gas flow rates) and a heating element around the cold tip. The Air Products helium Cryo-Tip (which we have successfully employed) has a useful refrigeration capacity for matrix Raman studies, of 500 mW at 4·4 K, 3 W at 20 K and 7 W at 50 K (see Chapter 3). The cooldown time, from room temperature, is approximately 30 minutes.

For the study of matrix Raman spectra of high temperature molecules we have used a Cryo-Tip and furnace set up. The arrangement basically consists of a copper or silver cold tip on which the matrix and molecular beam are condensed, and a Knudsen cell to generate the molecular beam. The matrix gas and the molecular beam are co-condensed simultaneously on the cold tip with the matrix gas sufficiently in excess to isolate the molecules effusing from the cell. Finally the Raman spectrum is recorded using specular reflection of the laser beam from the matrix. Two optical and sample arrangements are shown in Figure 9.4.

When working with dilute matrices (less than 1 per cent.) using the noble gases Ne, Ar, Kr and Xe as matrix supports, one has the initial problem of lining up the matrix with respect to the spectrometer on a sample band, in an attempt to obtain an optimum signal. This can present a major problem especially if, as is generally the case, the Raman signal is very weak. Thus in matrix Raman, it is advantageous, although not mandatory, to use a matrix support which itself has a Raman spectrum, so that one can optimize the matrix Raman signal by maximizing the spectrum of the matrix material. In such circumstances matrix materials such as N_2, O_2, CO, CO_2, SF_6, $SiMe_4$, etc., become extremely useful when viable.

9.5 Matrix Raman spectra—results and discussion

As matrix Raman spectroscopy is in a very early stage of its development there are few studies that may be profitably referred to. In an effort to discover the scope and limitations of laser Raman matrix studies our initial research programme was aimed at various general areas of investigation all of which involved highly reactive guest species trapped in host lattices at cryogenic temperatures. Thus we have divided this section into four sub-sections the contents of which have invariably been taken from our own research experiences.

9.5.1 Microwave discharge reaction products

The technique of matrix isolation IR spectroscopy has proved to be an extremely powerful tool for determining vibrational frequencies and molecular structures of highly reactive species. The need for reliable Raman data especially on centrosymmetric systems is obvious; the unstable species $XeCl_2$ presents such a situation. Using methods similar to those employed by Meinert,[21] Nelson and Pimentel,[22] various xenon-chlorine mixtures were passed through a microwave discharge (2450 MHz, Microtron 200) and condensed on to the cold tip (at 20 K) of a matrix Raman cell. We performed discharge experiments[23] using *identical* xenon-chlorine mixtures (100 : 1) to those used by Nelson and Pimentel[22] as well as experiments at higher concentrations (25 : 1). Typical laser power levels used were between 10–100 mW and spectral slit widths were usually 2–5 cm^{-1}. The laser wavelengths used in this study were the Kr 5682 Å and 6471 Å lines.

We observed identical matrix Raman spectra whether we used Xe : Cl_2 ratios of 100 : 1 or 25 : 1. In all experiments a strong new band appeared at 253 cm^{-1} which was not present in the matrix isolated spectrum of chlorine in xenon under identical conditions without the microwave discharge. The band at $253\ cm^{-1}$ was initially almost as intense as the Raman band of diatomic chlorine (Figure 9.5). Diffusion controlled warm up experiments were

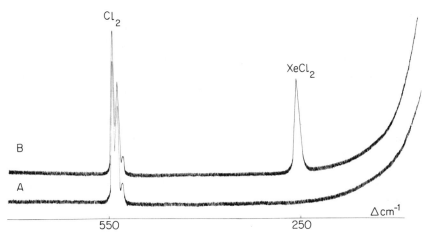

Figure 9.5 The Raman spectra of xenon-chlorine $25:1$ matrix at $4.2\ K$ (A) *without* microwave discharge of the gaseous mixture showing the presence of only Cl_2 (for which ^{35}Cl, ^{37}Cl isotope splitting was observed), (B) *with* microwave discharge of the gaseous mixture, showing the presence of both $XeCl_2$ and Cl_2

performed on the matrix isolated species at various temperatures, and the spectra then recorded after recooling the matrix at $4.2\ K$. At $30\ K$, the band at $253\ cm^{-1}$ showed a marked decrease in intensity relative to the band of diatomic chlorine which confirmed that we were detecting the matrix Raman spectrum of a highly reactive species.

In their matrix IR study Nelson and Pimentel observed only *one* band at $313\ cm^{-1}$ from discharging Xe/Cl_2 mixtures. No mention was made of any other IR band which could be assigned to Cl_3 ($370\ cm^{-1}$) and which was only observed on discharging Kr/Cl_2 mixtures.[24]

We performed Kr/Cl_2 and Ar/Cl_2 microwave discharge experiments under *identical* conditions to those of *our* Xe/Cl_2 experiments. The matrix Raman spectra of the condensed products of our Kr/Cl_2 and Ar/Cl_2 discharges were absolutely clear in the range $60-650\ cm^{-1}$ (apart from the stretching mode of diatomic chlorine). Hence we assign the scattering species in our Xe/Cl_2 experiments to a reactive species containing xenon and chlorine rather than to Cl_3.

As our spectral recordings were such that we had a sensitivity factor of about 100 times to spare, and that under such conditions *no* other Raman lines were observed in the region $60-650\ cm^{-1}$, the possible presence of $XeCl_4$ is excluded and $XeCl_2$ is favoured on several grounds. First, the dilute spectra

$(Xe : Cl_2 \equiv 100 : 1)$ were identical to the most concentrated runs $(Xe : Cl_2 \equiv 25 : 1)$, the latter being the conditions under which $XeCl_4$ would be most likely to form. Secondly, only *one* strong line was observed at $253 \, cm^{-1}$ characteristic of a totally symmetrical $Xe-Cl$ stretching mode of a *linear* and *symmetrical* $(D_{\infty h})$ species. If the species giving rise to the observed Raman scattering had been $XeCl_4$, then the Raman spectrum should have been typical and virtually identical to all known square planar chlorides, for example, the well documented ICl_4^-, $PdCl_4^{2-}$, $PtCl_4^{2-}$, $AuCl_4^{2-}$ ions.[25] These spectra are all very similar and definitive for square planar chlorides and all show three Raman lines of reasonable intensities in the region $300–60 \, cm^{-1}$. The intensity of the new band relative to the chlorine band (see Figure 9.5) suggests that under the conditions of our matrix isolation Raman experiment, about half of the chlorine in the gas mixture is converted to xenon dichloride.

Our results thus confirm Pimentel's IR study.[22] The asymmetric stretching frequency found in the matrix IR spectrum for $XeCl_2$ $(313 \, cm^{-1} \, \Sigma_u^+)$ when taken in conjunction with the corresponding symmetrical stretching frequency from matrix Raman $(253 \, cm^{-1} \, \Sigma_g^+)$ yields xenon-chlorine bond

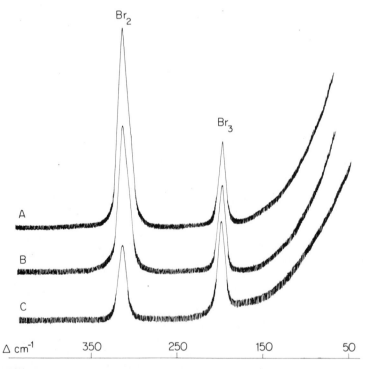

Figure 9.6 The matrix Raman spectrum of the products of a Kr/Br_2 $(50 : 1)$ microwave discharge (C) frozen at $4.2 \, K$, (B) and (A) allowed to diffuse at $35 \, K$ for 3 and 6 minutes respectively and then recooled to $4.2 \, K$

stretching (f_r) and stretch-stretch interaction (f_{rr}) force constants of 1·33 and 0·01 mdynÅ^{-1} respectively.

The tribromine radical (Br$_3$)

In this section we describe the matrix Raman spectra of the products of Ar/Br$_2$, Kr/Br$_2$ and Xe/Br$_2$ microwave discharge reactions.[26] The experiments were performed with noble gas to halogen ratios of 60–25 : 1. Each experiment was performed at least twice to ensure reproducibility. In the experiments with both Kr and Xe, a single Raman line was observed at 197 cm^{-1} (for Kr matrices) and 190 cm^{-1} (for Xe matrices) (apart from the Br$_2$ line at 305 cm^{-1}) corresponding to the Σ_g^+ mode. This band was not observed in Kr/Br$_2$ and Xe/Br$_2$ mixtures which had not been subjected to microwave discharge. (Figures 9.6 and 9.7). The sensitivity conditions were such that we

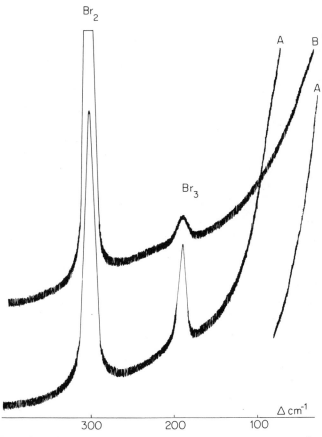

Figure 9.7 The matrix Raman spectrum of the products of a Xe/Br$_2$ (50 : 1) microwave discharge (A) frozen at 4·2 K, (B) allowed to diffuse at 35 K for 3 minutes and then recooled at 4·2 K

had at least a factor of ten to spare, but no other Raman lines could be observed in the range 300–50 cm^{-1}.

In diffusion controlled warm up experiments in the temperature range 4·2–50 K, the Raman lines at 197 cm^{-1} and 191 cm^{-1} showed a marked decrease in intensity while the Br_2 line increased in intensity, indicating that the species produced from the discharge was decomposing into bromine upon heating and diffusion. The matrix Raman spectrum of the products of three experiments involving the discharge of Ar/Br_2 mixtures (in the ratio of 50 : 1) showed extremely weak non-reproducible features in the region of 200 cm^{-1}.

We discount the possibility of the species being noble gas bromides since the species was produced in krypton discharges, whereas only the fluorides of krypton are known, and similar experiments we have performed with Kr/Cl_2 mixtures have failed to produce $KrCl_2$, which should be more stable than $KrBr_2$ or $KrBr$. We also discount the possibility of the compound being Br_4, since this molecule should exhibit two observable Br—Br stretches (both a_1 if a T-shaped molecule, proposed for other tetrahalogens[27] is assumed) in the region 150–300 cm^{-1}. Thus, we assign the band at c. 190 cm^{-1} to the totally symmetrical stretching mode (Σ_g^+) of the linear ($D_{\infty h}$) symmetrical Br_3 radical (note that v_1 of Br_3^- has been observed[28] in solution Raman experiments at 162 cm^{-1}).

As far as we can be certain, vibrational spectroscopic or structural data have not been previously available for Br_3, although its existence has been postulated from gas phase bromine atom recombination studies and from molecular beam kinetic data.[29] From the temperature dependence of the halogen atom recombination rates, the dissociation energy has been estimated as 12·6 kJmol^{-1} for Br_3 and 4 kJmol^{-1} for Cl_3.

The predicted instability of Cl_3 relative to Br_3 in the above results is reflected in our experiments. In four discharge experiments using Kr/Cl_2 ratios of 100–20 : 1 we could obtain no evidence for Cl_3. The only direct evidence for Cl_3 is that of Pimentel,[24] who *only* obtained Cl_3 when he simultaneously discharged chlorine and krypton. The structure that is proposed is also somewhat unusual, being linear but slightly asymmetric, which he says is a result of interactions of neighbouring chlorine molecules, despite the fact that the average Cl_3 radical is surrounded by at least 100 noble gas atoms. Although in our experiments, we obtain no evidence for Cl_3, it is possible that the local heating effect of the laser beam is enough to destroy it, a complication that would not occur in the infrared. We feel, however, that this is unlikely as we used as little as 10 mW of 5682 Å laser power, under which conditions we have previously obtained Raman spectra for Br_3 and $XeCl_2$.

The Raman frequency for matrix isolated Br_3 yields a value of 1·70 mdyn/Å for the force constant sum $f_r + f_{rr}$. This value when compared with $f_r =$

2·45 mdyn/Å for Br_2 and $f_r + f_{rr} = 1·23$ mdyn/Å for Br_3^- provides evidence that the bonding[30] in linear Br_3 resembles that in the negative trihalide ion Br_3^-, which has a linear geometry and vibrational force constants corresponding to 'half bonds'.

9.5.2 High temperature species

Selenium dioxide and its dimer (SeO_2 *and* $(SeO_2)_2$)

The room temperature form of selenium dioxide is a solid with the well documented[31] infinite chain structure A:

```
        O      O
   \   / \    / \   /
    Se    Se    Se              Se
    |     |     |              /  \
    O     O     O             O    O

           A                      B
```

Electron diffraction,[32] gas phase IR[33] and Raman[34] and matrix IR spectroscopy[35] are all consistent with the presence of a non-linear triatomic molecule B as the main constituent in the vapour of selenium dioxide. Mass spectroscopic data[36] has shown that the vapour of selenium dioxide is approximately 99 per cent. monomeric, where the remaining 1 per cent. of the gaseous species was essentially all dimer, $(SeO_2)_2$.

To date the data available specifically for the dimer $(SeO_2)_2$ has consisted of a single line observed in the IR spectrum[35] of matrix isolated SeO_2 where the monomer had been allowed to diffuse at 35 K in an argon matrix. With this very limited data it was not possible to define with any degree of certainty the vibrational assignments and molecular structure of the dimer, $(SeO_2)_2$.

The purpose of our matrix Raman study[37] was twofold. First, we were interested to know whether we could produce the monomer in a convenient manner and co-condense with a matrix gas into our Raman cell. Having produced the matrix isolated monomer, was it then possible to detect its Raman spectrum at low solute concentrations?

The second and, for us, the most interesting part of the investigation, was the detection by Raman spectroscopy of the diffusion controlled polymerization products of the monomer to the dimer and to higher aggregates. The matrix preparation involved the vaporization of selenium dioxide from a molybdenum Knudsen cell (0·05 cm diameter orifice, heated to 420 K in a tantalum-wound resistance furnace) and co-condensation with the matrix gas (CO_2) on to a copper block at 4·2 K. The matrix was deposited over a period of three hours and the matrix thickness was estimated to be about 0·1 mm. Table 9.1 and Figure 9.8b–d show our spectra of matrix isolated selenium dioxide before and after diffusion had been allowed to occur. Included for comparison (Table 9.1 and Figure 9.8a) are the Raman spectra of gaseous SeO_2 (and SO_2) monomer.[34]

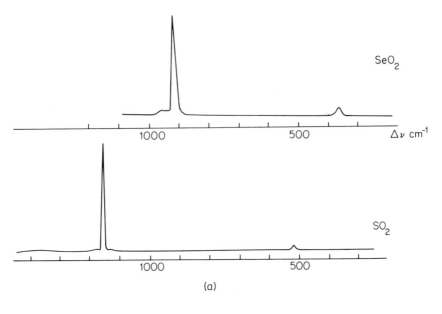

(a)

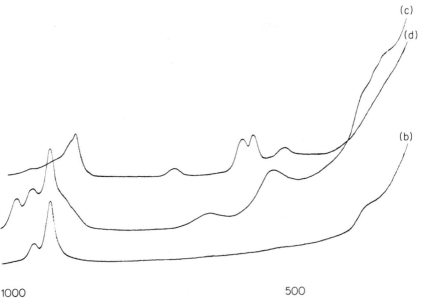

1000   500

Figure 9.8 The Raman spectrum of, (a) gaseous SeO_2 at 300°C and SO_2 at 25°C, (b) selenium dioxide monomer, isolated in a carbon dioxide matrix ($M : A \simeq 100\cdot1$) at 4·2 K, (c) matrix isolated selenium dioxide after diffusion at 50 K, showing both monomer and dimer bonds, (d) matrix isolated selenium dioxide after diffusion at 50–90 K, showing mainly polymer bands of $(SeO_2)_n$ where $n > 3$

Table 9.1 The Raman spectrum of selenium dioxide isolated in a carbon dioxide matrix[a]

Raman (vapour[34])	Matrix Raman			Raman (RT powder)
	Monomer	Dimer[b]	Trimer[b] and higher polymer	
		$1002\,s$		
$967\,w^c(v_3)$	$967\,mw(v_3)$			$940\,w$
$923\,sp(v_1)$	$933\,s(v_1)$			$908\,mw$
			$900\,msh$	
			$887\,s$	$884\,s$
			$710\,w$	
				$704\,mw$
		$660\,m$		
			$592\,m$	$594\,s$
			$572\,m$	
		$543\,s$		
			$520\,mw$	$521\,mw$
$368\,wp(v_2)$	$382\,mw(v_2)$			
		$363\,m$		$356\,w$
		$352\,m$		
				$299\,w$
				$285\,mw$
				$250\,s$
				$195\,m$
				$122\,m$

a. 1 per cent. matrix deposited at 4·2 K.
b. Observed by diffusion controlled experiments in the temperature range 30–90 K.
c. This value is quoted from the vapour phase infrared as the corresponding band in the Raman was very weak.

The results of Figure 9.8b show unambiguously that pure SeO_2 monomer is isolated in the low temperature matrix since the Raman spectrum is very similar to that of the vapour (Figure 9.8a) measured at 570 K.[34] Excellent spectra were obtained for concentrations of less than 1 per cent. The symmetrical stretching and deformational modes v_1 and v_2 show gas to matrix shifts of 10 cm^{-1} and 14 cm^{-1}, respectively. Matrix isolation Raman data for SeO_2 monomer also support the reported matrix IR data[35] and clearly identifies v_2, the matrix frequency which was previously in question.[35]

Spectacular changes in the Raman spectrum of the system occurred as the matrix was allowed to warm up. By measuring the spectrum at various matrix temperatures the progress and results of diffusion controlled reactions between SeO_2 monomers could be observed and the formation of the dimer $(SeO_2)_2$ and higher polymers could be detected (Figures 9.8c and 9.8d respectively). It is interesting to note the tendency of the matrix Raman spectra, after repeated diffusion controlled warm up experiments, to approach that of the room temperature powder Raman spectrum (Table 9.1).

Raman data for $(SeO_2)_2$ (Table 9.1) strongly favour a double oxygen bridged structure $(OSeO_2SeO)$. The lone pair of electrons on selenium are likely to be stereochemically active in a double oxygen bridged dimer $(OSeO_2SeO)$ (as they are in the SeO_2 chain polymer)[31] giving an essentially tetrahedral co-ordination around each selenium atom. Two structures seem likely (C and D).

C	D
C_{2h} symmetry	C_{2v} symmetry
$\Gamma_{vib} = 4a_g + 2b_g + 2a_u + 4b_u$	$\Gamma_{vib} = 5a_1 + 2a_2 + 3b_1 + 2b_2$
(6 Raman active modes)	(12 Raman active modes)

The matrix Raman spectrum of $(SeO_2)_2$ shows at least five bands which we shall show are most satisfactorily assigned to the vibrational modes of the *trans*-centrosymmetric double oxygen bridged structure C. It is interesting to note that this structure would have been predicted on stereochemical and electrostatic grounds since in C the lone-pair-lone-pair interactions and charge repulsion between the terminal oxygen atoms are minimized as compared to those in the *cis* structure D.

In order to be rigorous in our analysis of the dimer problem it is necessary, however, to also consider single oxygen bridge structures. Two such structures (E and F) should be seriously considered.

E	F

Both of these single oxygen bridge configurations take into account the lone pair of electrons on each selenium where one selenium atom is based on sp^2 and the other on sp^3 hybridization. Each of these dimer configurations (E and F) would allow ready conversion to the chain structure of the polymer A without the necessity of breaking SeO bridge bonds in the diffusion controlled polymerization process.

Normal co-ordinate calculations for $(SeO_2)_2$

To provide additional evidence for our vibrational and stereochemical assignments for the $(SeO_2)_2$ molecule we have computed the vibrational frequencies for the double and single oxygen bridged dimers in the chair (C_{2h}), boat (C_{2v}) and both single oxygen bridge (C_s) configurations respectively. Although we would not expect to be able unambiguously to distinguish the four stereochemistries C, D, E and F simply from normal co-ordinate calculations, we feel that a combination of data, that is, the *number, frequencies*, and *intensities* of the observed Raman active bands together with the *calculated frequencies* of vibration, does provide convincing evidence that (a) we have isolated $(SeO_2)_2$ and (b) that the centrosymmetric chair (C_{2h}) configuration is strongly favoured.

The force field for SeO_2 monomer was calculated from the gas phase vibrational frequencies ($v_1 = 923$, $v_2 = 368$, $v_3 = 967$ cm^{-1}). We transferred these force constants directly to the $(SeO_2)_2$ dimer. An identical force field was used for the four proposed configurations of the dimer and the vibrational frequencies were computed for each. The bond lengths and angles used in the dimers are generally similar to those reported for the SeO_2 chain polymer[31] which contains both terminal and bridge Se—O bonds. In our calculations we made the approximation that the bridge bond stretching force constant was 25 per cent. of the corresponding terminal force constant, as found previously for oxygen and fluorine bridged systems (see for example references 38, 39).

The calculated and observed frequencies are shown in Tables 9.2 and 9.3 together with the vibrational assignments and approximate description of the modes. The agreement is remarkably close for only the chair form; this is noticeably so in the SeO_b bridge stretching and the Se—O$_t$ deformation region. It is also reassuring to find that the two observed SeO_b bridge stretching modes (543 s and 660 mw cm^{-1}) are extremely close to those calculated for the chair configuration C (a_g 531 cm^{-1}; b_g 663 cm^{-1}) and that the most intense of these two lines corresponds to the totally symmetrical vibration.

Table 9.2 Computed vibrational frequencies (in cm^{-1}) for structures C, D, E, F of $(SeO_2)_2$

Observed	C		D		E		F	
1002	1001·5	A_g	1003·1	A_1	1017·0	A''	1017·6	A''
			1001·8	B_1	1006·9	A'	1006·7	A'
					981·1	A'	981·1	A'
660	662·5	B_g	655·53	A_2	671·4	A'	634·3	A'
			594·1	B_1				
543	522·3	A_g	523·6	A_1				
			466·4	A_1	420·2	A'	459·7	A'
					402·6	A''	402·6	A''
363	346·4	B_g	374·3	B_2	391·7	A'	418·1	A'
			339·3	B_1				
352	346·9	A_g	337·6	B_2	229·9	A'	228·0	A'
			214	A_2	192·7	A'	182·9	A'
	163·7	A_g	164·1	A_1	125·1	A'	113·8	A'

Table 9.3 Vibrational assignment for the matrix Raman
spectrum of $(SeO_2)_2$ in the chair form

		Approximate description of mode
1002 s	$\nu_1\,a_g$	νSeO_t
660 mw	$\nu_3\,b_g$	νSeO_b
543 s	$\nu_2\,a_g$	νSeO_b
363 mw	$\nu_5\,b_g$	δSeO_t
352 mw	$\nu_4\,a_g$	δSeO_t
not obs.	$\nu_6\,a_g$	$\nu SeSe$

ν_1 ν_2 ν_3

ν_4 ν_5 ν_6

We may summarize the evidence for $(SeO_2)_2$ as follows. The new spectrum obtained in the initial stages of the diffusion controlled warm up experiments for the isolated SeO_2 monomer are most likely to be associated with the $(SeO_2)_2$ dimer. Although asymmetrical single bridge models for $(SeO_2)_2$ cannot be entirely discounted as a possible structure, they are unlikely on the basis of Raman activities and calculations. Of the most likely symmetrical double oxygen bridged structures, the centrosymmetric model C was strongly favoured purely on the basis of activities and frequencies of the observed bands. The agreement between calculated and observed frequencies of vibration was remarkably close for the chair form.

9.5.3 Lithium atom co-condensation reaction products

Lithium superoxide (LiO_2)

The structure and bonding in LiO_2 has been the subject of a detailed investigation by matrix isolation IR spectroscopy[40] (§5.4). Three IR bands could be ascribed to the expected $(a_1 + b_2)\,\nu Li{-}O$ stretching and $(a_1)\,\nu O{-}O$ stretching modes of a triangular LiO_2 molecule with C_{2v} symmetry; an extremely important feature was the assignment of the $O{-}O$ stretching mode to a band at 1097 cm^{-1}. This frequency was virtually unshifted with respect to the corresponding mode in the free superoxide ion, the Raman frequency of which had been observed in alkali halide lattices.[41] This data was interpreted as an indication that the bonding in LiO_2 should *not* be considered to be that of a covalently bound triangular molecule, but rather that of a Li^+ cation bonded directly to an O_2^- anion by coulombic attractive forces. The IR frequency of the O—O stretching mode in LiO_2 implied the almost complete transference of the Li (2s) valence electron into an oxygen π^* molecular orbital.

Additional evidence to test the proposed structure and bonding in LiO_2 may be obtained from the Raman spectrum of matrix isolated LiO_2.[42] If the bonding of lithium to two equivalent oxygen atoms can best be rationalized as ionic bonding then the matrix Raman spectrum should display only *one* strong band, corresponding to the O—O stretching mode of the O_2^- ion, the frequency of which should be coincident with the band observed in the IR at $1097 \, cm^{-1}$. For the electrostatic model, the symmetric and asymmetric motions of the Li^+ cation relative to the O_2^- anion in the proposed C_{2v} triangular molecule, both involve a dipole change, and consequently finite and observable matrix IR absorption intensities. However, on the basis of the Woodward intensity theory for ion-pairs[43] the corresponding Li—O modes in the matrix Raman spectrum should be extremely weak, at least 100 times less intense than the stretching mode of the covalently bonded O_2^- anion and probably not observable.

In the present study we simultaneously co-condensed a molecular beam of lithium atoms from a stainless steel Knudsen cell, with a jet of oxygen gas, on to a cold tip at 4.2 K. The $Li : O_2$ ratios were calculated to be approximately $1 : 100$ in A and $1 : 1000$ in C. A blank run, using pure oxygen was also performed. The matrix Raman spectra together with diffusion controlled warm up experiments are shown in Table 9.4 and Figure 9.9. Our matrix

Table 9.4 The matrix Raman frequencies (in cm^{-1}) of the co-condensation products of atomic lithium and molecular oxygen[a]

4.2 K ($LiO_2 \simeq 1 : 1000$)	4.2 K ($Li : O_2 \simeq 1 : 100$)	Warm up to 30 K and recooled up to 4.2 K	Assignment
C	A	B	
1551 *vvs*	1551 *vvs*	1551 *vvs*	O_2
1504 *w*	1504 *w*	1504 *w*	O_2
	1148 *wsh* $\Big\}$[b]	1148 *w*[b]	
	1134 *mw*		
1097 *s*	1097 *s*	1097 *vvw*	νO–O(LiO_2)
		802 *s*[c]	
	~517 *w*[b]	~517 *w*[b]	
	~464 *w*[b]	~464 *w*[b]	

a. The IR frequencies[40] of LiO_2 are assigned to bands at

$$\nu_1 \; 1097; \qquad \nu_2 \; 744; \qquad \nu_3 \; 507 \, (^7LiO_2)$$
$$\nu_1 \; 1097; \qquad \nu_2 \; 699; \qquad \nu_3 \; 492 \, (^6LiO_2)$$

b. These bands are assigned to higher aggregates of LiO_2 (see Figure 9.9A).
c. Possibly ν(O—O) of Li_2O_2.

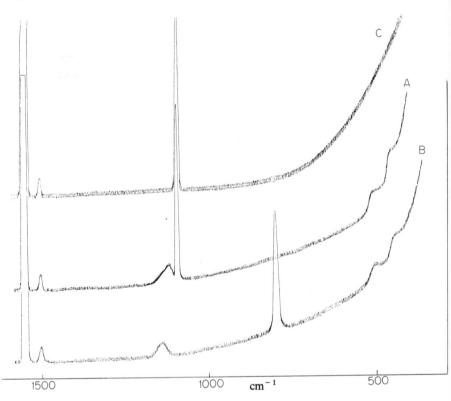

Figure 9.9 The matrix Raman spectrum at 4·2 K of (A) the co-condensation products of a molecular beam of lithium atoms with a jet of oxygen gas (Li : $O_2 \simeq 1 : 100$), (B) the matrix described in (A) but after controlled warm up to 30 K, (C) the same as in (A) but with Li : $O_2 \simeq 1 : 1000$

Raman spectra clearly show *one* major feature at 1097 cm^{-1} (Figure 9.9C). This Raman line is strong and sharp and is coincident with the O—O stretching mode observed in the infrared spectrum of matrix isolated LiO_2. Using conditions of extremely high sensitivity and warm up experiments in the range 4·2 → 35 K we were unable to observe or associate any bands in the range 800–300 cm^{-1} to LiO stretching modes of LiO_2. The 1097 cm^{-1} band showed a spectacular decrease in intensity during diffusion controlled warm up experiments (Figure 9.9B). If Raman lines associated with Li—O motion of LiO_2 were present in our matrix spectra, we estimate that they must be of the order or less than 1 per cent. of the intensity of the observed O—O stretching mode.

We conclude from these data that the assignment of the O—O stretching mode and the previously proposed model for the bonding in matrix isolated LiO_2 are both correct. The intense O—O stretching mode and apparent absence (or extreme weakness) of Li—O stretching modes in the matrix Raman spectrum of LiO_2, together with the previous matrix IR data *prove* that the bonding between the lithium and oxygen should be considered to be essentially electrostatic in nature.

9.5.4 Miscellaneous highly reactive and thermally unstable compounds

The PCl_3—NMe_3 system

Holmes[44] has reported that PCl_3 and NMe_3 form a single 1 : 1 addition complex at low temperatures. Evidence was not cited for any complexes of higher stoichiometries in this system. At room temperature the 1 : 1 complex is completely dissociated into its constituents. Neither vibrational nor structural data has been previously reported for $PCl_3 \cdot NMe_3$. The data for this adduct would be particularly interesting as it is isoelectronic with the thermally unstable addition compound $SiCl_4 \cdot NMe_3$, the data[45] for which has been interpreted in terms of a 5-co-ordinate trigonal bipyrimidal molecule with axial NMe_3 and C_{3v} symmetry. Thus for $PCl_3 \cdot NMe_3$ at low temperatures it is particularly relevant to discover if the complex is monomolecular and if this is the case, whether the data implies the presence of a pseudo-trigonal bipyramidal species with a stereochemically active lone pair of electrons. If this proves to be the base then four possibilities occur.

Thus our vibrational study may be divided into four parts:
- (i) The low temperature spectra of (*a*) the pure 1 : 1 complex; and the 1 : 1 complex in (*b*) a CO_2 matrix and (*c*) a NMe_3 matrix.
- (ii) Isotope substitution data for the completely deuterated complex $PCl_3 \cdot N(CD_3)_3$.
- (iii) Comparison of this data with the analogous system $PBr_3 \cdot NMe_3$.
- (iv) Normal co-ordinate analysis of $PCl_3 \cdot (CH_3)_3$ and $PCl_3 \cdot N(CD_3)_3$ in its various symmetries.

As the Raman data[46] is most relevant to this particular chapter, we shall discuss this aspect of the present study in greatest detail. The Raman spectrum of the pure 1 : 1 adduct is shown in Figure 9.10a and Table 9.5 (IR data are included in Table 9.5). Although the spectra are complex they are not incompatible with a $PCl_3 \cdot NMe_3$ compound with C_s symmetry. Figure 9.10b and Table 9.5 show the matrix Raman data of $PCl_3 \cdot NMe_3$ in a CO_2 matrix where the $M : A = 100 : 1$. We find that the main features of the matrix spectrum are essentially unchanged when compared to the pure 1 : 1 spectrum. However, when the matrix support was chosen to be excess ligand

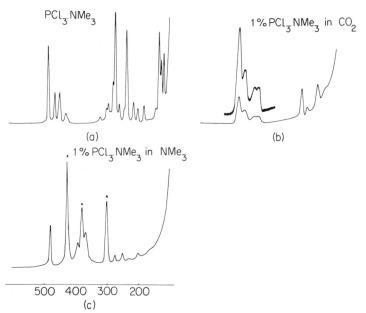

Figure 9.10 The Raman spectrum of (a) pure $PCl_3 . NMe_3$, (b) $PCl_3 . NMe_3$ isolated in a CO_2 matrix at $M : A = 100 : 1$, (c) $PCl_3 . NMe_3$ isolated in a NMe_3 matrix at $M : A = 100 : 1$ (asterisked lines indicate free NMe_3).

Table 9.5 Infrared and Raman data for the PCl_3—NMe_3 system[a]

PCl_3—NMe_3 (1 : 1) pure		PCl_3—$N(CD_3)_3$ (1 : 1) pure	PCl_3—NMe_3 (1 per cent. CO_2 matrix)	PCl_3—NMe_3 (1 per cent. NMe_3 matrix)	Assignment of 1 : 1 monomer
IR	Raman				
500 *wsh*	496 *s*	498 *vs*	505 *s*		a' $vPCl$
				486 *s*	
480 *s*		483 *ms*			a'' $vPCl$
	471 *m*	474 *ms*	480 *msh*		a' $vPCl$
	452 *m*		439 *m*		a' δNMe_3
				435 *m*	
431 *ms*	{429 *w*		426 *m*		a'' δNMe_3
	{421 *vwsh*				
				412 *wsh*	
				402 *m*	
		388 *ms*			a' $\delta N(CD_3)_3$
				373 *ms*	
		340 *w*			a'' $\delta N(CD_3)_3$

a. Frequencies in cm^{-1} and only the PCl stretching and δNMe_3 (and $\delta N(CD_3)_3$) deformational region are tabulated.

NMe_3, the spectra of Figures 9.10a and 9.10b undergo a dramatic change to yield the matrix Raman spectrum of Figure 9.10c (where $PCl_3 \cdot NMe_3$: $NMe_3 = 1 : 100$).

With a combination of vibrational data derived from isotope substitution using $PCl_3 \cdot N(CD_3)_3$ and halogen substitution using $PBr_3 \cdot N(CH_3)_3$ one can easily distinguish P—Cl, and P—Br skeletal modes from NMe_3 ligand modes. This assignment is particularly clear cut in the skeletal stretching and ligand deformational region of the spectra ($500–300 \text{ cm}^{-1}$) where the ligand modes experience isotope shifts ($40–60 \text{ cm}^{-1}$) far in excess of the P-halogen skeletal modes.

Thinking in terms of symmetry co-ordinates the strongest Raman band in B would be expected to be the symmetrical PCl_3 equatorial stretching mode; in D the strongest Raman band would be the symmetrical axial PCl_2 stretch. In neither of these 'modes' does the phosphorus move. Thus by analogy with *trans* $SiCl_4 \cdot 2py^{47}$; *trans*-$SiCl_4 \cdot 2PMe_3^{45}$ ($SiCl_4$ symmetric stretch at 320 cm^{-1} and 292 cm^{-1}) and $SiCl_4 \cdot NMe_3^{45}$ ($SiCl_3$ equatorial symmetric stretch at 385 cm^{-1}) we expect a strong band in the region $290–400 \text{ cm}^{-1}$ for either B or D. From the Raman data collected in Table 9.5 and Figures 9.10a–c it is immediately apparent that stereochemistries B and D are excluded.

Thus the data indicate a monomolecular form, either A or C. We are unable to distinguish between these two possibilities, both of which are expected to yield three high frequency $vPCl$ stretching modes (cf. PCl_3 shows $v_1 = 510 \text{ cm}^{-1}$ and $v_3 = 480 \text{ cm}^{-1}$ and our computed frequencies for A and C also yield high frequency $vPCl$ modes), but we note that *no* example of a non-equatorial lone pair is known so that shape A is the most likely. If this stereochemistry is correct then it is interesting to compare the known shapes of the isoelectronic species shown below:

| G^{46} | E^{45} | H^{48} | F^{45} |

The data thus imply that on replacing a chlorine atom in the silicon complexes E and F by a stereochemically active lone pair of electrons in the isoelectronic phosphorus complexes G and H respectively, the basic molecular framework remains unchanged.[48]

Of considerable interest is the new species formed with $PCl_3 \cdot NMe_3$ in a NMe_3 matrix. This species probably arises from the interaction of the 1 : 1

complex G with another ligand molecule in the matrix:

$$G \qquad\qquad J$$

The approach of a second ligand molecule along the least sterically crowded path would yield a $1:2$ complex J with the same stereochemistry as the known structure $cis\text{-}SbCl_3 \cdot 2NH_2\phi$.*[49] The matrix Raman spectrum for $PCl_3 \cdot 2NMe_3$ shows only high frequency νPCl stretching modes and a shift of all νPCl stretching modes to lower frequencies compared to those in $PCl_3 \cdot NMe_3$. These data are compatible with a pseudo octahedral cis-stereochemistry for $PCl_3 \cdot 2NMe_3$.

Octavalent xenon compound (XeO_3F_2)

The octavalent xenon compound XeO_3F_2 was characterized recently by observation of its mass spectrum.[50] The compound is extremely unstable and so it was studied by the matrix isolation technique to observe both the IR and the Raman spectra (§3.6.3).[51] Figures 9.11a and 9.11b and Table 9.6

Table 9.6 Observed fundamental frequencies (in cm^{-1}) of XeO_3F_2 and assignments[6,51]

Raman	IR	Assignment
806·7	—	$\nu_1\,(a_1')$
567·4	—	$\nu_2\,(a_1')$
—	631·7	$\nu_3\,(a_2'')$
—	375·4	$\nu_4\,(a_2'')$
892	895·8	$\nu_5\,(e')$
316	320·8	$\nu_6\,(e')$
190	n.o.	$\nu_7\,(e')$
361	—	$\nu_8\,(e'')$

show the best matrix IR and Raman spectra obtained. The data immediately suggest a high symmetry molecule, probably D_{3h} with two fluorine atoms on the threefold axis and the three oxygen atoms at 120° angles around the

* Note that the reported X-ray structure[49] for the $1:1$ complex $SbCl_3NH_2\phi$ is analogous to structure G shown above.

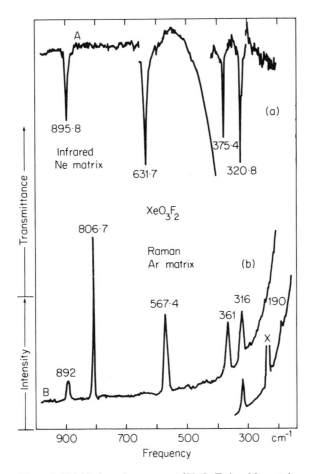

Figure 9.11 (a) Infrared spectrum of XeO_3F_2 in a Ne matrix at 4.2 K; $M/A \simeq 500:1$, deposition time 1 hour, (b) Raman spectrum of XeO_3F_2 in an Ar matrix at 4.2 K, M/A between $400:1$ and $100:1$, deposition time 45 minutes. The line marked with an X at 233 cm^{-1} on the lowest trace is due to a krypton emission line that was not masked for that particular run

equator. Such a structure for XeO_3F_2 was predicted by Gillespie[52] some years before the material was first synthesized.

Molecules of XeO_3F_2 with D_{3h} symmetry would have eight fundamental modes with symmetry designations $2a_1' + 2a_2'' + 3e' + e''$. Species a_1', e' and e'' are Raman active, and species a_2'' and e'' are IR active. The observed spectra fit these symmetry predictions and the assignments (Table 9.6) are obvious from the selection rules. The agreement of the observed spectra with the predictions based on D_{3h} symmetry is so good that no other symmetry needs to be considered.

The molecule may be considered as related to XeF_2, which has been described as a trigonal bipyramid with the fluorine atoms on the axis and three lone pairs of electrons on the equator.[52] The XeO_3F_2 molecule is similar, the lone pairs having become bonding pairs to oxygen atoms. The intermediate case XeO_2F_2 has been similarly described, with one equatorial lone pair and with a nearly linear $F—Xe—F$ axis.[53]

9.6 Matrix Raman Depolarization Measurements

An approximation, termed the 'cold gas model', considers the guest species to be an infinitely dilute cold gas where the host species is assumed absent.[54] Thus all intramolecular processes are assumed to occur without environmental influence. The cold gas model may be applied to matrix Raman spectra where the guests are assumed to be a perfectly random collection of non-interacting species in a weakly interacting host.

In this treatment of Raman depolarization measurements on matrix isolated species, the isotropic optical behaviour of a transparent medium (for example, the fcc lattice of the inert gases and N_2) facilitates the observations. Provided the matrix support is transparent, depolarization of the incident laser beam and the scattered Raman light should be minimal and experimental depolarization ratios should be meaningful for matrix isolated species. Although perfectly transparent matrices are difficult to obtain in practice and non-ideal depolarization ratios will inevitably result from frosty matrices, the data permits symmetry assignments.

Referencing the classical expressions for randomly oriented molecules in a matrix (or fluid), we are not surprised to find that for ideal Raman scattering from matrices, the same equations apply as for fluids where a band may be assigned to a totally symmetrical mode if its depolarization ratio is less than 0·75.

Raman depolarization measurements have already been realized in practice for both stable[55] and unstable[56] matrix isolated species. Nibbler[55] reported depolarization measurements of various stable molecules such as, CCl_4, CH_4, CO_2, CS_2, COS in N_2 matrices (at $M/A \simeq 500$). Although the measured depolarization ratios were not ideal they were sufficiently clear cut to distinguish totally symmetrical from depolarized asymmetrical modes. Other examples which demonstrate the usefulness of these measurements are the unstable monomeric dihalides of group IV.[56]

9.7 Conclusions

As with any new technique in its early stages of development, matrix Raman spectroscopy is certainly amenable to refinement:

(i) Increasing signal-to-noise and throughput by (a) silvered rather than aluminized optics for higher reflectivity, (b) finer ruled gratings to permit wider slits without sacrificing bandpass, (c) improved illumination-collection optics to gather more light, (d) dye lasers and the possible utilization of resonance conditions.

(ii) Optimizing detection by improved, high quantum efficiency photomultiplier tubes, for example the RCA developmental C-31034.

(iii) Closer approach to the excitation line by (a) a third monochromator, (b) a single moded laser with iodine filter.

In the light of the successful studies reported by several groups of workers during the time this chapter was in press, it is now clear that Raman matrix isolation spectroscopy is a viable technique.

There are many people and institutions to whom I shall always be indebted for making the matrix Raman technique and this introductory chapter a reality.

I am sincerely grateful to Dr. Gordon Briggs, Mr. David Boal, Mr. Helmut Huber, Dr. Michael Menzinger, and Mr. Anthony Vander Voet for their technical assistance, many helpful discussions, enthusiasm and patience which they have given towards the development of the various aspects of matrix isolation laser Raman spectroscopy at the University of Toronto. I thank Dr. H. H. Claassen and Dr. James Shirk for helpful discussions and for making available preliminary matrix Raman data for SF_6 and XeO_3F_2.

I would like to thank Professor A. G. Brook and Professor E. A. Robinson for their financial assistance and continued interest in the matrix technique and the National Research Council of Canada for financial support to G.A.O. and a N.R.C. scholarship to D.H.B.

9.8 Addendum

Since the original drafting of this chapter considerable advances have been made in matrix Raman sampling techniques and a number of important chemical applications are beginning to appear in the open literature.

In order to update this rapidly growing area the following results obtained mainly in our laboratory, are discussed which illustrate the usefulness and desirability of combining the matrix IR and Raman experiments for the study of some transition metal atom–molecule co-condensation reactions.

The reader is also referred to the elegant matrix Raman experiments of Andrews (alkali metal atom–oxygen co-condensation reactions;[57] and the detection of the free radical OF by laser induced photodetachment of F atoms from OF_2) and Nibler (matrix isolated rare earth trifluoride monomers, for example PrF_3).[58]

The reaction of transition metals with neutral ligand molecules such as CO, CS, CS_2, COS, N_2 and O_2 to form molecular species (with the metals usually in the zero valent oxidation state) is presently providing chemists with simple binary species of low co-ordination numbers for the study of the metal ligand bond without the interference of other ligands. Dinitrogen and molecular dioxygen compounds are of special significance due to their relationship to biological problems and information obtained on the metal ligand bond in these matrix stabilized compounds should be useful in further elucidating the larger inorganic and biochemical systems. Although the majority of these binary species are stable only under matrix isolation conditions, some are stable enough to be prepared synthetically, thus enabling greater elucidation of their chemistry.

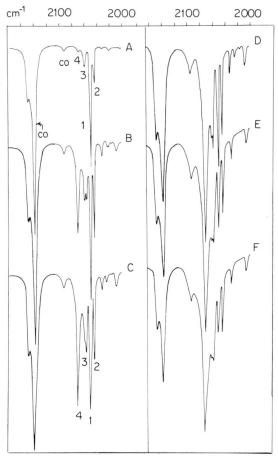

Figure 9.12 The matrix infrared spectra of the products of the co-condensation reaction between Pd atoms and CO in Ar (where CO : Ar = 1 : 1000); (A) on deposition at 10 K; (B–F) after successive diffusion warm up experiments at 20–30 K[62,63]

In order to obtain compounds analogous to the nickel carbonyls, Ogden[59] and independently Ozin[60] investigated the previously unknown Pd/CO and Pd/CO, Pt/CO systems respectively. Ogden reported the matrix infrared spectrum of the palladium tetracarbonyl which on the basis of $^{12}C^{18}O/^{12}C^{16}O$ isotopic splittings could be assigned to tetrahedral symmetry. Ozin and his co-workers reported both the matrix IR and Raman evidence for the formation not only of the $M(CO)_4$ species, but in dilute matrices, of the intermediates $M(CO)$, $M(CO)_2$ and $M(CO)_3$ (see Figures 9.12 and 9.13). The stoichiometries of the species were confirmed by $^{12}C^{18}O$

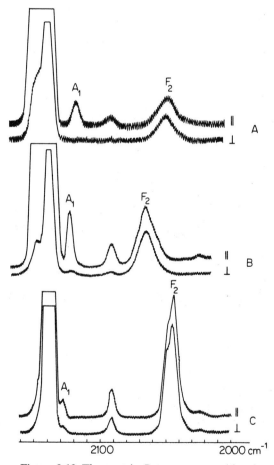

Figure 9.13 The matrix Raman spectra (showing parallel and crossed polarization measurements) of the products of the co-condensation reaction (C) between nickel atoms and carbon monoxide, (B) palladium atoms and carbon monoxide, and (A) platinum atoms and carbon monoxide[60]

isotopic substitution and by the kinetic studies of the diffusion of the compounds during controlled warm up experiments of the matrices.[62,63]

The matrix IR and Raman spectra of the products of the co-condensation reaction of cobalt atoms with pure CO have been recorded[64] including depolarization measurements. The data are consistent with the presence of the tetracarbonyl cobalt free radical species $Co(CO)_4$ having symmetry D_{2d} which is to be expected for a d^9 tetrahedral molecule subject to a Jahn–Teller distortion (Figure 9.14).

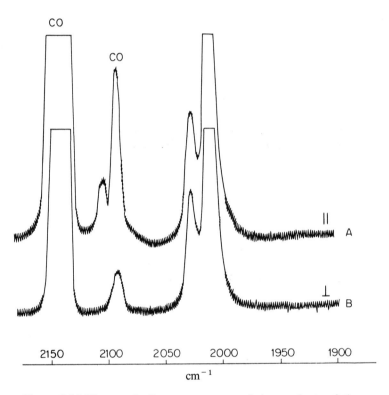

Figure 9.14 The matrix Raman spectrum of the products of the co-condensation reaction of cobalt atoms and CO at 4·2 K (A) parallel and (B) crossed polarizations showing the presence of $Co(CO)_4$[64]

Complexes of nickel with N_2 had been reported by Burdett and Turner[65] from co-condensation reactions and the bands observed were assigned (erroneously as later experiments were to indicate) to the species NiN_2 and $Ni(N_2)_2$ as well as to cluster type compounds $Ni_x(N_2)_y$. Huber et al.[66] have studied the matrix IR and Raman spectra of the nickel dinitrogen system in

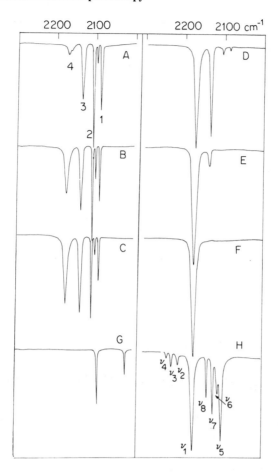

Figure 9.15 The matrix infrared spectra of the co-condensation reaction of Ni atoms with (A) a $^{14}N_2 : Ar = 1 : 10$ mixture at 10 K (B–F) after successive diffusion controlled warm up experiments at 20, 20, 25, 30, and 35 K and then recooled to 10 K (G) a $^{14}N_2 : ^{15}N_2 : Ar = 1 : 1 : 2000$ mixture at 10 K (H) a $^{14}N_2 : ^{15}N_2 : Ar = 1 : 1 : 20$ mixture deposited at 10 K, allowed to diffuse at 35 K and recooled to 10 K

detail including concentration dependence studies, and isotopic substitution using $^{14}N_2$, $^{15}N_2$ and $^{14}N^{15}N$ (Figures 9.15, 9.16 and 9.17). Results, which proved to be entirely analogous to the $Ni(CO)_x$ system,[67] were obtained and indicated the presence of the compounds $Ni(N_2)_x$ ($x = 1, 2, 3, 4$). The

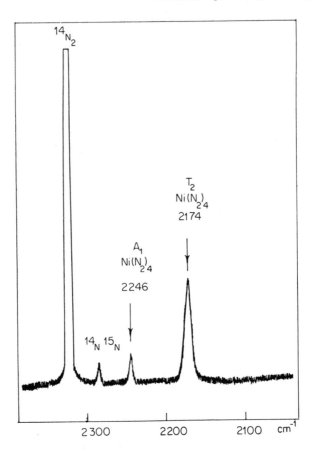

Figure 9.16 The matrix Raman spectra of the co-condensation reaction of Ni atoms with a $^{14}N_2$: Ar (1 : 10) mixture at 4·2 K

$^{14}N^{15}N$ experiments for NiN_2 (Figure 9.18) clearly indicated the presence of 'end-on bonded' molecular dinitrogen. The isotopic substitution experiments, the applicability of the Cotton–Kraihanzel force field approximations[68] and isotope intensity sum rules, with the resultant agreement of observed and calculated frequencies and relative intensities combined to lead to the assignment of structures of the Ni/N_2 species similar to the nickel carbonyl series $Ni(CO)_n$ where $n = 1, 2, 3,$ and 4. It can be noted here that in dilute N_2/Ar matrices the tetrakis-dinitrogen compound, $Ni(N_2)_4$ is a regular tetrahedron, but that in solid α-nitrogen the symmetry is distorted to C_2 due to the low substitutional site symmetry in the N_2

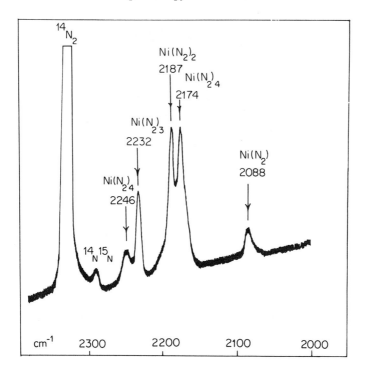

Figure 9.17 After diffusion controlled warm up at 35 K showing $Ni(N_2)_4$[66]

crystal. The mixed $^{14}N^{15}N$ experiments were shown to be relevant to N_2 chemisorption studies.[69] According to Eischens,[70] although chemisorption of nitrogen is most often assumed end-on, the $^{14}N^{15}N$ experiments had failed to indicate evidence for band splitting of broadening.

The Pd/N_2[66] and Pt/N_2 (Figures 9.19 and 9.20) systems were also studied by Ozin and his co-workers, and they found the species of highest co-ordination number to be the tris-dinitrogen compounds $Pd(N_2)_3$ and $Pt(N_2)_3$ respectively. Concentration dependence studies and isotopic substitution experiments in both the IR and Raman confirmed the linear $C_{\infty v}$, linear $D_{\infty h}$ and trigonal planar D_{3h} stereochemistry for the three $M(N_2)_x$ $(x = 1–3)$ compounds formed.

As a continuation of their work on the isoelectronic and isostructural binary carbonyls and dinitrogen complexes of nickel, $Ni(CO)_n$ and $Ni(N_2)_n$ where $n = 1–4$, Kündig and Ozin[64] co-condensed nickel atoms with

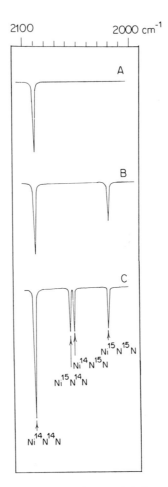

Figure 9.18 The matrix infrared spectra of the products of the co-condensation reaction of Ni atoms with:
(A) $^{14}N_2 : Ar \sim 1 : 1000$;
(B) $^{14}N_2 : {}^{15}N_2 : Ar \sim$
        $2 : 1 : 2000$;
(C) $^{14}N_2 : {}^{14}N^{15}N : {}^{15}N_2Ar \sim$
        $3 : 2 : 1 : 6000$
showing that NiN_2 has N_2 bonded in an 'end-on' fashion[69]

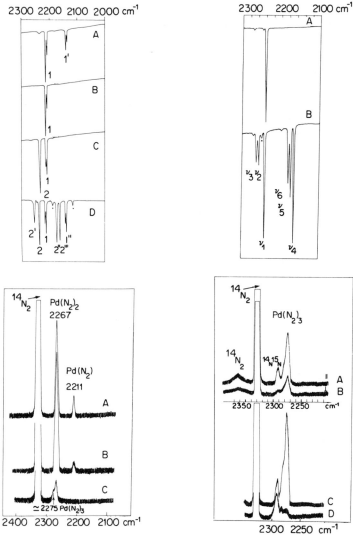

Figure 9.19 *Top left*: The matrix infrared spectra of the co-condensation reaction of Pd atoms with (A) a $^{14}N_2 : ^{15}N_2 : Ar \sim 1 : 1 : 2000$ mixture at 10 K; (B) a $^{14}N_2 : Ar \sim 1 : 1000$ mixture; (C) a $^{14}N_2 : Ar \sim 1 : 10$ mixture; (D) a $^{14}N_2 : ^{15}N_2 : Ar \sim 1 : 1 : 20$ mixture. *Top* right: The matrix infrared spectrum of the co-condensation reaction of Pd atoms with (A) pure $^{14}N_2$ and (B) with an equimolar mixture of $^{14}N_2/^{15}N_2$ at 10 K. *Bottom left*: The matrix Raman spectrum of the co-condensation reaction of Pd atoms with a $^{14}N_2 : Ar \sim 1 : 10$ mixture (A) at 4·2 K; (B–C) after successive diffusion controlled warm up experiments at 20 K and 34 K and recooled to 4·2 K. *Bottom right*: The matrix Raman spectrum of the co-condensation reaction of Pd atoms with pure $^{14}N_2$ at 4·2 K (A) parallel polarization; (B) crossed polarization; (C) after diffusion controlled warm up at 30 K and recooled to 4·2 K[66]

$CO/N_2/Ar$ mixtures to obtain the binary mixed complexes $Ni(N_2)_n(CO)_{m-n}$. In pure CO/N_2 and $CO/N_2/Ar = 1/1/20$ matrices at 10 K, evidence for all the possible mixed species (co-ordination numbers 4, 3 and 2) was obtained. Diffusion controlled warm up experiments to 35–40 K yielded the five possible tetraco-ordinate species (Figure 9.21) $Ni(N_2)_n(CO)_{4-n}$ where $n = 0$–4. The matrix IR data obtained in the co-condensation experiments are in close agreement with the data previously obtained by Rest[71] for $Ni(CO)_3(N_2)$ formed from the photolysis of $Ni(CO)_4$ in N_2 matrices. The N—N and C—O frequency data for the mixed nickel complexes is shown in Table 9.7 where it is found that all IR active stretching frequencies lie *above* that of $Ni(N_2)_4$ (2173 cm^{-1}) with the NN frequencies decreasing in the order $Ni(CO)_3(N_2) > Ni(CO)_2(N_2)_2 > Ni(CO)(N_2)_3 > Ni(N_2)_4$. This is the trend expected when the weak π-acceptor N_2 is successively replaced by the strong π-acceptor CO. An inverse effect is observed for the CO stretching frequencies where all IR active CO stretching frequencies for the mixed species lie *below* that of $Ni(CO)_4$ (2048 cm^{-1}) with the order $Ni(CO)_4 > Ni(CO)_3(N_2) > Ni(CO)(N_2)_3 > Ni(CO)_2(N_2)_2$.

From the amount of work that has been done in the last few years with metal atom/gas reactions both in matrices and on the preparative scale, it is seen that there is not only a great versatility in this area of chemical research but also a very promising future.

Table 9.7 Vibrational frequencies for the mixed dinitrogen-carbonyl intermediates of nickel $Ni(CO)_n(N_2)_{4-n}$ (where $n = 0-4$)[a]

Symmetry	Molecule	νCO assignment		νNN assignment	
T_d	$Ni(N_2)_4$			2248 A_1	2173 T_2
C_{3v}	$Ni(N_2)_3(CO)$		2006 A_1	2293[b] A_1	2211 E
C_{2v}	$Ni(N_2)_2(CO)_2$	2084 A_1	1996 B_2	2285[b] A_1	2243 B_2
C_{3v}	$Ni(N_2)(CO)_3$	2098 A_1	2030 E		2263 A_1
T_d	$Ni(CO)_4$	2130 A_1	2043 T_2		

a. See reference 64.
b. In the matrix infrared spectra the A_1 νNN stretching modes of $Ni(N)_3(CO)$ and $Ni(N_2)_2(CO)_2$ were too weak to observe.

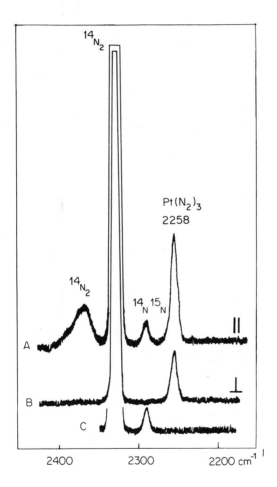

Figure 9.20 The same as Figure 9.19 *bottom right* except using Pt atoms[66]

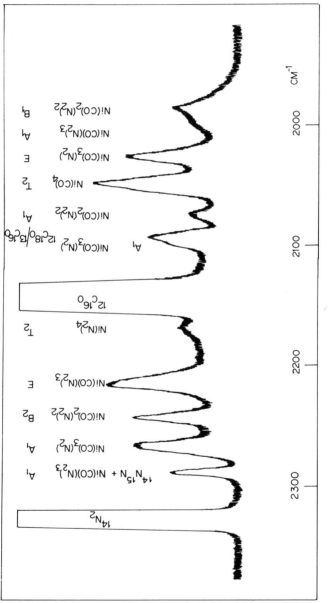

Figure 9.21 The matrix Raman spectrum of the co-condensation reaction of Ni atoms with a CO/N_2 matrix at 4.2 K[64] showing all of the binary mixed dinitrogen-carbonyl species $Ni(N_2)_n(CO)_{4-n}$ where $n = 0$–4

9.9 References

1. A. Cabana, A. Anderson and R. Savoie, *J. Chem. Phys.*, **42**, 1122, (1965).
2. E. Nixon, *Raman Newsletter* 1969.
3. I. R. Beattie and R. Collis, *J. Chem. Soc.* (**A**), 2960, (1969).
4. P. A. Giguere, K. Herman and X. Deglise, 'Paper JII', *Abstracts of the 25th Symposium on Molecular Structure and Spectroscopy*, The Ohio State University, 1970.
5. D. O'Hare and B. Leroi, *2nd International Raman Spectroscopy Conference*, Oxford University, 1970.
6. J. S. Shirk and H. H. Claassen, *J. Chem. Phys.*, **54**, 3237, (1971).
7. J. Dewar, *Proc. Roy. Soc.*, **68**, 360, (1901).
8. L. Vergard, *Nature*, **113**, 716, (1924).
9. G. N. Lewis and D. Lipkin, *J. Amer. Chem. Soc.*, **64**, 2801, (1942).
10. M. Freamo and F. O. Rice, *J. Amer. Chem. Soc.*, **73**, 5529, (1951).
11. A. M. Bass and H. P. Broida, *Formation and Trapping of Free Radicals*, Academic Press, New York and London (and references therein), 1960.
12. I. Norman and G. Porter, *Nature*, **174**, 508, (1954).
13. E. Whittle, D. A. Dows and G. C. Pimentel, *J. Chem. Phys.*, **22**, 1943, (1954).
14. G. C. Pimentel, G. E. Ewing and W. E. Thompson, *J. Chem. Phys.*, **32**, 927, (1960).
15. E. D. Becker and G. C. Pimentel, *J. Chem. Phys.*, **25**, 224, (1956).
16. M. J. Linevsky, *J. Chem. Phys.*, **34**, 587, (1961).
17. W. Weltner, *Proc. Int. Symp. Condensation and Evaporation of Solids*, p. 243, 1962.
18. See previous Chapters.
19. M. Hass and D. L. Griscom, *J. Chem. Phys.*, **51**, 5185, (1969).
20. G. A. Ozin (unpublished design).
21. H. Meinert, *Z. Chem.*, **6**, 71, (1966).
22. L. Y. Nelson and G. C. Pimentel, *Inorg. Chem.*, **6**, 1758, (1967).
23. D. H. Boal and G. A. Ozin, *Inorg. Chem.* (in press).
24. L. Y. Nelson and G. C. Pimentel, *J. Chem. Phys.*, **47**, 3671, (1967).
25. H. Stammereich and R. Forneris, *Spectrochim. Acta*, **16**, 363, (1960); J. D. S. Goulden, A. Maccoll and D. J. Millen, *J. Chem. Soc.*, 1635, (1950).
26. D. H. Boal and G. A. Ozin, *J. Chem. Phys.*, **55**, 3598, (1971).
27. L. Y. Nelson and G. C. Pimentel, *Inorg. Chem.*, **7**, 1695, (1968).
28. W. B. Person, A. R. Anderson, J. N. Fordemwalt, H. Stammreich and R. Forneris, *J. Chem. Phys.*, **35**, 908, (1961).
29. Y. T. Lee, P. R. leBreton, J. D. McDonald and D. R. Herschback, *J. Chem. Phys.*, **51**, 455 (and references therein), (1969).
30. G. C. Pimentel, *J. Chem. Phys.*, **19**, 446, (1951).
31. J. D. McCullough, *J. Amer. Chem. Soc.*, **59**, 789, (1937).
32. K. J. Palmer and W. Elliot, *J. Amer. Chem. Soc.*, **60**, 1852, (1938).
33. P. A. Giguére and M. Falk, *Spectrochim. Acta.*, **16**, 1, (1960).
34. I. R. Beattie and J. Horder, (private communication).
35. J. W. Hastie, R. Hauge and J. L. Margrave, *J. Inorg. Nucl. Chem.*, **31**, 281, (1969).
36. P. J. Ficalora, J. L. Margrave and J. C. Thompson, *J. Inorg. Nucl. Chem.*, **31**, 3771, (1969).
37. D. Boal, G. Briggs, H. Huber, G. A. Ozin, E. A. Robinson and A. Vander Voet, *Nature*, **231**, 174, (1971); G. A. Ozin and A. Vander Voet, *J. Mol. Struct.*, **10**, 173, (1971).
38. I. R. Beattie and G. A. Ozin, *J. Chem. Soc.*, **A**, 2615, (1969).
39. I. R. Beattie, K. M. Livingston, G. A. Ozin and D. J. Reynolds, *J. Chem. Soc.*, **A**, 958, (1969).
40. L. Andrews, *J. Chem. Phys.*, **50**, 4288, (1969).

41. J. Rolfe, W. Holzer, W. F. Murphy and H. J. Bernstein, *J. Chem. Phys.*, **49**, 963, (1968).
42. H. Huber and G. A. Ozin, *Mol. Spectroscopy*, **41**, 595, (1972).
43. J. H. B. George, J. A. Rolfe and L. A. Woodward, *Trans. Faraday Soc.*, **49**, 375, (1953).
44. R. R. Holmes, *J. Phys. Chem.*, **64**, 1295, (1960).
45. I. R. Beattie and G. A. Ozin, *J. Chem. Soc.*, A, 370, (1970).
46. D. H. Boal and G. A. Ozin, *J. Chem. Soc. Dalton*, **1972**, 1824.
47. I. R. Beattie, T. R. Gilson and G. A. Ozin, *J. Chem. Soc.*, A, 2772, (1968).
48. D. Friesen and G. A. Ozin (unpublished work).
49. R. Hulme and D. Mullen, *Acta Cryst.*, **A25**, 171, (1969).
50. J. L. Huston, *Inorg. Nucl. Chem. Lett.*, **4**, 29, (1968).
51. H. H. Claassen and J. L. Huston, *J. Chem. Phys.*, **55**, 1505, (1971).
52. R. J. Gillespie in *Noble-Gas Compounds*, ed. H. H. Hyman, p. 334, University of Chicago Press, 1963.
53. P. Tsao, C. C. Cobb and H. H. Claassen, *J. Chem. Phys.*, **54**, 5247, (1971).
54. B. Meyer, *Low Temperature Spectroscopy*, Elsevier, New York, 1971.
55. J. W. Nibler and D. A. Coe, *J. Chem. Phys.*, **55**, 5133, (1971).
56. H. Huber, G. A. Ozin and A. Vander Voet, *Nature*, **232**, 166, (1971); and G. A. Ozin and A. Vander Voet, *J. Chem. Phys.*, **56**, 4768, (1972).
57. R. R. Smardzewski and L. Andrews, *J. Chem. Phys.*, **57**, 1327, (1972); L. Andrews, A. Hatzenbuhler, ibid, **56**, 3398, (1972); L. Andrews, ibid, **57**, 51, (1972).
58. M. Lesiecki, J. W. Nibler and C. W. DeKock, *J. Chem. Phys.*, **57**, 1352, (1972).
59. J. H. Darling and J. S. Ogden, *Inorg. Chem.*, **11**, 666, (1972).
60. H. Huber, E. P. Kündig, M. Moskovits and G. A. Ozin, *Nature Physical Science*, **235**, 98, (1972).
61. E. P. Küdig, M. Moskovits and G. A. Ozin, *J. Mol. Structure*, (in press).
62. E. P. Kündig, M. Moskovits and G. A. Ozin, *Can. J. Chem.*, **50**, 3587, (1972).
63. E. P. Kündig, M. Moskovits and G. A. Ozin, unpublished results, 1972.
64. E. P. Kündig and G. A. Ozin, unpublished results, 1972.
65. J. K. Burdett and J. J. Turner, *J. Chem. Soc.*, **D**, 885, (1971).
66. H. Huber, E. P. Kündig, M. Moskovits and G. A. Ozin, *J. Am. Chem. Soc.*, **95**, 332, (1973).
67. R. L. DeKock, *Inorg. Chem.*, **10**, 1205, (1971).
68. F. A. Cotton and C. S. Kraihanzel, *J. Am. Chem. Soc.*, **84**, 4432, (1962).
69. M. Moskovits and G. A. Ozin, *J. Chem. Phys.*, **58**, 125, (1973).
70. R. P. Eischens, *Accounts Chem. Res.*, **5**, 75, (1972).
71. A. J. Rest, *J. Organometal. Chem.*, **40**, C76 (1972).

Author Index

Only references to matrix isolation and directly related work are included

Subject Index